AF249077

Ecology and Management of
Invasive Plants

Ecology and Management of Invasive Plants

Congyan Wang
Hongli Li

Basel • Beijing • Wuhan • Barcelona • Belgrade • Novi Sad • Cluj • Manchester

Editors

Congyan Wang
School of Environment
and Safety Engineering
Jiangsu University
Zhenjiang
China

Hongli Li
School of Ecology and
Nature Conservation
Beijing Forestry University
Beijing
China

Editorial Office
MDPI AG
Grosspeteranlage 5
4052 Basel, Switzerland

This is a reprint of articles from the Special Issue published online in the open access journal *Plants* (ISSN 2223-7747) (available at: www.mdpi.com/journal/plants/special_issues/FVE66LN620).

For citation purposes, cite each article independently as indicated on the article page online and as indicated below:

Lastname, A.A.; Lastname, B.B. Article Title. *Journal Name* **Year**, *Volume Number*, Page Range.

ISBN 978-3-7258-2436-6 (Hbk)
ISBN 978-3-7258-2435-9 (PDF)
doi.org/10.3390/books978-3-7258-2435-9

Contents

Article

Factors Influencing the Variation of Plants' Cardinal Temperature: A Case Study in Iran

Sima Sohrabi [1,*], Javid Gherekhloo [2], Saeid Hassanpour-bourkheili [2], Afshin Soltani [2] and Jose L. Gonzalez-Andujar [3]

1 Department of Agronomy, Ferdowsi University of Mashhad, Iran and Leader of Iranian Invasive Plants Working Group, Gorgan 4917739001, Iran
2 Department of Agronomy, Gorgan University of Agricultural Sciences and Natural Resources, Gorgan 4913815739, Iran; gherekhloo@gau.ac.ir (J.G.); s.hassanpour.b@gmail.com (S.H.-b.); afshin.soltani@gmail.com (A.S.)
3 Instituto de Agricultura Sostenible (IAS-CSIC), 14004 Cordoba, Spain; andujar@csic.es
* Correspondence: simsoh@gmail.com

Abstract: The establishment and spread of plants in their native or alien geographical ranges are determined by their germination. This study investigated the impact of different factors on variations in cardinal temperatures. We used the lm procedure and measured the effect size by the Eta-square approach to find the association of different factors (species, ecotypes, origin (native/alien), year, and life cycle) with the cardinal temperatures of 31 species. Our results showed that the base, optimum, and maximum temperatures responded differently to these factors. The base temperature was less impacted by ecotypes compared with the optimum and maximum temperatures, whereas the species had a higher impact on the variation in the base temperature. The effect of the origin of weedy plants on the base temperature was higher than the optimum and maximum temperatures. The effect of the year on the optimum temperature was more prominent than that on the base and maximum temperatures. The results confirmed that weedy alien plants preferred high and narrow ranges of base, optimum, and maximum temperatures and probably will be more problematic in summer crops. The results indicate that alien plants can benefit from warmer conditions in invaded areas at the germination stage. These findings lay the foundation for further studies to elucidate which factors are more important.

Keywords: alien plants; base temperature; establishment; germination; life cycle

Citation: Sohrabi, S.; Gherekhloo, J.; Hassanpour-bourkheili, S.; Soltani, A.; Gonzalez-Andujar, J.L. Factors Influencing the Variation of Plants' Cardinal Temperature: A Case Study in Iran. *Plants* **2024**, *13*, 2848. https://doi.org/10.3390/plants13202848

Academic Editors: Congyan Wang and Hongli Li

Received: 2 September 2024
Revised: 8 October 2024
Accepted: 8 October 2024
Published: 11 October 2024

1. Introduction

Seed germination is a critically important stage of life in plants and the consequent success or failure of the plant's establishment heavily depends on this process [1–3]. The timing of germination also plays a significant role in the inter- and intra-specific competition between plants [4,5]. The variation in seed response between and within populations is one of the key factors responsible for the establishment and persistence of alien plant species [6–8]. The invasiveness and pre-adapting ability of an alien plant are associated with certain attributes, including rapid and prolific germination, rapid growth and high fecundity, and great environmental tolerance [9–11]. Despite increased scientific efforts to study biological invasions of certain plant species, there is still a lack of adequate understanding of how alien vs. native plant species respond to environmental cues in different regions. The responses of alien and native species will vary depending on a number of factors, including the habitat, species, environmental parameters, distribution range, invasion status, and the intensity of management measures [12–14]. It is more probable that alien species with wider distribution ranges have a higher chance of adapting to various conditions compared to alien plants with a limited distribution [15]. Therefore, it is essential to implement rigorous management measures to prevent alien plants from adapting to environmental stresses [16,17]. Environmental factors are mentioned as one of the main barriers to invasion by alien plants [18]. The decision on which alien species should be prioritized for management will depend on the potential for an alien species to invade and its negative impact on the region's agricultural diversity and biodiversity [18,19].

Each plant species has its specific range of cardinal temperatures with base (T_b), optimum (T_{opt}), and maximum (T_{max}) temperatures that determine the geographical limits for growth [20]. The maximum growth and development rate occurs around the optimum temperature (T_{opt}) range [21–23]. The base and maximum temperatures are the lowest and highest temperatures, respectively, at which a plant is able to grow [24] and germinate [25]. Cardinal temperatures can be estimated from data on plant development—which is primarily a temperature-dependent process—by conducting germination tests under experimental conditions within a range of constant temperatures. Various non-linear functions are applied to describe specific ranges of cardinal temperatures, which vary in terms of simplicity and realistic description [26]. Among these functions are Dent, segmented and beta models, which have been frequently used to estimate the cardinal temperatures of germination in numerous crops and weeds such as safflower (*Carthamus tinctorius* L.) [27], Asian spiderflower (*Cleome viscosa* L.) [28]; purple nutsedge (*Cyperus rotundus* L.) [29]; sea barley (*Hordeum marinum* Huds.) [30]; etc.

The response of germination to temperature depends on various factors such as the plant's species, variety, growth environment, or origin [26,31,32]. Understanding how alien plants respond to temperature is crucial for detecting their competitive ability and their response to climate change [33], which may differ from that of native species [34,35]. A substantial body of research has demonstrated that the cardinal temperature exhibits considerable variation across different populations. However, there is currently a paucity of information regarding the full extent of this variation. Moreover, no research has been conducted to ascertain the most influential factors responsible for fluctuations within the cardinal temperatures of germination. Consequently, there is a need to determine the impact of various factors on the three cardinal temperature components. Thus, the objective of this study was to evaluate the effect of the species, ecotypes, origin (native/alien), year, and life cycle of plant species on the variation in the base, optimum, and maximum seed temperatures.

2. Results

2.1. Description of Dataset

From 84 selected records, 46 ecotypes and 31 species were obtained, which belonged to 14 families. Most of the records were associated with the Poaceae, Asteraceae, Brassicaceae, and Plantaginaceae families. Of the 31 different species studied, 8 species were alien (19 records) and belonged to five families (Supplementary Data). The species studied were distributed almost all over Iran, with most of them occurring in the northern and northeastern parts of the country (Figure 1).

2.2. The lm Procedure Output

The lm procedure (a two-way ANOVA linear model) showed that some factors are more important than others with regard to the variation in cardinal temperatures. Being of native or alien origin had a significant impact on the variation in base, optimum, and maximum temperatures. All the factors, apart from the life cycle, significantly affected the base and optimum temperatures. Our results revealed that the base temperature is less influenced by ecotypes than by the optimum and maximum temperatures. The maximum temperature was not affected by year despite the base and optimum temperatures (Table 1). Alien species had higher base temperatures (mean = 10.6 °C) than the native ones (mean = 5.51 °C), and the average optimum and maximum temperatures were around 31 and 45 °C, respectively. Alien plants were also subject to narrow optimum and maximum temperatures in comparison with the native species (Figure 2).

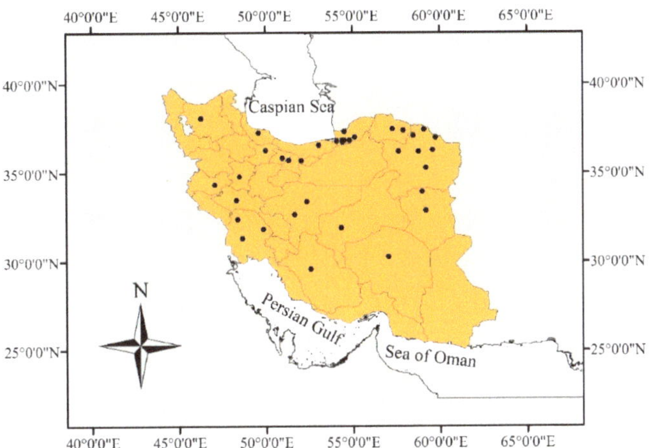

Figure 1. The dots show distribution of the examined populations in Iran.

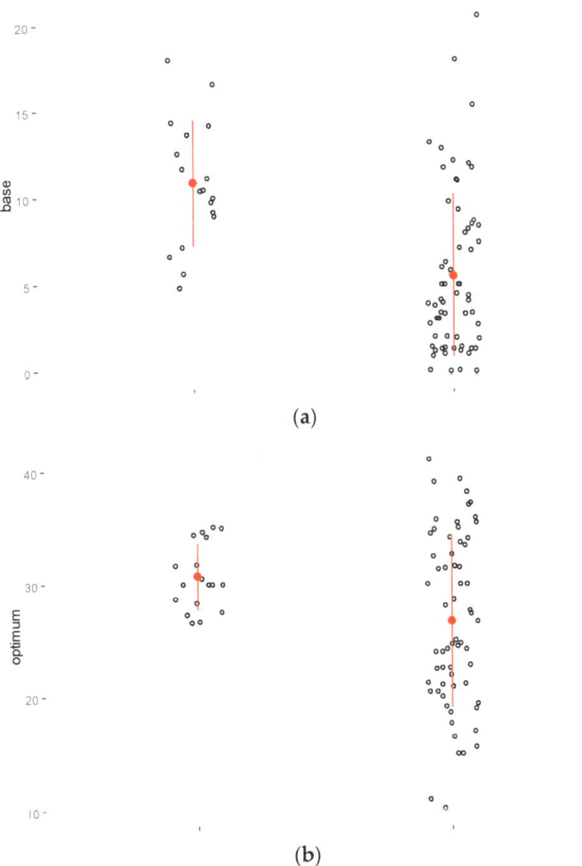

Figure 2. *Cont.*

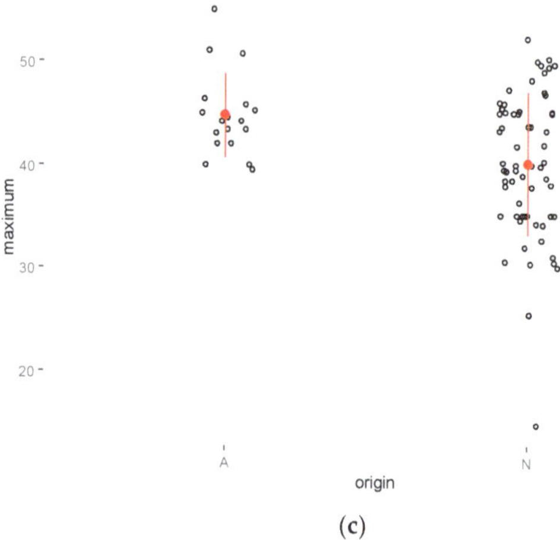

(c)

Figure 2. The effect of origin on the variation in the base (**a**) (mean = 5.51 for native plants (N) and 10.6 for alien plants; *p*-value: <0.001), optimum (**b**) (mean = 26.7 for native plants and 30.9 for alien plants; *p*-value: 0.003), and maximum (**c**) temperature (mean = 40 for native plants and 44.9 for alien plants; *p*-value: <0.001).

Table 1. The lm procedure (a two-way ANOVA linear model) of the effect of different factors on three components of cardinal temperature.

Cardinal Temperature	Parameter	Df	Sum_Squares	Mean_Square	*p*-Value
	Origin	1	416.11	416.11	<0.001 **
	Life cycle	2	37.13	18.56	0.1
Base	Populations	45	703.29	15.63	0.063 ·
	Species	24	705.81	4.64	0.008 **
	Year	1	89.04	89.04	0.004 **
	Origin	1	276.15	276.15	0.003 **
	Life cycle	2	106.78	53.39	0.098 *
Optimum	Populations	45	2064.29	45.87	0.058 ·
	Species	24	1132.21	47.18	0.058 ·
	Year	1	326.2	326.2	0.002 **
	Origin	1	320.81	320.81	<0.001 **
	Life cycle	2	215.53	107.76	0.009 **
Maximum	Populations	45	1594.85	35.44	0.054 ·
	Species	24	1361.32	56.72	0.012 *
	Year	1	40.92	40.92	0.115

**, *, and · are significant at 0.001, 0.01, and 0.05, respectively.

2.3. Effect Size of Factors

The effect size in ANOVA measured the degree of association between the effect and the dependent variable. The interaction between the species and base, optimum, and maximum temperatures accounted for 40, 31, and 38% of the total variability, respectively. The effect of ecotypes on the variation in the base, optimum, and maximum temperatures explained 32, 51, and 46% of the total variability, respectively. Being native or alien had the most effect on the base temperature (21%), whereas its effect on the optimum and maximum temperatures was around 7 to 9%, respectively. The effect of the year of study on the optimum temperature was higher than that on the base and maximum temperatures (Figure 3). In this study, the base and maximum temperatures were more affected by the species type and not by the ecotypes. The latter had a greater effect on the variation in the optimum temperature. In general, the year and life cycle had a lesser influence on the variation in cardinal temperature (Figure 3).

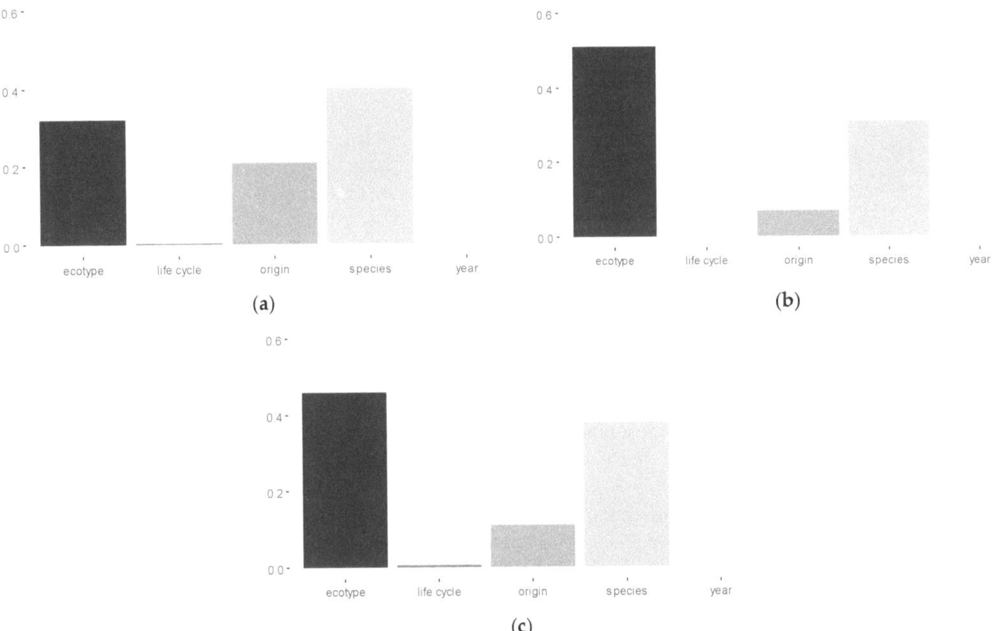

Figure 3. Relative effect sizes (Eta-squared) for the base (**a**), optimum (**b**), and maximum (**c**) temperatures by different factors.

3. Discussion

The species most represented in this study was related to two families with the largest number of plant species. Asteraceae and Poaceae contain the highest number of weedy species of all the flora of many other countries (https://powo.science.kew.org) Our result showed that the base temperature was affected by all the factors. Other studies have reported positive correlations between the base temperature and other germination traits [36]. Plant species can alter the base, optimum, and maximum temperature to cope with unpredictable conditions, in harmony with efficient adaptation strategies [37]. The genetic diversity of species, biodiversity level of the region, seed age, and climate conditions of the maternal plant have an effective role in their germination time and in the consequences of shaping a community assembly [38–41]. Species that detect and respond more quickly to environmental changes at ecological and genetic levels through adaptation are expected to have a selective advantage over species that respond more slowly [3]. Plant species' response to environmental conditions changes over time and, being more acknowledged for the range of their response changes, will be necessary to predict their distribution and impact on different ecosystem levels [42–44]. The duration of the growing season and seed size had a considerable impact on the optimum temperature of *Nigella sativa* L. [45] and *Trigonella foenum-graecum* L. [32]. The effect of the year on the variation in cardinal temperatures can be attributed to the effect of the region's annual rainfall, temperature, and amount of sunshine. Plant species that grow under optimum conditions will have a larger seed size and germinate faster, which influences cardinal temperatures [46–48]. The seed mass and germination were improved under a favorable habitat´s quality in different populations of *Scorzonera hispanica* L. [49]. The tolerance ranges (i.e., ecological valencies) to various environmental factors from the natural selection of plants will be important to predict their response in different locations, especially in introduced plants and their subsequent distribution [15,50]. More studies about different populations' responses will be helpful to draw accurate conclusions about plants' performance.

The wider base, optimum, and maximum temperature range in native species can be attributed to their greater genetic diversity in their native area. The lesser trait similarity between native and alien plants could lead to significant impacts of alien plants on the communities that they invade [51,52]. Alien plants had a higher base temperature than native plants, which suggests that growing in new regions with warmer temperatures may be more beneficial for alien species. Some alien plants are more competitive under climate change conditions due to their rapid establishment and growth in warmer temperatures [53]. In general, the variation in the cardinal temperature of native plants is higher than that of alien ones. The future flora of plant communities could be modified by the different responses of alien and native species to their germination stage [33,35]. Accordingly, Trotta et al. [33] reported that warmer temperatures favor the germination of alien plants rather than that of native species, so they may be more prone to emerge in summer crops as weeds. In Iran, 50% of alien plants grow in agricultural habitats, so that awareness of their response could help to develop robust management tactics [54]. Widespread invasive plants have been able to sense the changes in the climate and respond to them more rapidly via plastic and/or adaptive changes. Therefore, invasive species are predicted to have an advantage over slow-responding plants [55–57]. The prolific growth and worldwide distribution of *Amaranthus retroflexus* L. is related to its high invasion potential [6]. As the establishment and spread of plant species in their native or alien geographic areas are determined by germination as a key process and mechanism [58], finding out the timing of germination will be essential to estimate the growth, maturity, and seed production of the plants [59]. The increasing pressure from biological invasions on ecosystems, intensified by the effects of climate change, requires the swift development of robust and effective management strategies to control invasive species [60,61]. Invasive species, particularly weeds, represent a significant risk to agricultural productivity, biodiversity, and ecosystem services. It is therefore essential to adopt a comprehensive and adaptive approach to weed management, integrating multiple control techniques [62]. Preventive management is a vital tool in reducing the risk of new invasive species introductions, as well as in mitigating further detrimental impacts once they are established [63]. Preventative measures may include the introduction of stricter phytosanitary regulations, early detection and rapid response systems, and the use of certified weed-free seeds and planting materials. Furthermore, public awareness campaigns and education on the risks associated with invasive species are essential for reducing the likelihood of unintentional introductions through human activities.

New weed control tools are being developed that focus on species-specific characteristics and the ecological dynamics of invasion. These tools are particularly valuable for addressing alien species in the early "introduction" phase of invasion, where containment and localized eradication efforts are more feasible [64,65]. For example, early intervention techniques can prevent invasive weeds from becoming established and spreading across larger areas, thereby reducing long-term management costs and minimizing ecosystem disruption.

Control practices targeting weed seedbanks can be highly effective in limiting the persistence and spread of invasive species. One such method involves manipulating soil conditions, such as by lowering soil temperature, to delay seed germination. By extending the dormancy period, seeds are exposed to a longer period of vulnerability, increasing the likelihood of predation by natural enemies such as ants, beetles, and other seed predators [66]. This strategy not only increases seed mortality but also reduces the overall seedbank density, thereby limiting future weed infestations. Another successful approach is the use of weed emergence models [67]. These models are based on environmental factors, such as soil temperature, which influence weed seed dormancy and germination. Gaining insight into the factors influencing variations in cardinal temperatures allows the creation of more precise models and the implementation of timely control interventions, such as herbicide application or mechanical control. This significantly improves the efficiency of invasive species management. By accurately identifying the ideal environmental conditions for germination and growth, control measures can be better timed, ensuring maximum effectiveness and reducing the impact of invasive species on ecosystems and crops.

The effect size result demonstrated the strength of the relationship between plant origin and base temperature. Likewise, the ecotypes influenced the optimum temperature and the species influenced the base temperature. Effect sizes are the most important outcome of empirical studies and can show the magnitude of the reported effects [68,69]. Differential associations of factors in plant response to water stress were represented by effect size [70]. The current evaluations of the variation in cardinal temperature have focused on germination changes; however, increases in minimum temperature may be more significant in their effect on growth and phenology [71]. These findings could be further explored with different settings and more species in different communities and countries, adding depth to our understanding of the alien plant's response to temperature.

We hope the aforementioned alien weeds garner significant attention from policy-makers and prompt the necessary management actions in Iran, particularly given their advantage in thriving under warmer temperatures. Climate change is creating increasingly favorable conditions for these invasive species, allowing them to expand their range, out-compete native flora, and disrupt agricultural systems. The rise in temperatures not only accelerates the growth and reproduction of these weeds but also extends their growing seasons, making them more resilient to traditional control methods.

4. Materials and Methods

4.1. Data Collection

The database employed in this study was taken from peer-reviewed publications in the English and Persian languages (Supplementary Data) via the Web of Science, Google Scholar, Iranian journals, and congress proceedings. The search encompassed literature related to Iranian plant species (both native and alien) and cardinal temperature from 2020 to 2024, using the following terms: "cardinal temperature" or "weed germination", or "plant germination" or "germination temperature" or "germination range" or "base temperature" or "optimum temperature". The following criteria were applied for this investigation: (1) the outputs of the segmented model, including base, optimum, and maximum temperature, (2) the year (of seed collection), (3) the location of seed collection, (4) the origin of species selected (native or alien) The criteria mentioned above were extracted when there were at least two studies on the selected species (Table 2). We screened 84 records (31 different species) by using the above criteria (Supplementary Data). The determination of the origin (native or alien) and nomenclature of species were based on the POWO (Plants of the World Online) database (https://powo.science.kew.org). The location of the study was used to determine the ecotype as a population (or subspecies or race) that is adapted to local environmental conditions. We used the segmented model for the greater availability of information on the species in Iran [72]. A list of the species studied in this paper is presented in Table 2.

Table 2. The list of the species studied in this paper and their main characteristics.

Scientific Name	Family	Common Name	Life Cycle	Number of Ecotypes	Native Geographical Distribution
Abutilon theophrasti Medik.	Malvaceae	Velvetleaf	Annual	2	Central Asia to China
Amaranthus blitoides S.Wats.	Amaranthaceae	Prostrate pigweed	Annual	2	Central and E. Central U.S.A.
Amaranthus retroflexus L.	Amaranthaceae	Redroot pigweed	Annual	3	Mexico
Amaranthus viridis L.	Amaranthaceae	Slender amaranth	Annual	2	SE. Mexico to Tropical America
Bassia scoparia (L.) Beck	Amaranthaceae	Mexican fireweed	Annual	2	E. Europe to Temp. Asia
Carthamus tinctorius L.	Asteraceae	Safflower	Annual	3	Central and E. Türkiye to Iran
Cleome viscosa L.	Cleomaceae	Asian spiderflower	Annual	2	Tropical and Subtropical Old World
Cucumis melo L. subsp. *agrestis* var. *agrestis* (Naudin) Pangalo	Cucurbitaceae	Wild melon	Annual	2	Ethiopia to S. Africa, SW. Syria to Arabian Peninsula and Indian Subcontinent, New Guinea to N. and Central Australia.
Cynanchum acutum L.	Apocynaceae	Stranglewort	Perennial	3	Canary Islands (Lanzarote), Medit. to Siberia and N. China.
Echinochloa crus-galli (L.) P.Beauv.	Poaceae	Barnyard grass	Annual	2	S. and E. Europe to Asia, W., E., and S. Tropical Africa to S. Africa, Madagascar.
Eruca sativa (L.) Cav.	Brassicaceae	Rocket	Annual	2	Medit. to China and Arabian Peninsula
Euphorbia maculata L.	Euphorbiaceae	Spotted spurge	Annual	2	SE. Canada to Belize, Cuba, Bahamas
Hordeum murinum L.	Poaceae	Wall barley	Annual	2	Macaronesia, Europe, Medit. to Central Asia and W. Himalaya
Hordeum vulgare subsp. *spontaneum* (K.Koch) Asch. and Graebn.	Poaceae	Spontaneous barley	Annual	2	E. Medit. to Central Asia and China (Sichuan, Yunnan)
Ipomoea nil (L.) Roth	Convolvulaceae	Ivy morning glory	Annual	2	Tropical and Subtropical America
Ipomoea purpurea (L.) Roth	Convolvulaceae	Common morning glory	Annual	2	Tropical and Subtropical America
Nigella sativa L.	Ranunculaceae	Black cumin	Annual	3	Romania to W. and SW. Iran
Papaver somniferum L.	Papaveraceae	Opium poppy	Annual	2	Macaronesia, W. and Central Medit.
Phalaris minor Retz.	Poaceae	Little seed canary grass	Annual	3	Macaronesia, Medit. to Himalaya and Eritrea
Plantago major L.	Plantaginaceae	Greater plantain	Perennial	6	Temp. Eurasia to Arabian Peninsula, Macaronesia, N. and S. Africa
Plantago ovata Forssk.	Plantaginaceae	Blond plantain	Annual	3	SE. Spain, N. Africa to India and Somalia, SW. and S. Central U.S.A. to N. Mexico
Polygonum aviculare L.	Polygonaceae	Prostrate knotweed	Annual	2	Temp. Northern Hemisphere, Macaronesia to Eritrea
Portulaca oleracea L.	Portulacaceae	Common purslane	Annual	2	Macaronesia, Tropical Africa, Medit. to Pakistan and Arabian Peninsula.
Prosopis farcta (Banks and Sol.) J.F.Macbr.	Caesalpinioideae	Syrian mesquite	Perennial	4	N. Africa to Central Asia and India
Rapistrum rugosum (L.) All.	Brassicaceae	Turnipweed	Annual	2	Macaronesia, Medit. to Central Asia and Iran, NE. Tropical Africa to Arabian Peninsula

Table 2. *Cont.*

Scientific Name	Family	Common Name	Life Cycle	Number of Ecotypes	Native Geographical Distribution
Setaria italica (L.) P. Beauvois	Poaceae	Foxtail millet	Annual	2	China
Silybum marianum (L.) Gaertn.	Asteraceae	Milk thistle	Biennial	5	Macaronesia, Medit. to Central Asia and India, Ethiopia
Sinapis arvensis L.	Brassicaceae	Wild mustard	Annual	3	Temp. Eurasia, N. Africa to Arabian Peninsula
Sisymbrium irio L.	Brassicaceae	London rocket	Annual	2	Europe to N. China and Himalaya, Sahara to N. and NE. Tropical Africa, Arabian Peninsula
Trigonella foenum graecum L.	Fabaceae	Fenugreek	Perennial	7	Iraq to N. Pakistan
Xanthium strumarium L.	Asteraceae	Rough cocklebur	Annual	3	S. Central and S. Europe to China and Indochina, Taiwan, NW. Africa

4.2. Statistical Analyses

The statistical analysis involved the linear model (lm) procedure and the calculation of effect size using Eta-squared (η^2). We employed two-way analysis of variance (ANOVA) at a 0.05 significance level to assess the differences among factors concerning the three key temperature parameters. In the linear model, we focused on explaining the variance attributed to each model term, which facilitated the prediction and elucidation of variability among the factors analyzed through ANOVA.

To quantify the proportion of total variance in the dependent variable associated with the factors, we calculated effect sizes for the ANOVAs. The effect size reflects the strength of the association between the factors and the dependent variable, measured specifically by Eta-squared (η^2), defined as $\eta^2 = SSeffect/SStotal$. This calculation was implemented using the eta_squared function from the effectsize library [69]. For each factor, we utilized the model to examine interactions with reference to the base, optimum, and maximum temperatures. Data processing and statistical analyses were conducted using R version 4.3.0 beta (R Core Team), with the effect size and ggplot2 (v. 3.4.2) packages employed for enhanced data visualization.

5. Conclusions

This study provided the first comparison of the effect of different factors on three components of cardinal temperature and showed that the factors examined had significant effects apart from the life cycle of the species. Variations in base temperature were more affected by the plant's species, while optimum temperature was more influenced by its ecotype. Our results also demonstrated that the base temperature was more affected by the plant's origin, so that alien plants preferred a higher and narrower range of base, optimum, and maximum temperatures than the native ones. From this evidence, it was elucidated that alien plants can benefit more under warmer conditions in invaded areas and may be more problematic in summer crops. These findings lay the foundation for carrying out subsequent studies with broader species ranges of native and alien plants. The comparison of the response of these plants to environmental conditions will be important for predicting their impact on plant communities and improving management programs, as well as how they relate to climate patterns.

Supplementary Materials: The following supporting information can be downloaded at https://www.mdpi.com/article/10.3390/plants13202848/s1.

Author Contributions: Conceptualization: S.S., J.G. and A.S.; methodology: S.S. and A.S.; formal analysis: S.S.; writing–review and editing: S.S., J.G., A.S., S.H.-b. and J.L.G.-A. All authors have read and agreed to the published version of the manuscript.

Funding: This work was funded by the Gorgan University of Agricultural Sciences and Natural Resources (GUASNR), Iran (project no. 02-490-49).

Data Availability Statement: The original contributions presented in this study are included in the Supplementary Materials; further inquiries can be directed to the corresponding author.

Conflicts of Interest: The authors declare no conflicts of interest.

References

1. Dubey, P.S.; Mall, L.P. Ecology of germination of weed seeds. *Oecologia* **1972**, *10*, 105–110. [CrossRef] [PubMed]
2. Rajjou, L.; Duval, M.; Gallardo, K.; Catusse, J.; Bally, J.; Job, C.; Job, D. Seed germination and vigor. *Annu. Rev. Plant Biol.* **2012**, *63*, 507–533. [CrossRef] [PubMed]
3. Gioria, M.; Pyšek, P. Early bird catches the worm: Germination as a critical step in plant invasion. *Biol. Invasions* **2017**, *19*, 1055–1080. [CrossRef]
4. Kader, M.A.; Jutzi, S.C. Effects of thermal and salt treatments during imbibition on germination and seedling growth of sorghum at 42/19 °C. *J. Agron. Crop Sci.* **2004**, *190*, 35–38. [CrossRef]
5. Schlaepfer, D.R.; Glättli, M.; Fischer, M.; van Kleunen, M. A multi-species experiment in their native range indicates pre-adaptation of invasive alien plant species. *New Phytol.* **2010**, *185*, 1087–1099. [CrossRef]

6. Hao, J.-H.; Lv, S.-S.; Bhattacharya, S.; Fu, J.-G. Germination response of four alien congeneric Amaranthus species to environmental factors. *PLoS ONE* **2017**, *12*, e0170297. [CrossRef]
7. Ozaslan, C.; Farooq, S.; Onen, H.; Ozcan, S.; Bukun, B.; Gunal, H. Germination biology of two invasive *Physalis* species and implications for their management in arid and semi-arid regions. *Sci. Rep.* **2017**, *7*, 16960. [CrossRef]
8. Šerá, B.; Žarnovičan, H.; Hodálová, I.; Litavský, J. Reproductive capacity and seed germination after various storage of the invasive alien plant *Amorpha fruticosa* L.—A case study from Bratislava. *Biologia* **2024**, *79*, 1–9. [CrossRef]
9. Baker, H.G. The evolution of weeds. *Annu. Rev. Ecol. Evol. Syst.* **1974**, *5*, 1–24. [CrossRef]
10. Zhang, Q.; Wang, Y.; Weng, Z.; Chen, G.; Peng, C. Adaptation of the invasive plant *Sphagneticola trilobata* (L.) Pruski to drought stress. *Plants* **2024**, *13*, 2207. [CrossRef]
11. Wang, C.; Liu, Y.; Li, C.; Li, Y.; Du, D. The invasive plant *Amaranthus spinosus* L. exhibits a stronger sesistance to drought than the native plant *A. tricolor* L. under co-cultivation conditions when treated with light drought. *Plants* **2024**, *13*, 2251. [CrossRef]
12. Hejda, M.; Pyšek, P.; Jarošík, V. Impact of invasive plants on the species richness, diversity and composition of invaded communities. *J. Ecol.* **2009**, *97*, 393–403. [CrossRef]
13. Parepa, M.; Fischer, M.; Bossdorf, O. Environmental variability promotes plant invasion. *Nat. Commun.* **2013**, *4*, 1604. [CrossRef] [PubMed]
14. Hunter, D.M.; DeBerry, D.A. Environmental drivers of plant invasion in wetland mitigation. *Wetlands* **2023**, *43*, 81. [CrossRef]
15. Holm, L.; Doll, J.; Holm, E.; Pancho, J.; Herberger, J. *World Weeds: Natural Histories and Distribution*; John Wiley and Sons: New York, NY, USA, 1977; p. 1152.
16. Huston, M.A. Management strategies for plant invasions: Manipulating productivity, disturbance, and competition. *Divers. Distrib.* **2004**, *10*, 167–178. [CrossRef]
17. Johnson, L.R.; Handel, S.N. Management intensity steers the long-term fate of ecological restoration in urban woodlands. *Urban For. Urban Green.* **2019**, *41*, 85–92. [CrossRef]
18. Kumschick, S.; Gaertner, M.; Vilà, M.; Essl, F.; Jeschke, J.M.; Pyšek, P.; Ricciardi, A.; Bacher, S.; Blackburn, T.M.; Dick, J.T.A.; et al. Ecological impacts of alien species: Quantification, scope, caveats, and recommendations. *BioScience* **2014**, *65*, 55–63. [CrossRef]
19. Sohrabi, S.; Gherekhloo, J.; Zand, E.; Nezamabadi, N. The necessity of monitoring and assessing alien plants in Iran. *Iran Nat.* **2023**, *8*, 81–90.
20. Soltani, E.; Baskin, C.C.; Gonzalez-Andujar, J.L. An overview of environmental cues that affect germination of nondormant seeds. *Seeds* **2022**, *1*, 146–151. [CrossRef]
21. Yan, W.; Hunt, L.A. An equation for modelling the temperature response of plants using only the cardinal temperatures. *Ann. Bot.* **1999**, *84*, 607–614. [CrossRef]
22. Phartyal, S.S.; Thapliyal, R.C.; Nayal, J.S.; Rawat, M.M.S.; Joshi, G. The influences of temperatures on seed germination rate in Himalayan elm (*Ulmus wallichiana*). *Seed Sci. Technol.* **2003**, *31*, 83–93. [CrossRef]
23. Hatfield, J.L.; Prueger, J.H. Temperature extremes: Effect on plant growth and development. *Weather Clim. Extrem.* **2015**, *10*, 4–10. [CrossRef]
24. Garcia-Huidobro, J.; Monteith, J.L.; Squire, G.R. Time, temperature and germination of pearl millet (*Pennisetum typhoides* S. & H.): I. Constant temperature. *J. Exp. Bot.* **1982**, *33*, 288–296. [CrossRef]
25. Gherekhloo, J.; Sohrabi, S.; Ansari, O.; Bagherani, N.; Prado, R.D. Study of dormancy breaking and factors affecting germination of *Parapholis incurva* (L.) C.E.Hubb. as annual grass. *Taiwania* **2023**, *68*, 185–192. [CrossRef]
26. Andreucci, M.P.; Moot, D.J.; Black, A.D.; Sedcole, R. A comparison of cardinal temperatures estimated by linear and nonlinear models for germination and bulb growth of forage brassicas. *Eur. J. Agron.* **2016**, *81*, 52–63. [CrossRef]
27. Torabi, B.; Attarzadeh, M.; Soltani, A. Germination response to temperature in different safflower (*Carthamus tinctorius*) cultivars. *Seed Technol.* **2013**, *35*, 47–59.
28. Elahifard, E.; Derakhshan, A. Asian spiderflower (*Cleome viscosa*) germination ecology in southern Iran. *Weed Biol. Manag.* **2018**, *18*, 110–117. [CrossRef]
29. Mijani, S.; Rastgoo, M.; Ghanbari, A.; Nassiri Mahallati, M.; González-Andújar, J.L. Development of a new thermal time model for describing tuber sprouting of purple nutsedge (*Cyperus rotundus* L.). *Weed Res.* **2021**, *61*, 431–442. [CrossRef]
30. Taheri, M.; Gherekhloo, J.; Sohrabi, S.; Siahmarguee, A.; Hassanpour-bourkheili, S. Sea barley (*Hordeum Marinum*) seed germination ecology and seedling emergence. *Acta Bot. Hung.* **2024**, *66*, 119–134. [CrossRef]
31. Sohrabi Kertabad, S.; Rashed Mohassel, M.H.; Nasiri Mahalati, M.; Gherekhloo, J. Some biological aspects of the weed Lesser celandine (*Ranunculus ficaria*). *Planta Daninha* **2013**, *31*, 577–585. [CrossRef]
32. Solouki, H.; Kafi, M.; Nabati, J.; Ahmadi-Lahijani, M.J.; Nezami, A.; Ahmady, R.S. Quantifying cardinal temperatures of fenugreek (*Trigonella foenum graecum* L.) ecotypes using non-linear regression models. *J. Appl. Res. Med. Aromat. Plants* **2022**, *31*, 100401. [CrossRef]
33. Trotta, G.; Vuerich, M.; Petrussa, E.; Hay, F.R.; Assolari, S.; Boscutti, F. Germination performance of alien and native species could shape community assembly of temperate grasslands under different temperature scenarios. *Plant Ecol.* **2023**, *224*, 1097–1111. [CrossRef]
34. Gioria, M.; Pyšek, P.; Osborne, B.A. Timing is everything: Does early and late germination favor invasions by herbaceous alien plants? *J. Plant Ecol.* **2016**, *11*, 4–16. [CrossRef]

35. Nešić, M.; Obratov-Petković, D.; Skočajić, D.; Bjedov, I.; Čule, N. Factors affecting seed germination of the invasive species *Symphyotrichum lanceolatum* and their implication for invasion success. *Plants* **2022**, *11*, 969. [CrossRef] [PubMed]
36. Maleki, K.; Soltani, E.; Seal, C.E.; Colville, L.; Pritchard, H.W.; Lamichhane, J.R. The seed germination spectrum of 486 plant species: A global meta-regression and phylogenetic pattern in relation to temperature and water potential. *Agric. For. Meteorol.* **2024**, *346*, 109865. [CrossRef]
37. Donohue, K.; Rubio de Casas, R.; Burghardt, L.; Kovach, K.; Willis, C.G. Germination, postgermination adaptation, and species ecological ranges. *Annu. Rev. Ecol. Evol. Syst.* **2010**, *41*, 293–319. [CrossRef]
38. Colautti, R.I.; Lau, J.A. Contemporary evolution during invasion: Evidence for differentiation, natural selection, and local adaptation. *Mol. Ecol.* **2015**, *24*, 1999–2017. [CrossRef]
39. Dlugosch, K.M.; Anderson, S.R.; Braasch, J.; Cang, F.A.; Gillette, H.D. The devil is in the details: Genetic variation in introduced populations and its contributions to invasion. *Mol. Ecol.* **2015**, *24*, 2095–2111. [CrossRef]
40. Lu, J.J.; Tan, D.Y.; Baskin, C.C.; Baskin, J.M. Effects of germination season on life history traits and on transgenerational plasticity in seed dormancy in a cold desert annual. *Sci. Rep.* **2016**, *6*, 25076. [CrossRef]
41. Moyano, J. Origins of successful invasions. *Nat. Ecol. Evol.* **2023**, *7*, 1583–1584. [CrossRef]
42. Tardieu, F. Plant response to environmental conditions: Assessing potential production, water demand, and negative effects of water deficit. *Front. Physiol.* **2013**, *4*, 17. [CrossRef] [PubMed]
43. Gray, S.B.; Brady, S.M. Plant developmental responses to climate change. *Dev. Biol.* **2016**, *419*, 64–77. [CrossRef] [PubMed]
44. Afuye, G.A.; Kalumba, A.M.; Orimoloye, I.R. Characterisation of vegetation response to climate change: A review. *Sustainability* **2021**, *13*, 7265. [CrossRef]
45. Kiani, M.; Alahdadi, I.; Soltani, E.; Benakashani, F. Changes of seed quality and germination of some black cumin ecotypes (*Nigella sativa* L.) during development and maturity. *J. Plant. Prod. Res.* **2021**, *27*, 227–240. [CrossRef]
46. Peco, B.; Rico, L.; Azcárate, F.M. Seed size and response to rainfall patterns in annual grasslands: 16 years of permanent plot data. *J. Veg. Sci.* **2009**, *20*, 8–16. [CrossRef]
47. Kuniyal, C.P.; Purohit, V.; Butola, J.S.; Sundriyal, R.C. Seed size correlates seedling emergence in *Terminalia bellerica*. *S. Afr. J. Bot.* **2013**, *87*, 92–94. [CrossRef]
48. Martins, A.A.; Opedal, Ø.H.; Armbruster, W.S.; Pélabon, C. Rainfall seasonality predicts the germination behavior of a tropical dry-forest vine. *Ecol. Evol.* **2019**, *9*, 5196–5205. [CrossRef]
49. Münzbergová, Z.; Plačková, I. Seed mass and population characteristics interact to determine performance of Scorzonera hispanica under common garden conditions. *Flora Morphol. Distrib. Funct. Ecol. Plants* **2010**, *205*, 552–559. [CrossRef]
50. Mickelbart, M.V.; Hasegawa, P.M.; Bailey-Serres, J. Genetic mechanisms of abiotic stress tolerance that translate to crop yield stability. *Nat. Rev. Genet.* **2015**, *16*, 237–251. [CrossRef]
51. Skálová, H.; Moravcová, L.; Pyšek, P. Germination dynamics and seedling frost resistance of invasive and native Impatiens species reflect local climatic conditions. *Perspect. Plant Ecol. Evol. Syst.* **2011**, *13*, 173–180. [CrossRef]
52. Hulme, P.E.; Bernard-Verdier, M. Comparing traits of native and alien plants: Can we do better? *Funct. Ecol.* **2018**, *32*, 117–125. [CrossRef]
53. Haeuser, E.; Dawson, W.; van Kleunen, M. Introduced garden plants are strong competitors of native and alien residents under simulated climate change. *J. Ecol.* **2019**, *107*, 1328–1342. [CrossRef]
54. Sohrabi, S.; Vilà, M.; Zand, E.; Gherekhloo, J.; Hassanpour-bourkheili, S. Alien plants of Iran: Impacts, distribution and managements. *Biol. Invasions* **2023**, *25*, 97–114. [CrossRef]
55. Dickson, T.L.; Hopwood, J.L.; Wilsey, B.J. Do priority effects benefit invasive plants more than native plants? An experiment with six grassland species. *Biol. Invasions* **2012**, *14*, 2617–2624. [CrossRef]
56. Fridley, J.D. Extended leaf phenology and the autumn niche in deciduous forest invasions. *Nature* **2012**, *485*, 359–362. [CrossRef]
57. Wainwright, C.E.; Wolkovich, E.M.; Cleland, E.E. Seasonal priority effects: Implications for invasion and restoration in a semi-arid system. *J. Appl. Ecol.* **2012**, *49*, 234–241. [CrossRef]
58. Verdú, M.; Traveset, A. Early emergence enhances plant fitness: A phylogenetically controlled meta-analysis. *Ecology* **2005**, *86*, 1385–1394. [CrossRef]
59. Roché, C.T.; Thill, D.C.; Shafii, B. Reproductive phenology in yellow starthistle (*Centaurea solstitialis*). *Weed Sci.* **2017**, *45*, 763–770. [CrossRef]
60. Sun, Y.; Kaleibar, B.P.; Oveisi, M.; Müller-Schärer, H. Addressing climate change: What can plant invasion science and weed science learn from each other? *Front. Agron.* **2021**, *2*, 626005. [CrossRef]
61. Soler, J.; Izquierdo, J. The invasive *Ailanthus altissima*: A biology, ecology, and control review. *Plants* **2024**, *13*, 931. [CrossRef]
62. Gonzalez-Andujar, J.L. Integrated weed management: A shift towards more sustainable and holistic practices. *Agronomy* **2023**, *13*, 2646. [CrossRef]
63. Gentili, R.; Schaffner, U.; Martinoli, A.; Citterio, S. Invasive alien species and biodiversity: Impacts and management. *Biodiversity* **2021**, *22*, 1–3. [CrossRef]
64. Rask, A.M.; Kristoffersen, P. A review of non-chemical weed control on hard surfaces. *Weed Res.* **2007**, *47*, 370–380. [CrossRef]
65. Udugamasuriyage, D.; Kahandawa, G.; Tennakoon, K.U. Nonchemical aquatic weed control methods: Exploring the efficacy of UV-C radiation as a novel weed control tool. *Plants* **2024**, *13*, 1052. [CrossRef]

6. Bussière, F.; Cellier, P. Modification of the soil temperature and water content regimes by a crop residue mulch: Experiment and modelling. *Agric. For. Meteorol.* **1994**, *68*, 1–28. [CrossRef]
7. Gonzalez-Andujar, J.L.; Chantre, G.R.; Morvillo, C.; Blanco, A.M.; Forcella, F. Predicting field weed emergence with empirical models and soft computing techniques. *Weed Res.* **2016**, *56*, 415–423. [CrossRef]
8. Lakens, D. Calculating and reporting effect sizes to facilitate cumulative science: A practical primer for *t*-tests and ANOVAs. *Front. Psychol.* **2013**, *4*, 863. [CrossRef]
9. Ben-Shachar, M.S.; Lüdecke, D.; Makowski, D. Estimation of effect size indices and standardized parameters. *J. Open Source Softw.* **2020**, *5*, 2815. [CrossRef]
0. Sun, Y.; Wang, C.; Chen, H.Y.H.; Ruan, H. Response of plants to water stress: A meta-analysis. *Front. Plant Sci.* **2020**, *11*, 978. [CrossRef]
1. Khodapanah, G.; Gherekhloo, J.; Sohrabi, S.; Ghaderi-Far, F.; Golmohammadzadeh, S. Phenological response patterns and productive ability of Fallopia convolvulus to weather variability in Iran. *Braz. J. Agric. Sci.* **2023**, *18*, e3211. [CrossRef]
2. Ghaderi-Far, F.; Gorzin, M. *Applied Research in Seed Technology*; Gorgan University of Agricultural Science and Natural Resources Press: Gorgan, Iran, 2019.

Review

The Invasive *Ailanthus altissima*: A Biology, Ecology, and Control Review

Jordi Soler * and Jordi Izquierdo

Department of Agri-Food Engineering and Biotechnology, Universitat Politècnica de Catalunya, 08860 Castelldefels, Spain; jordi.izquierdo@upc.edu
* Correspondence: soleraaa@hotmail.com; Tel.: +34-689460905

Abstract: Tree of Heaven (*Ailanthus altissima* (Mill.) Swingle) is a tree native to China which has invaded disturbed areas in many regions worldwide. Its presence endangers natural ecosystems by displacing native species, modifying habitats, changing community structures, and affecting ecosystem processes. Its invasive nature is enhanced by its high ability to reproduce both vegetatively through root regrowth and sexually through seeds. Seeds, which are wind dispersed, are the main mechanism by which this species reaches new habitats. When they germinate and develop the root system, roots emit new shoots that contribute to a rapid increase in the tree density and the subsequent expansion of the population nearby. The contradictory results about the ecological requirements for seeds to germinate and their degree of dormancy and longevity indicate the complexity and difficulty of understanding the mechanisms that govern the biology and adaptability of this plant. The management of this weed aims at its eradication, with programs based on herbicide applications carried out by injecting the active ingredient directly to the trunk. But, not many active ingredients have shown total control, so new ones should be tested in order to increase the range of available herbicides. During the last few decades, some biological agents have been identified, but their efficacy in controlling the tree and their safety for the local flora have not yet been determined. A correct management strategy should take into account all these aspects in order to contain the expansion of this species and, ultimately, allow its eradication.

Keywords: Tree of Heaven; biological invasion; germination requirements; biological control; chemical control; management

Citation: Soler, J.; Izquierdo, J. The Invasive *Ailanthus altissima*: A Biology, Ecology, and Control Review. *Plants* **2024**, *13*, 931. https://doi.org/10.3390/plants13070931

Academic Editors: Hongli Li and Congyan Wang

Received: 19 February 2024
Revised: 8 March 2024
Accepted: 11 March 2024
Published: 23 March 2024

1. Introduction

Tree of Heaven (*Ailanthus altissima* (Mill.) Swingle is one of the most invasive weeds in temperate climates of the world [1]. It is a dioecious species that belongs to the family Simaroubaceae and to the genus *Ailanthus*, which comprises up to 15 different species, as found in the extensive review by Kowarik [2]. Adults can reach a height of up to 27 m and 1 m diameter at breast height [3]. The leaves are sparse and pinnaticomposite and they can be glabrous or pubescent, with leaf length measuring up to 90 cm [4]. The cotyledons are rounded and with epigeal germination [5]. Male flowers contain 10 stamens, while female flowers have 10 non-functional stamens and a pistil with 5–6 free carpels. The fruits are winged and elongated up to 5 cm in length, samara type, and grouped in leafy and hanging clusters [4]. The blooming period in Mediterranean latitudes occurs during May and July [6]. The diameter at breast height is a good indicator of the amount of seed production [7], with a female adult tree producing up to 300,000 seeds per year [8].

Seeds are mainly dispersed by wind [9], but they can remain in the canopy throughout the winter [5]. The species has a root system with high potential to expand by producing new shoots that contribute to its vegetative dispersion. Most of the roots are found in the first 46 cm of soil [3], with roots sprouts found up to 27 m from the parent tree [10]. More information about the biology of this species can be found in [11].

A. altissima is present in all continents except Antarctica and prefers areas with high human disturbance [2]. The initial ornamental use of this tree in many gardens and backyards favored its spread; however, because of its invasive traits, ornamental uses have decreased, and the origin of new infestations nowadays lies with existing natural populations [11]. Infestations seem to be dependent on the level of disturbance of the invaded area [12], as a positive correlation is found between the presence of *A. altissima* along different parts of the Danube river and areas with dense human population [13]. Additionally, it is not affected by urban pollution [14]. It adapts well to a wide range of soils and prefers warm climes [3], although it is able to establish itself in many different climatic conditions [11]. It is well established in the Mediterranean basin, in which many eradication programs have been carried out [15].

As an invasive plant, *A. altissima* has all the detrimental ecological effects on local ecosystems that have been described for these types of species elsewhere [16,17]. However, it has several particular traits that contribute to generating further damage. The rapid entry into flowering (fourth year), the huge amount of seed production, the wind dispersal, and the vegetative reproduction favor its expansion and enhance its persistency [7]. Its ecological preference for altered and degraded ecosystems and its well-developed root system make this species frequent in communication corridors such as railways, freeways, or walkways [2], in which it damages constructions and pavements and reduces visibility [18], in rural areas along fencerows, woodland edges, or forest openings [2], and in heritage areas such as archaeological monuments, degrading them [19–22]. But, apart from physical damage, this species has allelopathic effects. Ailanthone, a quassinoid compound, has exhibited pre- and postemergence herbicidal activity against other species [11]. It can be found in different parts of the tree and negatively affects the growth of different native species [23], allowing an increase in the presence of other non-native species [24]. Assays with concentrated extracts of ailanthone have shown different effects towards different weeds [25], with dicots being the more affected species [26]. Perhaps this fact may explain why the removal of *A. altissima* in natural areas does not allow native plants to recover until two years later [27]. The modification of bacterial colonies from soil and retardations in the growth of plants under *A. altissima* canopies have also been reported [28].

2. Seed Dormancy

Although *A. altissima* was classified as a species having seeds with non-dormant embryos [29], some years later, dormancy was described [30,31]. The environmental requirements for *A. altissima* seeds to germinate have been shown to be greatly variable according to the bibliography, which suggests that several intrinsic factors, such as embryo immaturity or the presence of inhibitors, must have some influence on the level of dormancy of the seeds [32]. Additionally, the level of dormancy of seeds seems to also be related to the environmental conditions suffered by the parent plant, as great variability was found among the germination rates of seeds collected at the same time from different individuals [7]. In fact, the seeds have to break both the physiological dormancy produced by the presence of inhibiting hormones and the mechanical barrier of the coat that contains the embryo [33].

In order to break the dormancy of *A. altissima* seeds, different methods have been tested. Gibberellic acid [34,35], cold stratification [3,33,36–41], wet or dry stratifications [42,43], and sulfuric acid or boiling water [33] have been applied to the seeds over different periods of time. From the results reported, cold stratification seems to give the highest germination rates. A summary of these methods and the results obtained can be found in Table 1.

Table 1. Summary of germination requirements of *Ailanthus altissima* seeds according to different authors. * ≈ the number is an approximation from a graphic.

Seed Harvest Date	Germination Place	Pre-Treatment	Light/Dark (h)	Temperature (°C)	Seeds per Treatment	Germination (%)	Author
March	Growth chamber	No	12/12	20.5	100	67 to 97	[44]
September	Growth chamber	No	12/12	20.5	150	50 to 90	[44]
November	Greenhouse	No	Natural sunlight	21–24	10	66.1	[45]
December	Greenhouse	No	Natural sunlight	Ambient	25	90.67 to 91.38	[46]
October	Laboratory	Seeds floating on water up to 5 months	Fluorescent lights	22–25	50	94.4	[47]
October	Laboratory	Seeds in leaf litter under forest canopy up to 5 months + 4 °C for 5 weeks in moist sand	Fluorescent lights	22–25	50	78.9	[47]
January	Greenhouse	No	Natural sunlight	15–20	250	52.7	[48]
January	Greenhouse	Floating seeds in water for 3 days 10 days 20 days	Natural sunlight	15–20	250	86.8 58.8 32.4	[48]
January	Greenhouse	Submerged seeds in water for 3 days 10 days 20 days	Natural sunlight	15–20	250	67.7 35.3 30.7	[48]
Not defined	Field	Yes	Sunlight limited at 100, 65, 35, 7%	12.9–46.9	96	* ≈ 80	[49]
Not defined	Greenhouse	0–1.5 °C (no time defined) + water soaked for 48 h + 3–5 °C moist vermiculite for 3 months + Bare soil or low leaf litter or high leaf litter	12/12	20–30	45	≈ 70 ≈ 80 ≈ 50	[50]
October	Growth chamber	Stratification for 8 weeks	Not defined	25	20	60.2	[51]
October	Growth chamber	Stratification for 8 weeks	Not defined	25	20	58.3	[52]
	Growth chamber	Gibberellic acid 40 ppm	Light/dark	30		good	[34]
Not defined	Growth chamber	Stored from 2 to 4 years + 4 °C for one month + Gibberellic acid 500 ppm	Dark	40 °C for 24 h	Not defined	good	[35]

Table 1. *Cont.*

Seed Harvest Date	Germination Place	Pre-Treatment	Light/Dark (h)	Temperature (°C)	Seeds per Treatment	Germination (%)	Author
October	Greenhouse	Seeds incubated under field conditions for 1 to 5 years at: soil depth = 10 cm	Natural sunlight	26.5	30–50	1.9 to 81	[38]
		soil depth = 0 cm				79.4 to 83	
	Greenhouse	1–4 °C for 88 days	Natural sunlight	25/20	50	87	
	Greenhouse	Stored 5 years in lab conditions	Natural sunlight	Not defined	50	83.5	
Late summer	Greenhouse	1.7 °C for 28 days	Natural sunlight	Ambient	depending on source: 40, 43, or 64	0 to 78.1	[7]
December	Laboratory	Seeds incubated under litter and duff layers in field conditions	12/12 Fluorescent lights	18-20	100	28 and 79	[32]
December	Field	1 year with cold moist sand	High flux of sunlight Low flux of sunlight	Ambient	50	≈ 25 ≈ 21	[43]
		3 months with cold moist sand	High flux of sunlight Low flux of sunlight			≈ 15 ≈ 8	
October	Growth chamber	2 months at 17–20 °C	16/8	15 20 30	80	≈ 55 ≈ 25 ≈ 18	[37]
	Growth chamber	Stored 1 year (no treatment) Stored 2 years (no treatment) Stored 3 years (no treatment)	16/8	20	100	12 19 20	
	Field	no	Natural sunlight	Ambient	792	≈25	
Dispersal season	Growth chamber	4 °C more than 1 year	16/8	24/16	125	50.8	[41]
Fall	Growth chamber	4 °C during winter	Not defined	Not defined	Not defined	87	[36]
	Field		Natural sunlight	Ambient	100	≈ 6 to 9	
					80	≈ 6 to 13	

Table 1. *Cont.*

Seed Harvest Date	Germination Place	Pre-Treatment	Light/Dark (h)	Temperature (°C)	Seeds per Treatment	Germination (%)	Author
October	Growth chamber	No	12/12	15/6 20/10 30/20	25	0 71 87	[39]
			24 Dark	15/6 20/10 30/20		0 75 84	
	Growth chamber	4 °C for one month	12/12	15/6 20/10 30/20		19 51 82	
			24 Dark	15/6 20/10 30/20		7 37 89	
October	Growth chamber	Moist at 5 °C for 12 days Dry at 5 °C for 12 days Moist at 25 °C for 12 days Dry at 25 °C for 12 days Control 5 °C for 4 days 5 °C for 12 days	Dark	20(16 h)/ 30(8 h)	30	95 76 84 75 70 77 96	[42]
November	Growth chamber	4 °C for 1 month	Dark	25 25/30 30 40	100 naked embryos	40 73 94 51	[40]
Not defined	Growth chamber	Control Sulfuric acid 95% for 10 min Sulfuric acid 50% for 10 min Hot water 95 °C for 15 min 3 °C for 10 days 3 °C for 15 days 3 °C for 20 days	Not defined	20	100	26 20 60 40 29 32 52	[33]

3. Seed Germination

3.1. Temperature Requirements

The variability of the results found makes setting an optimal temperature for seed germination difficult. In growth chambers under constant temperature, the highest germination rates were obtained at 15 °C and the lowest at 30 °C [37]. Simulating different heat treatments, as forest fires would do, germination decreased with increasing heat temperatures [41]. However, with an alternating temperature regime of 15/6 °C, seeds germinated almost four times less than at an alternant higher temperature of 30/20 °C [39]. If the coat of the seed is removed, the best germinating temperatures for naked embryos were found by alternating 25/30 °C or by a constant 30 °C [40]. In growth chambers at 20.5 °C, germination reached 90 and 97% depending on the place the seeds were collected [44]. At a constant 20 °C, germination rates ranged from 44.4 to 26.2% depending on the intensity of the chamber flux light [43]. In greenhouse conditions with natural light and a temperature ranging between 21 and 24 °C, germination rates were 66.1% [45] or 90.7% and 91.4% depending on the natural stands from which the seeds were collected [46]. A pre-treatment with gibberellic acid showed 30 °C as the optimal temperature [34].

3.2. Light Requirement for Germination and Growth

Although *A. altissima* is considered to be a shade intolerant species [3], it can germinate and grow under a natural forest canopy with low light conditions [53] or be competitive in a closed-canopy forest [54]. Simulating different leaf litter layers over *A. altissima* seeds in the greenhouse, no differences in germination were found, meaning that a lack of direct light it is not the main condition for its germination [50]. Furthermore, measurements of leaf water potential found no differences between trees growing under high irradiance conditions with shaded ones. However, it seems that germination rates are affected by light exposure, because the average time needed for seedlings to emerge was longer when the flux of natural light was reduced, for example by using plastic nets to mimic shadow conditions [49] or positioning under a dense forest canopy [43]. The inhibition effect of the coat was also deduced from the study of [40], in which they found that naked embryos in dark conditions, which do not promote the germination of the seeds, were able to achieve a germination rate of 94%.

Additionally, longer photoperiods allowed seedlings to more quickly develop their vascular system than seedlings growing in completely dark conditions, although the increment was non-significant [55].

3.3. Water Requirements for Germination and Growth

In laboratory conditions and using Polyethylene glycol (PEG) to simulate different water stress conditions, the germination rate of *A. altissima* seeds decreased when reducing the water potential. Germination significantly decreased when water potential decreased from −2 to −4 bar, with almost no germination found at -6 bar and none at all at −8 bar [51,52]. Similar decreasing trends when lowering the water potential were observed by [56], although *A. altissima* seedlings supported better water stress than *Phytolacca americana* and *Robinia pseudoacacia* [57].

Greenhouse experiments on the effect of water availability on plant growth showed that decreasing irrigation regimes (1, 0.25, 0.1, and 0.05 L per week) reduced the leaf and root area of the plants, although the results were not statistically different [58]. With a similar water regime experiment (0.3 and 0.03 L per week) a positive correlation between drought and growth was found, with seedlings having a more reduced growth, height, and dry weight at low water availability [45]. The differences between both studies probably lie in the fact that the first experiment lasted two years while the second lasted only one, suggesting that plants have mechanisms to adapt to a water-scarce environment. In adult plants, the ability of this species to cope with drought may also be related to its ability to take water mainly from deeper soil layers (more than 75 cm) than from the first 25 cm of soil layers [59]. Other authors pointed out that *A. altissima* was more efficient in terms of root-to-leaf water transport capacity than native species [60].

Water exposure seems to affect seed germination. Some studies report increases in the germination rate of seeds exposed to 3 days of water (floating or submerged) compared to 20 days [48], although the opposite trend was found, with no descending rate observed for seeds floating for 5 months [47].

4. Seed Longevity

For species that are reproduced by seeds, the longevity of the seeds is a key factor for determining the persistency of the species in the habitat. *A. altissima* seeds have very low level of predation [12,44] and although some authors found that the longevity of the *A. altissima* seed bank was not significant [8], or not enough to form a long-term seed bank [2,43], more recent studies have shown that the viability of stored seeds can be as long as three [37], five [38], or nine years [7].

5. Seedlings Survival

The survival of *A. altissima* seedlings depends primarily on soil water availability and the competition with the native flora [14]. If the native flora has a very dense canopy, seedlings will hardly survive in such a shady environment, but if the level of disturbance of the canopy forest is significant, more light will reach the understory, more seeds will germinate, and more seedlings will survive [37,43]. Shade and cold conditions act as limiting factors for seedlings' establishment [11].

6. Dispersion

The main aim of any biological dispersal process is to allow the reproductive structures of the plant to reach long distances from the mother plant [48]. The distance is more related to the height of the plant than to the seed mass [54]. In the case of *A. altissima*, long distances from the seed source are mainly achieved by means of the wind [9,47,61] and secondarily by water [3] or animals such as birds or ants [54]. Wind is the primary source of dispersion as the fruit is a samara well-adapted to wind dispersal [3] and fruits may reach distances up to 200 m from the mother plant [62]. The final distance will depend on the wind speed and the orography, because some studies have found shorter distances [12,63,64]. Water is a secondary source of dispersion. Samaras are also adapted to float on water [48,65,66] and can be scattered downstream [47,54,67]. Seeds can remain viable in water for a long time (94.4% germinability after five months) [47].

7. Vegetative Reproduction

Asexual reproduction is an important trait to consider for this species, as new shoots from roots act as a dispersion mechanism [11]. The absence of a taproot is common [3] and a root system that has an asymmetrical shape, adapted to the soil characteristics [2]. New shoots can appear from stumps or roots, and the shapes of leaves vary if they appear from root sprouts (from unifoliolate to pentafoliolatemore or others) or seedlings (trifoliolate) [5]. When aerial parts suffer damage or die, new shoots from the root system appear [3].

8. Management

Different strategies such as chemical, mechanical, and biological or a combination of them have been applied for the management of *A. altissima* trees in natural ecosystems. Many of these strategies showed a good efficacy; however, it is important to point out that due to the high capacity of the plant to resprout from the root system as explained before, any actuation on well-established individuals will need long-term supervision to check the efficacy of the measurement [2,68,69], particularly when female trees are present.

8.1. Mechanical

Mechanical control can be performed by hand or with any tool, but it is only effective against seedlings, because once the root system is established, cutting or breaking the roots will promote resprouting [5], and successive cuts will increase the number of shoots [11]. Mechanical control on established trees showed very weak control of the populations. For example, when comparing mechanical versus chemical, it was found a mortality of 21.3% by manual cutting versus near total control with different herbicides [70]. Other authors have demonstrated that herbicides such as glyphosate, imazapyr, picloram, triclopyr, or 2.4-D had better control than mechanical methods alone [21,71,72]. When comparing cutting versus herbicide application over the cut stump with glyphosate, imazapyr, or triclopyr, trees without herbicide produced more resprouts than trees with herbicide [73]. Similar results were obtained, with a mortality of 52% by only cutting compared to near 90%, when herbicides were applied over the cut stump [74]. A combination of mechanical actuations plus chemical control seems to be the best procedure rather than chemical control alone [68,71].

8.2. Chemical

Herbicides are the most popular method to manage *A. altissima* populations [75]. Systemic herbicides are the most efficient particularly when applied at the end of the growing season because they are transported to the root system via the phloem with the descending movement of the sap [69].

The application of the herbicides on the trees is performed by means of different techniques: stem injection, basal bark, or cut stump. Stem injection is performed by making holes with a drill into the trunk and filling each hole with herbicide, by the E-Z-ject Lance system (injecting into the trunk solid capsules containing herbicide) or by hack-and-squirt which is spraying into cuts performed with a hack along the stem. Basal bark consists of spraying herbicides into the lower part of an uncut trunk. Cut stump consists of spraying or injecting the herbicide on the cut surface of the trunk.

Herbicides have been applied diluted and undiluted, with different results. The best results with the stem injection technique were observed using undiluted glyphosate and making holes [76]. However, undiluted triclopyr applied by hack-and-squirt showed no total effectiveness over the trees [73]. The E-Z-ject Lance system has been tested with triclopyr and imazapyr [77,78], and glyphosate [71,78,79] with varied efficacy.

The efficacy of basal bark applications depended on the diameter of the tree. Diluted triclopyr showed good control in most cases [70,74,77,79], but diluted mixtures of triclopyr + fluroxypyr, aminopyralid + fluroxypyr or glyphosate alone did not show total mortality when applied to trees with bigger diameters [21].

For cut stump applications, mortality seems to depend on the concentration of the active ingredient. Spraying diluted active ingredients alone (i.e., triclopyr or glyphosate) or mixed (i.e., triclopyr + fluroxypyr, aminopyralid + fluroxypyr or glyphosate) did not achieve total mortality of the trees [21,73], while the same active ingredients undiluted reached total control [76]. Another case is granular herbicides like Metsulfuron methyl, where undiluted applications are not possible, which had a great mortality but not all trees died [74].

8.3. Biological

Although *A. altissima* tissues contain chemical compounds that likely act as a natural defense against pests [2], during the last decades many different organisms have been identified as biological agents of *A. altissima* trees, some of them with high specificity. These natural enemies are arthropods and fungus and most of them have been reported in Chinese ecosystems, although lately they have also been reported in the places where *A. altissima* has been introduced, probably due to accidental introductions [80]. Mites have been reported to attack leaves, with the genus *Aculus* spp. being the most mentioned ones [81–84]. Coleoptera such as *Eucryptorrhynchus brandti* and *Eucryptorrhynchus chinensis* have shown good specificity over *A. altissima* in China [80]. These coleoptera also showed good specificity in quarantine trials, preferring this tree over others when feeding at the larval stage and for oviposition [75,85,86] making them a good option for biological control. In Italy, the orange whitefly (*Leurocanthus spiniferus*) has been reported on *A. altissima* for the first time, but this insect cannot be considered a biological agent because the trees tolerated the infestation [87].

Some generalist insects have a range of hosts that include *A. altissima* leaves in their diet, such as the butterflies *Atteva punctella* and *Samia cynthia*, whose host range includes trees from the genus Simarouba [80,88] and the beetle *Maladera castanea* [3]. Regarding biological control, *A. altissima* acts as a host for the invasive pest *Lycorma delicatula* in North America [89].

Fungus has an important role in *A. altissima* biocontrol. Some *Verticilium*, *Alternaria* and *Cercospora* species have shown good results as biocontrol agents [80]. The *Verticilium* species are the most important and they have been identified in many countries like the USA [3,90–95], Austria [96], Italy [97], and Spain [98]. Common symptoms of trees affected by *Verticillium* are premature defoliation, yellowish vascular discoloration, and final mortality [91]. Different strains of *Verticillium* may act depending on the climate, with *V. dahliae* being the most common in warm areas and *V. nonalfae* in cooler regions [96]. Trees can be infected during winter and show the first symptoms the next growing season. Under laboratory conditions, *V. nonalfalfae* was transmitted by *E. brandti*, which carried the propagules of the fungi externally and internally [99].

9. Challenges

A. altissima has a very extensive bibliography involving many other topics such as medicinal properties or phytosanitary activity of some of its components. All this information has not been cited in this review because it is not relative to the invasive aspect of this weed in natural ecosystems. However, from the information reviewed, it appears that there is a need for further study of the behavior of this prolific species. Understanding the mechanisms of seed dormancy, determining the ecological requirements for seeds to germinate, or finding the best herbicide combination to control this weed are some of the aspects that are not well known. Additionally, some challenges derived from its control still have to be addressed. The use of herbicides may provoke soil/water contamination for drift or root exudates. When managing its populations, the vegetal residues of the trees generated should be properly treated to avoid the negative effects of their allelopathic compounds, by converting the residues into mulching in a secure way. The biological control with the fungus *Verticillum* in the invaded areas faces the challenge of the possible effect on native flora. As this tree has been shown to have growth limitations when living in closed-canopy forests, it would be interesting to determine the best planting density of native species in order to deter its establishment. New research studies are needed in order to properly develop successfully management programs aimed to eradicate this weed from our natural systems.

Author Contributions: Conceptualization, J.S.; writing—original draft preparation, J.S.; writing—review and editing, J.S. and J.I.; visualization, J.S. and J.I.; supervision, J.I.; project administration, J.I. All authors have read and agreed to the published version of the manuscript.

Funding: This research received no external funding.

Data Availability Statement: The data presented in this study are available on request from the corresponding author.

Conflicts of Interest: The authors declare no conflicts of interest.

References

1. Weber, E.; Gut, D. Assessing the risk of potentially invasive plant species in central Europe. *J. Nat. Conserv.* **2004**, *12*, 171–179. [CrossRef]

2. Kowarik, I.; Säumel, I. Biological Flora of Central Europe: *Ailanthus altissima* (Mill.) Swingle. *Perspect. Plant Ecol. Evol. Syst.* **2007**, *8*, 207–237. [CrossRef]
3. Miller, J.H. *Ailanthus altissima* (Mill.) Swingle. Ailanthus. *Silv. N. Am.* **1990**, *2*, 101–104.
4. Sànchez-Cuxart, A.; Llistosella, J. *Guia il·Lustrada per a Conèixer els Arbres*; Publicacions i Edicions de la Universitat Barcelona: Barcelona, Spain, 2015.
5. Hu, S.Y. Ailanthus. *Arnoldia* **1979**, *39*, 29–50.
6. Sanz e Lorza, M.; Dana Sánchez, E.D.; Sobrino Vesperinas, E. *Atlas de las Plantas Alóctonas Invasoras en España*; Dirección General para la Biodiversidad: Madrid, Spain, 2004.
7. Wickert, K.L.; O'Neal, E.S.; Davis, D.D.; Asson, M.T. Seed production, viability, and reproductive limits of the invasive *Ailanthus altissima* (Tree-of-Heaven) within invaded environments. *Forests* **2017**, *8*, 226. [CrossRef]
8. Evans, C.W.; Moorhead, D.J.; Bargeron, C.T.; Douce, G.K. *Invasive Plant Responses to Silvicultural Practices in the South*; the University of Georgia Bugwood Network: Tifton, GA, USA, 2006.
9. Bossard, C.C.; Randall, M.J.; Hoshovsky, C.M. *Invasive Plants of California's Wildlands*; Univ of California Press: California, UK, 1957.
10. Howard, J.L. Ailanthus altissima. In *Fire Effects Information System*; US Department of Agriculture, Forest Service, Rocky Mountain Research Station, Fire Sciences Laboratory (Producer): Fort Collins, CO, USA, 2004.
11. Sladonja, B.; Sušek, M.; Guillermic, J. Review on invasive Tree of Heaven (*Ailanthus altissima* (Mill.) Swingle) conflicting values: Assessment of its ecosystem services and potential biological threat. *Environ. Manag.* **2015**, *56*, 1009–1034. [CrossRef] [PubMed]
12. Martin, P.; Canham, C. Dispersal and recruitment limitation in native versus exotic tree species: Life-history strategies and janzen-connell effects. *Oikos* **2010**, *119*, 807–824. [CrossRef]
13. Wagner, S.G.K.; Moser, D.G.K.; Franz Essl, F. Urban rivers as dispersal corridors: Which factors are important for the spread of alien woody species along the Danube? *Sustainability* **2020**, *12*, 2185. [CrossRef]
14. Feret, P.O. Ailanthus: Variation, cultivation, and frustration. *J. Arboric.* **1985**, *11*, 361–368. [CrossRef]
15. Brunel, S.; Brundu, G.; Fried, G. Eradication and control of Invasive Alien Plants in the Mediterranean Basin: Towards Better Coordination to Enhance Existing Initiatives. *Bull. OEPP/EPPO Bull.* **2013**, *43*, 290–308. [CrossRef]
16. Andersen, M.C.; Adams, H.; Hope, B.; Powell, M. Risk assessment for invasive species. *Risk Anal.* **2004**, *24*, 787–793. [CrossRef] [PubMed]
17. Vilà, M.; Ibáñez, I. Plant invasions in the landscape. *Landsc. Ecol.* **2011**, *26*, 461–472. [CrossRef]
18. Casella, F.; Vurro, M. *Ailanthus altissima* (Tree of heaven): Spread and harmfulness in a case-study urban area. *Arboric. J.* **2013**, *35*, 172–181. [CrossRef]
19. Celesti-Grapow, L.; Blasi, C. The role of alien and native weeds in the deterioration of archaeological remains in Italy. *Weed Technol.* **2004**, *18*, 1508–1513.
20. Trotta, G.; Savo, V.; Cicinelli, E.; Carboni, M.; Caneva, G. Colonization and damages of *Ailanthus altissima* (Mill.) Swingle on archaeological structures: Evidence from the Aurelian Walls in Rome (Italy). *Int. Biodeterior. Biodegrad.* **2020**, *153*, 105054. [CrossRef]
21. Fogliatto, S.; Milan, M.; Vidotto, F. Control of *Ailanthus altissima* using cut stump and basal bark herbicide applications in an eighteenth-century fortress. *Weed Res.* **2020**, *60*, 125–434. [CrossRef]
22. Celesti-Grapow, L.; Ricotta, C. Plant invasion as an emerging challenge for the conservation of heritage sites: The spread of ornamental trees on ancient monuments in Rome, Italy. *Biol. Invasions* **2021**, *23*, 1191–1206. [CrossRef]
23. Gómez-Aparicio, L.; Canham, C.D. Neighbourhood analyses of the allelopathic effects of the invasive Tree *Ailanthus altissima* in temperate forests. *J. Ecol.* **2008**, *96*, 447–458. [CrossRef]
24. Brooks, R.K.; Barney, J.N.; Salom, S. The Invasive Tree, *Ailanthus altissima*, impacts understory nativity, not seedbank nativity. *For. Ecol. Manag.* **2021**, *489*, 119025. [CrossRef]
25. Novak, M.; Novak, N.; Milinović, B. Differences in allelopathic effect of tree of heaven root extracts and isolated ailanthone on test-species. *J. Cent. Eur. Agric.* **2021**, *22*, 611–622. [CrossRef]
26. Heisey, R.M. Identification of an allelopathic compound from *Ailanthus altissima* (Simaroubaceae) and characterization of its herbicidal activity. *Am. J. Bot.* **1996**, *83*, 192–200. [CrossRef]
27. Terzi, M.; Fontaneto, D.; Casella, F. Effects of *Ailanthus altissima* invasion and removal on high-biodiversity Mediterranean grasslands. *Environ. Manag.* **2021**, *68*, 914–927. [CrossRef]
28. Medina-Villar, S. Impactos ecológicos de los árboles exóticos invasores en la estructura y funcionamiento de los ecosistemas fluviales y de ribera. *Ecosistemas* **2016**, *25*, 116–120. [CrossRef]
29. Barton, L.V. Experiments at boyce Thompson institute on germination and dormancy in seeds. *Sci. Hortic.* **1939**, *7*, 186–193.
30. Little, S. *Ailanthus altissima* Mill. Swingle: Ailanthus. In *Seeds of Woody Plants in the United States*; Schopmeyer, C.S., Ed.; Tech. Coord.; USDA Forest Service: Washington, DC, USA, 1973; pp. 201–202.
31. Dirr, M.A.; Heuser, C.W., Jr. *The Reference Manual of Woody Plant Propagation*; Varsity Press: Athens, GA, USA, 1987.
32. Redwood, M.E.; Matlack, G.R.; Huebner, C.D. Seed longevity and dormancy state in an invasive tree species: *Ailanthus altissima* (Simaroubaceae). *J. Torrey Bot. Soc.* **2019**, *146*, 79–86. [CrossRef]
33. Deltalab, B.; Naziri Moghaddam, N.; Khorrami Raad, M.; Kaviani, B. The effect of cold and acid scarification on seed germination of three green space tree species. *J. Ornam. Plants* **2023**, *13*, 85–97.

34. Jian, Z.; Shouhua, G.; Yu, S.; Yong-qi, Z.; Xiao-yan, Y.; Lin, Y. The operational seed germination conditions of *Ailanthus altissima*. *Acta Bot. Boreali-Occident. Sin.* **2007**, *5*, 1030–1034.

35. Bao, Z.; Nilsen, E.T. Interactions between seedlings of the invasive tree *Ailanthus altissima* and the native tree *Robinia pseudoacacia* under low nutrient conditions. *J. Plant Interact.* **2015**, *10*, 173–184. [CrossRef]

36. Facelli, J.M. Multiple indirect effects of plant litter affect the establishment of woody seedlings in old fields. *Ecology* **1994**, *75*, 1727–1735. [CrossRef]

37. Constán-Nava, S.; Bonet, A. Genetic Variability Modulates the effect of habitat type and environmental conditions on early invasion success of *Ailanthus altissima* in Mediterranean ecosystems. *Biol. Invasions* **2012**, *14*, 2379–2392. [CrossRef]

38. Rebbeck, J.; Jolliff, J. How long do seeds of the invasive tree, *Ailanthus altissima* remain viable? *For. Ecol. Manag.* **2018**, *429*, 175–179. [CrossRef]

39. Pepe, M.; Gratani, L.; Fabrini, G.; Arone, L. Seed germination traits of *Ailanthus altissima*, *Phytolacca americana* and *Robinia pseudoacacia* in response to different thermal and light requirements. *Plant Species Biol.* **2020**, *35*, 300–314. [CrossRef]

40. Kheloufi, A.; Mansouri, L.M.; Zerrouni, R.; Abdelhamid, O. Effect of temperature and salinity on germination and seedling establishment of *Ailanthus altissima* (Mill.) Swingle (*Simaroubaceae*). *Reforesta* **2020**, *9*, 44–53. [CrossRef]

41. Cruz, O.; Riveiro, S.F.; Arán, D.; Bernal, J.; Casal, M.; Reyes, O. Germinative behaviour of *Acacia dealbata* Link, *Ailanthus altissima* (Mill.) Swingle and *Robinia pseudoacacia* L. in relation to fire and exploration of the regenerative niche of native species for the control of invaders. *Glob. Ecol. Conserv.* **2021**, *31*, e01811. [CrossRef]

42. Graves, W.R. Stratification not required for Tree-of-Heaven seed germination. *Tree Plant. Notes* **1990**, *41*, 1012.

43. Kota, N.L.; Landenberger, R.E.; McGraw, J.B. Germination and early growth of Ailanthus and tulip poplar in three levels of forest disturbance. *Biol. Invasions* **2007**, *9*, 197–211. [CrossRef]

44. Cabra-Rivas, I.; Castro-Díez, P. Potential Germination success of exotic and native trees coexisting in central Spain riparian forests. *Int. J. Ecol.* **2016**, *2016*, 7614683.

45. Stevens, M.T.; Roush, C.D.; Chaney, L. Summer Drought Reduces the Growth of Invasive Tree-of-Heaven (*Ailanthus altissima*) seedlings. *Nat. Areas J.* **2018**, *38*, 230–236. [CrossRef]

46. Delgado, J.A.; Jimenez, M.D.; Gomez, A. Samara size versus dispersal and seedling establishment in *Ailanthus altissima* (Miller) Swingle. *J. Environ.* **2009**, *30*, 183–186.

47. Kaproth, M.; McGraw, J. Seed viability and dispersal of the wind-dispersed invasive *Ailanthus altissima* in aqueous environments. *For. Sci.* **2008**, *54*, 490–496.

48. Kowarik, I.; Säumel, I. Water dispersal as an additional pathway to invasions by the primarily wind-dispersed tree *Ailanthus altissima*. *Plant Ecol.* **2008**, *198*, 241–252. [CrossRef]

49. González-Muñoz, N.; Castro-Díez, P.; Fierro-Brunnenmeister, N. Establishment success of coexisting native and exotic trees under an experimental gradient of irradiance and soil moisture. *Environ. Manag.* **2011**, *48*, 764–773. [CrossRef] [PubMed]

50. Kostel-Hughes, F.; Young, T.P.; Wehr, J.D. Effects of leaf litter depth on the emergence and seedling growth of deciduous forest tree species in relation to seed size. *J. Torrey Bot. Soc.* **2005**, *132*, 50–61. [CrossRef]

51. Yigit, N.; Sevik, H.; Cetin, M.; Kaya, N. Determination of the Effect of Drought Stress on the Seed Germination in Some Plant Species. Doctoral Dissertation, InTech, London, UK, 2016.

52. Sevik, H.; Cetin, M. Effects of water stress on seed germination for select landscape plants. *Pol. J. Environ. Stud.* **2015**, *24*, 689–693. [CrossRef] [PubMed]

53. Knüsel, S.; De Boni, A.; Conedera, M.; Schleppi, P.; Thormann, J.J.; Frehner, M.; Wunder, J. Shade tolerance of *Ailanthus altissima* revisited: Novel insights from southern Switzerland. *Biol. Invasions* **2017**, *19*, 455–461. [CrossRef]

54. Thomson, F.J.; Moles, A.T.; Auld, T.D. Kingsford, R.T. Seed dispersal distance is more strongly correlated with plant height than with seed mass. *J. Ecol.* **2011**, *99*, 1299–1307. [CrossRef]

55. Borger, G.A.; Kozlowski, T.T. Effects of photoperiod on early periderm and xylem development in *Fraxinus pennsylvanica*, *Robinia pseudoacacia* and *Ailanthus altissima* seedlings. *New Phytol.* **1972**, *71*, 703–708.

56. Song, L.; Wenwen, L.; Shufen, C. Effect of PEG on seed germination of *Ailanthus altissima*. *J. Ningxia Agric. Coll.* **2005**, *4*, 25–29.

57. Pepe, M.; Crescente, M.F.; Varone, L. Effect of Water Stress on Physiological and Morphological Leaf Traits: A Comparison among the Three Widely-Spread Invasive Alien Species *Ailanthus altissima*, *Phytolacca americana*, and *Robinia pseudoacacia*. *Plants* **2022**, *11*, 899. [CrossRef]

58. Trifilò, P.; Raimondo, F.; Nardini, A.; Lo Gullo, M.A.; and Salleo, S. Drought resistance of *Ailanthus altissima*: Root hydraulics and water relations. *Tree Physiol.* **2004**, *24*, 107–114. [CrossRef]

59. Granda, E.; Antunes, C.; Máguas, C.; Castro-Díez, P. Water use partitioning of native and non-native tree species in riparian ecosystems under contrasting climatic conditions. *Funct. Ecol.* **2022**, *36*, 2480–2492. [CrossRef]

60. Petruzzellis, F.; Nardini, A.; Savi, T.; Tonet, V.; Castello, M.; and Bacaro, G. Less safety for more efficiency: Water relations and hydraulics of the invasive tree *Ailanthus altissima* (Mill.) Swingle compared with native *Fraxinus ornus* L. *Tree Physiol.* **2019**, *39*, 76–87. [CrossRef] [PubMed]

61. Planchuelo, G.; Catalán, P.; Delgado, J.A. Gone with the wind and the stream: Dispersal in the invasive species *Ailanthus Altissima*. *Acta Oecologica* **2016**, *73*, 31–37. [CrossRef]

62. Kota, N.L. Comparative Seed Dispersal, Seedling Establishment and Growth of Exotic, Invasive Ailanthus Altissima (Mill.) Swingle and Native Liriodendron tulipifera (L.). Graduate Theses, West Virginia University, Morgantown, WV, USA, 2005.

63. Cho, C.W.; Lee, K.J. Seed dispersion and seedling spatial distribution of the tree of heaven in urban environments. *Korean J. Environ. Ecol.* **2002**, *16*, 87–93.

64. Landenberger, R.E.; Kota, N.L.; McGraw, J.B. Seed dispersal of the non-native invasive tree *Ailanthus altissima* into contrasting environments. *Plant Ecol.* **2007**, *192*, 55–70. [CrossRef]

65. Säumel, I.; Kowarik, I. Urban rivers as dispersal corridors for primarily wind-dispersed invasive tree species. *Landsc. Urban Plan.* **2010**, *94*, 244–249. [CrossRef]

66. Säumel, I.; Kowarik, I. Propagule morphology and river characteristics shape secondary water dispersal in tree species. *Plant Ecol.* **2013**, *214*, 1257–1272. [CrossRef]

67. Cabra-Rivas, I.; Alonso, A.; Castro-Díez, P. Does stream structure affect dispersal by water? A case study of the invasive tree *Ailanthus altissima* in Spain. *Manag. Biol. Invasions* **2014**, *5*, 179–186. [CrossRef]

68. Constán-Nava, S.; Bonet, A.; Pastor, E.; Lledó, J. Long-term control of the invasive tree *Ailanthus altissima: Insights* from Mediterranean protected forests. *For. Ecol. Manag.* **2010**, *260*, 1058–1064. [CrossRef]

69. EPPO. PM 9/29 *Ailanthus altissima*. *OEPP/EPPO Bull.* **2020**, *50*, 148–155. [CrossRef]

70. Burch, P.; Zedaker, S. Removing the invasive tree *Ailanthus altissima* and restoring natural cover. *J. Arboric.* **2003**, *29*, 18–24. [CrossRef]

71. Meloche, C.; Murphy, S.D. Managing Tree-of-Heaven (*Ailanthus altissima*) in Parks and Protected Areas: A Case Study of Rondeau Provincial Park (Ontario, Canada). *Environ. Manag.* **2006**, *37*, 764–772. [CrossRef]

72. Young, C.; Bell, J.; Morrison, L. Long-term treatment leads to reduction of tree-of-heaven (*Ailanthus altissima*) populations in the Buffalo National River. *Invasive Plant Sci. Manag.* **2020**, *13*, 276–281. [CrossRef]

73. DiTomaso, J.; Kyser, G. Control of *Ailanthus altissima* using stem herbicide application techniques. *Arboric. Urban For.* **2007**, *33*, 55–63. [CrossRef]

74. Johnson, J.M. An Evaluation of Application Timing and Herbicides to Control *Ailanthus altissima*. Master's Thesis, The Pennsylvania State University, The Graduate School College of Agricultural Sciences, University Park, PA, USA, 2011.

75. Kok, L.T.; Salom, S.M.; Yan, S.; Herrick, N.J.; McAvoy, T.J. Quarantine evaluation of *Eucryptorrhynchus brandti* (Harold) (Coleoptera: Curculionidae), a potential biological control agent of tree of heaven, *Ailanthus altissima* in Virginia, USA. In Proceedings of the XII International Symposium on Biological Control of Weeds, La Grande Motte, France, 22–27 April 2008.

76. Venegas, T.J.; Pérez, P.C. Análisis y optimización de técnicas de eliminación de especies vegetales invasoras en medios forestales de Andalucía. In Proceedings of the V Congreso Forestal Español, Ávila, Spain, 21–25 September 2009. S.E.C.F-Junta de Castilla y León, Ed..

77. Eck, W.; McGill, D. *Testing the Efficacy of Triclopyr and Imazapir Using Two Application Methods for Controlling Tree-of-Heaven along a West Virginia Highway*; e-Gen. Tech. Rep. SRS-101; U.S. Department of Agriculture, Forest Service, Southern Research Station: Asheville, NC, USA, 2007; pp. 163–168.

78. Lewis, K.; McCarthy, B. Nontarget Tree Mortality after Tree-of-Heaven (*Ailanthus altissima*) injection with Imazapyr. *North. J. Appl. For.* **2007**, *25*, 66–72. [CrossRef]

79. Bowker, D.; Stringer, J. Efficacy of herbicide treatments for controlling residual sprouting of tree-of-heaven. In Proceedings of the 17th Central Hardwood Forest Conference, Lexington, KY, USA, 5–7 April 2010; Fei, S., Lhotka, J., Stringer, J., Gottschalk, K., Miller, G., Eds.; Gen. Tech. Rep. NRS-P-78. U.S. Department of Agriculture, Forest Service, Northern Research Station: Newtown Square, PA, USA, 2011; pp. 128–133.

80. Ding, J.; Wu, Y.; Zheng, H.; Fu, W.; Reardon, R.; Liu, M. Assessing potential biological control of the invasive plant, tree-of-heaven, *Ailanthus altissima*. *Biocontrol Sci. Technol.* **2006**, *16*, 547–566. [CrossRef]

81. Marini, F.; Profeta, E.; Vidović, B.; Petanović, R.; de Lillo, E.; Weyl, P.; Hinz, H.L.; Moffat, C.E.; Bon, M.-C.; Cvrković, T.; et al. Field Assessment of the host range of *Aculus mosoniensis* (Acari: Eriophyidae), a biological control agent of the Tree of Heaven (*Ailanthus altissima*). *Insects* **2021**, *12*, 637. [CrossRef]

82. Skvarla, M.J.; Ochoa, R.; Ulsamer, A.; Amrine, J. The eriophyid mite *Aculops ailanthic* Lin, Jin, and Kuang, 1997 (Acariformes: Prostigmata: Eriophyidae) from tree of heaven in the United States new state records and morphological observations. *Acarologia* **2021**, *61*, 121–127. [CrossRef]

83. De Lillo, E.; Marini, F.; Cristofaro, M.; Valenzano, D.; Petanović, R.; Vidović, B.; Cvrković, T.; Bon, M.C. Integrative taxonomy and synonymization of *Aculus mosoniensis* (Acari: Eriophyidae), a potential biological control agent for Tree of Heaven (*Ailanthus altissima*). *Insects* **2022**, *13*, 489. [CrossRef]

84. Kashefi, J.; Vidović, B.; Guermache, F.; Cristofaro, C. Occurrence of *Aculus mosoniensis* (Ripka, 2014) (Acari; Prostigmata; Eriophyoidea) on Tree of heaven (*Ailanthus altissima* Mill.) is expanding across europe. first record in France confirmed by barcoding. *Phytoparasitica* **2022**, *50*, 391–398. [CrossRef]

85. Herrick, N.J.; McAvoy, T.J.; Zedaker, S.M.; Salom, S.M.; Kok, L.T. Site characteristics of *Leitneria floridana* (Leitneriaceae) as related to potential biological control of the invasive Tree-of-heaven, *Ailanthus altissima*. *Phytoneuron* **2011**, *27*, 1–10.

86. Herrick, N.J.; Mcavoy, T.J.; Snyder, A.L.; Salom, S.M.; Kok, L.T. Host-range testing of *Eucryptorrhynchus brandti* (Coleoptera: Curculionidae), a candidate for biological control of Tree-of-heaven, *Ailanthus altissima*. *Environ. Entomol.* **2012**, *41*, 118–124. [CrossRef] [PubMed]

87. Bubici, G.; Prigigallo, M.I.; Garganese, F.; Nugnes, F.; Jansen, M.; Porcelli, F. First report of *Aleurocanthus spiniferus* on *Ailanthus altissima*: Profiling of the insect microbiome and micrornas. *Insects* **2020**, *11*, 161. [CrossRef] [PubMed]

88. Wilson, J.; Landry, J.-F.; Janzen, D.; Hallwachs, W.; Nazari, V.; Hajibabaei, M.; Hebert, P. D-N. Identity of the Ailanthus webworm moth (Lepidoptera: Yponomeutidae), a complex of two species: Evidence from DNA barcoding, morphology and ecology. *ZooKeys* **2010**, *46*, 41–60. [CrossRef]

89. Kreitman, D.; Keena, M.A.; Nielsen, A.L.; Hamilton, G. Effects of temperature on development and survival of Nymphal *Lycorma delicatula* (Hemiptera: Fulgoridae). *Environ. Entomol.* **2021**, *50*, 183–191. [CrossRef]

90. Snyder, A.L.; Salom, S.M.; Kok, L.T. Survey of *Verticillium nonalfalfae* (Plectosphaerellaceae) on Tree-of-heaven in the southeastern United States. *Biocontrol Sci. Technol.* **2014**, *24*, 303–314. [CrossRef]

91. Rebbeck, J.; Malone, M.A.; Short, D.P.G.; Kasson, M.T.; O'Neal, E.S.; Davis, D.D. First report of Verticillium wilt caused by *Verticillium nonalfalfae* on Tree-of-Heaven (*Ailanthus altissima*) in Ohio. *Plant Dis.* **2013**, *97*, 999. [CrossRef] [PubMed]

92. O'Neal, S.E. Biological Control of *Ailanthus altissima*: Transmission, Formulation, and Risk Assessment of *Verticillium nonalfalfae*. Master's Thesis, The Pennsylvania State University. The Graduate School College of Agricultural Sciences, University Park, PA USA, 2014.

93. Kasson, M.T.; Short, D.P.; O'Neal, E.S.; Subbarao, K.V.; Davis, D.D. Comparative pathogenicity, biocontrol efficacy, and multilocus sequence typing of *Verticillium nonalfalfae* from the invasive *Ailanthus altissima* and other hosts. *Phytopathology* **2014**, *104*, 282–292. [CrossRef]

94. Kasson, M.T.; O'Neal, E.S.; Davis, D.D. Expanded Host Range Testing for *Verticillium nonalfalfae*: Potential Biocontrol Agent Against the Invasive *Ailanthus altissima*. *Plant Dis.* **2015**, *99*, 823–835. [CrossRef]

95. Brooks, R.K.; Wickert, K.; Baudoin, A.; Kasson, M.T.; Salom, S. Field-inoculated *Ailanthus altissima* stands reveal the biological control potential of *Verticillium nonalfalfae* in the mid-Atlantic region of the United States. *Biol. Control.* **2020**, *148*, 104298 [CrossRef]

96. Maschek, O.; Halmschlager, E. Natural distribution of Verticillium wilt on invasive *Ailanthus altissima* in eastern Austria and its potential for biocontrol. *For. Pathol.* **2017**, *47*, e12356. [CrossRef]

97. Pisuttu, C.; Marchica, A.; Bernardi, R.; Calzone, A.; Cotrozzi, L.; Nali, C.; Pellegrini, E.; Lorenzini, G. Verticillium wilt of *Ailanthus altissima* in Italy caused by *V. dahliae*: New outbreaks from Tuscany. *iForest* **2020**, *13*, 238–245.

98. Moragrega, C.; Carol, J.; Bisbe, E.; Fabregas, E.; Lorente, I. First report of Verticillium Wilt and mortality of *Ailanthus altissima* caused by *Verticillium dahliae* and *V. albo-atrum sensulato* in Spain. *Plant Dis.* **2021**, *105*, 3754. [CrossRef]

99. Snyder, A.L.; Salom, S.M.; Kok, L.T.; Griffin, G.J.; Davis, D.D. Assessing *Eucryptorrhynchus brandti* (Coleoptera: Curculionidae) as a potential carrier for *Verticillium nonalfalfae* (Phyllachorales). *Biocontrol Sci. Technol.* **2012**, *22*, 1005–1019. [CrossRef]

Article

Fitness and Hard Seededness of F$_2$ and F$_3$ Descendants of Hybridization between Herbicide-Resistant *Glycine max* and *G. soja*

Rong Liang [1], Jia-Li Liu [1], Xue-Qin Ji [1], Kenneth M. Olsen [2], Sheng Qiang [1] and Xiao-Ling Song [1,*]

1 Weed Research Laboratory, College of Life Sciences, Nanjing Agricultural University, Nanjing 210095, China; 2020216029@stu.njau.edu.cn (R.L.); 2021116017@stu.njau.edu.cn (J.-L.L.); 2020116016@stu.njau.edu.cn (X.-Q.J.); wrl@njau.edu.cn (S.Q.)
2 Department of Biology, Washington University in St. Louis, St. Louis, MO 63130, USA; kolsen@wustl.edu
* Correspondence: sxl@njau.edu.cn

Abstract: The commercial cultivation of herbicide-resistant (HR) transgenic soybeans (*Glycine max* L. Merr.) raises great concern that transgenes may introgress into wild soybeans (*Glycine soja* Sieb. et Zucc.) via pollen-mediated gene flow, which could increase the ecological risks of transgenic weed populations and threaten the genetic diversity of wild soybean. To assess the fitness of hybrids derived from transgenic HR soybean and wild soybean, the F$_2$ and F$_3$ descendants of crosses of the HR soybean line T14R1251-70 and two wild soybeans (LNTL and JLBC, which were collected from LiaoNing TieLing and JiLin BaiCheng, respectively), were planted along with their parents in wasteland or farmland soil, with or without weed competition. The fitness of F$_2$ and F$_3$ was significantly increased compared to the wild soybeans under all test conditions, and they also showed a greater competitive ability against weeds. Seeds produced by F$_2$ and F$_3$ were superficially similar to wild soybeans in having a hard seed coat; however, closer morphological examination revealed that the hard-seededness was lower due to the seed coat structure, specifically the presence of thicker hourglass cells in seed coat layers and lower Ca content in palisade epidermis. Hybrid descendants containing the *cp4-epsps* HR allele were able to complete their life cycle and produce a large number of seeds in the test conditions, which suggests that they would be able to survive in the soil beyond a single growing season, germinate, and grow under suitable conditions. Our findings indicate that the hybrid descendants of HR soybean and wild soybean may pose potential ecological risks in regions of soybean cultivation where wild soybean occurs.

Keywords: wild soybean (*Glycine soja*); transgenic soybean; plant invasion; weed management; seed bank

Citation: Liang, R.; Liu, J.-L.; Ji, X.-Q.; Olsen, K.M.; Qiang, S.; Song, X.-L. Fitness and Hard Seededness of F$_2$ and F$_3$ Descendants of Hybridization between Herbicide-Resistant *Glycine max* and *G. soja*. *Plants* **2023**, *12*, 3671. https://doi.org/10.3390/plants12213671

Academic Editors: Congyan Wang and Hongli Li

Received: 8 September 2023
Revised: 20 October 2023
Accepted: 21 October 2023
Published: 25 October 2023

1. Introduction

Genetically modified (GM) soybean (*Glycine max* Linn. Merr.) is one of the world's four most widely cultivated GM crops, with a planting area that increased from 500,000 hectares in 1996 to 91.9 million hectares in 2019, accounting for 48% of the global GM crop planting area [1,2]. Among GM soybean traits, herbicide resistance (HR) is the most important. In China, three of the four safety certificates for GM soybeans issued by the Ministry of Agriculture and Rural Affairs are for varieties with HR traits. Soybean's domestication origin is in eastern Asia, and one of the possible ecological risks posed by the commercial cultivation of HR soybeans in China is the potential for pollen-mediated gene flow to wild soybeans (*Glycine soja* Sieb. et Zucc.), creating GM hybrids whose descendants could persist indefinitely in the wild.

Wild soybean is the direct ancestor of cultivated soybean, and both *Glycine* species have the same chromosome number (2n = 40). Wild soybean, which occurs in all of China and neighboring regions of eastern Asia, is of great value in studying the origin and evolution of soybean [3–5]. While both species are predominantly self-pollinating, occasional cross-pollination can lead to gene flow from cultivated soybean to wild soybean since there is no reproductive isolation between them [5–8]. Pollen flow and hybridization between HR soybean and wild soybean have been widely reported [4,9–11]. However, whether HR soybean genes can introgress into the wild population also depends on the fitness of hybrids and descendants. Fitness is considered to value the adaptation of individuals or populations with specific genotypes under different ecological conditions [12]. Wild soybean is characterized by high genetic diversity [3,13], which could make the hybrids difficult to control, and the hybrids could potentially contaminate wild germplasm resources. Therefore, before introducing widespread commercial planting of HR soybean in regions where wild soybean occurs, it is important to evaluate the multigenerational fitness of hybrid descendants resulting from gene flow from HR soybeans to wild soybeans.

Seeds of wild soybean are characterized by a hard, impermeable seed coat, a trait referred to as hardseededness [5,14,15]. Hard seededness is one of the dormancy traits of wild soybean that inhibit germination until favorable conditions appear [16]. In contrast, the seeds of cultivated soybean are protein rich and perishable, which prevents the domesticated species from overwintering and persisting outside of cultivation [17]. Previous studies have established that the hybrids of wild soybean (as the seed parent) and HR soybean (as the pollen donor) were more similar to wild soybeans in seed morphology due to the segregation distortion, and the hard seed coat of hybrids needed to be scarified to break dormancy [14,18]. Studies have also shown that hybrids can complete the entire life cycle and that their fitness in soybean fields is comparable to or higher than that of wild soybean [19–21]. However, it is unknown to what extent crop-wild hybrids and their descendants consistently show the hard seededness that would be required for survival and long-term persistence outside of cultivation.

In a previous work, we evaluated the sexual compatibility of 10 populations of wild soybean with HR soybeans [22]; for 9 of the F_1 created, we determined that the fitness of hybrids was significantly lower than that of the corresponding wild soybean parent [14]. However, that study did not examine fitness past the F_1. In order to further explore the continuous impact of HR soybean gene flow on wild populations and the environment, the fitness of F_2 and F_3 derived from crosses of HR soybean line T14R1251-70 and two wild soybeans, LNTL and JLBC, which were collected from LiaoNing TieLing and JiLin BaiCheng, respectively, was investigated under two soil conditions and with or without weed competition in the current study. In addition, the hard seededness of hybrid seeds was assessed by observing the seed coat structure and determining the emergence rate after burying in different soil depths for different lengths of time. Our results on seed hardness and fitness of the F_2 and F_3 suggest that cultivation of HR soybean may pose risks for transgene escape to wild soybean and persistence of crop-wild hybrid descendants in the wild.

2. Results

2.1. Emergence Rate

For JLBC F_2, the mean emergence rate was 90.8%, which was significantly higher than the mean value of its wild parent grown in the same experiment (79.2%) ($p < 0.05$); no significant difference in mean emergence rates was observed for JLBC F_3 compared to its wild parent (Figure 1). In contrast, the mean emergence rate value of LNTL F_2 (77.5%) was significantly lower than that of its wild parent (88.1%) ($p < 0.01$); for LNTL F_3 and its wild parent, no significant difference was observed. Thus, variation in emergence rate differed in opposite directions at the F_2 between the two wild populations, and they were not consistent between the F_2 and F_3 generations for either population.

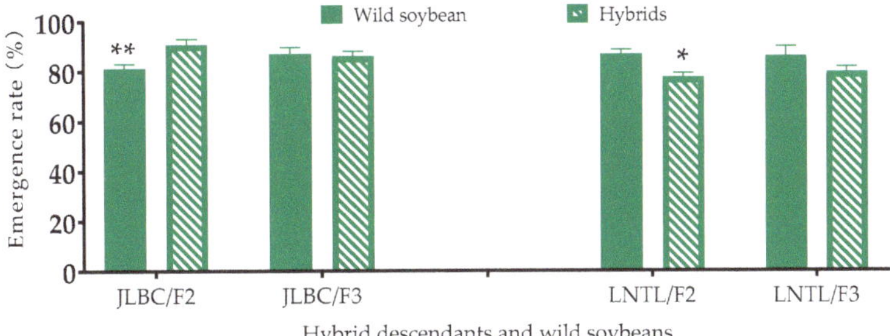

Figure 1. Emergence rate of F_2, F_3 and its wild soybean JLBC and LNTL. Note: * and ** indicates significant difference ($p < 0.05$) and extremely significant difference ($p < 0.01$) between hybrid descendants and its wild soybean.

2.2. True Leaf and Cotyledon Size

For the JLBC F_2 and F_3, both generations had statistically greater mean values of true leaf length than their wild soybean parent (15.8% and 13.2% longer, respectively). For JLBC F_3 only, true leaf width was statistically smaller than JLBC. Similarly, for the LNTL, F_2 and F_3 had significantly greater mean values of true leaf length compared to the LNTL (7.79% and 29.7% longer, respectively). However, the mean leaf width of the LNTL F_3 was also significantly greater than that of the wild parent.

No clear pattern was apparent for cotyledon size data. JLBC F_3 were significantly smaller than those of JLBC. Mean cotyledon width of LNTL F_2 was significantly smaller than that of LNTL, whereas for LNTL F_3, the mean values of both cotyledon length and width were significantly greater than those of their wild soybean parent (Figure 2).

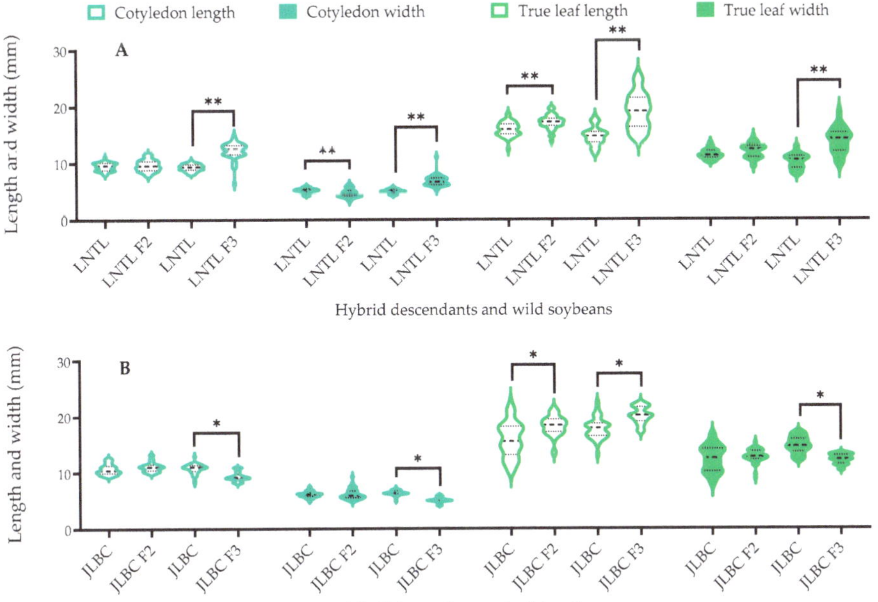

Figure 2. Size of cotyledon and true leaf of F_2, F_3 and wild soybeans ((**A**): LNTL; (**B**): JLBC). Note: * and ** indicates significant difference ($p < 0.05$) and extremely significant difference ($p < 0.01$) of the same trait between hybrid descendants and its wild soybean.

2.3. Plant Height at Third Trifoliolate Leaf Stage

There were differences in the mean plant height of JLBC, JLBC F_2, and F_3 among the four planting conditions. The mean values of JLBC F_2 and F_3 were 3.9–11.7% higher than JLBC. There was no significant difference in plant height of LNTL or F_2 among the four planting conditions, and the mean plant height of LNTL and F_3 was significantly higher when pure planted in farmland soil than that in wasteland soil. Under the same planting conditions, the mean plant heights of F_2 were 16.70–20.30% higher and F_3 were 36.98–44.63% higher than those of LNTL, respectively (Figure 3).

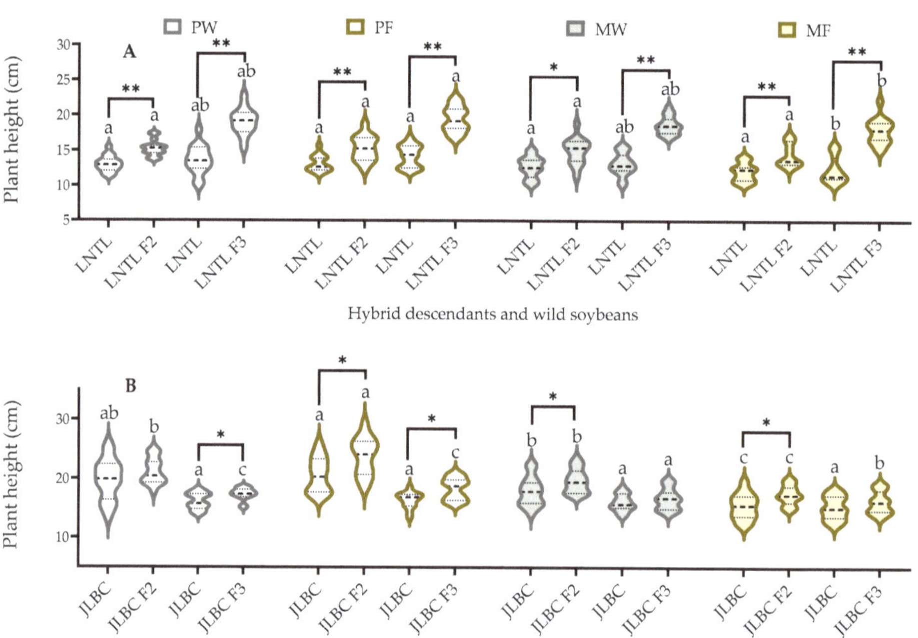

Figure 3. Plant height of F_2, F_3, and wild soybean ((**A**): LNTL; (**B**): JLBC) under four planting conditions (the third trifoliolate leaf stage). Note: PW: pure planting in wasteland soil; PF: pure planting in farmland soil; MW: mixed planting with weeds in wasteland soil; MF: mixed planting in farmland; * and ** indicates significant difference ($p < 0.05$) and extremely significant difference ($p < 0.01$) between hybrid descendants and its wild soybean. Different lowercase letters indicate significant difference ($p < 0.05$) of hybrid descendants or wild soybeans among four planting conditions.

2.4. Aboveground Dry Biomass

The mean aboveground dry biomass of JLBC F_2 was higher than that of their wild parent under pure planting, and that of JLBC F_3 was higher in farmland soil. JLBC F_2 and F_3 had 1.99–3.71 times greater mean aboveground dry biomass than JLBC under the same planting condition. The mean aboveground dry biomass of LNTL F_2 under mixed planting in farmland soil was significantly higher than that under the other three conditions, and that of LNTL in the same year was not significantly different among planting conditions. The aboveground dry biomass of LNTL F_3 and LNTL was significantly higher under pure planting in farmland soil and significantly lower under mixed planting in farmland soil than those under the other two conditions. Under the same planting conditions, the mean aboveground dry biomass of LNTL F_2 and F_3 was significantly higher than that of LNTL; F_2 was 1.3–1.59 times higher, while F_3 was 1.59–1.77 times higher than LNTL (Figure 4A,B).

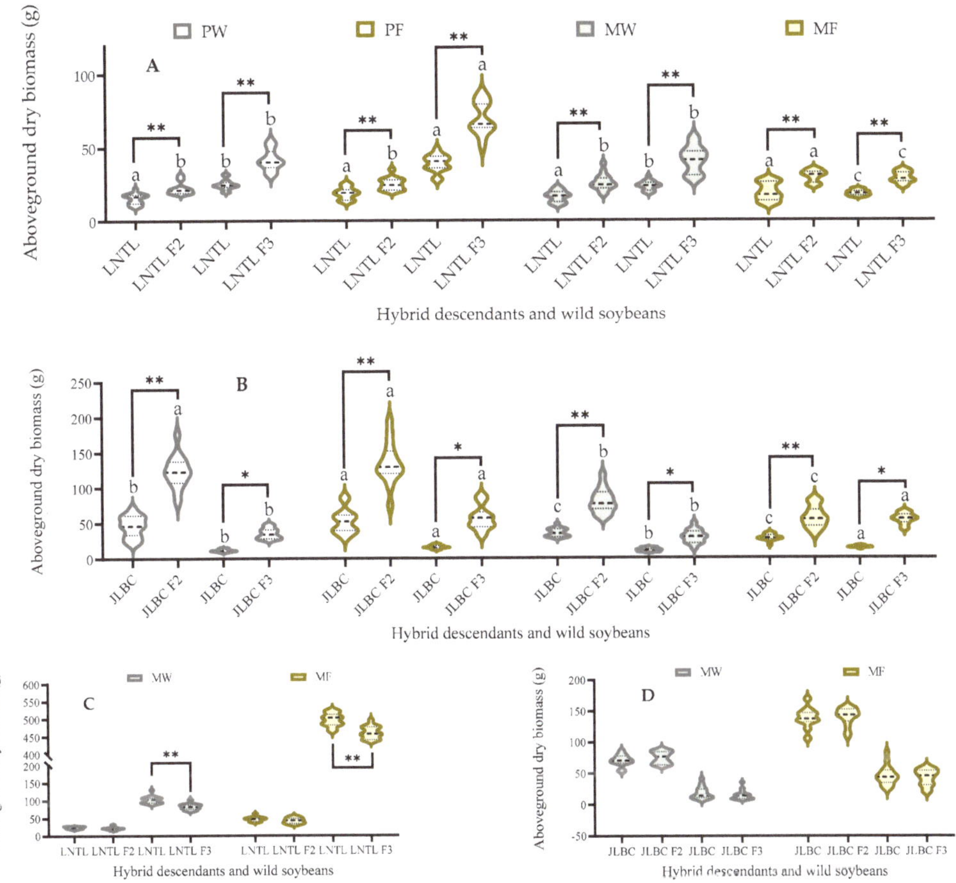

Figure 4. Aboveground dry biomass of F_2, F_3, wild soybeans (**A,B**) and weeds (**C,D**) under four planting condition. Note: PW: pure planting in wasteland soil; PF: pure planting in farmland soil; MW: mixed planting with weeds in wasteland soil; MF: mixed planting in farmland; * and ** indicates significant difference ($p < 0.05$) and extremely significant difference ($p < 0.01$) of the same trait between hybrid descendants and its wild soybean. Different lowercase letters indicate significant difference ($p < 0.05$) of hybrid descendants or wild soybeans among four planting conditions.

Aboveground dry biomass of weeds in farmland soil was always higher than that in wasteland soil (Figure 4C,D). There were no significant differences between the weed biomass with JLBC F_2, F_3, and JLBC. There was no significant difference between the weed biomass with LNTL F_2 and with LNTL, while that of LNTL F_3 was significantly lower than LNTL in both farmland and wasteland soil.

2.5. Vitro Pollen Germination Rate

The pollen germination rates of JLBC F_2 were higher when pure planted than when mixed planted, while those of JLBC F_3 were higher in farmland soil than in wasteland soil. The mean pollen germination rates of JLBC F_2 were higher than or comparable to JLBC, and those of JLBC F_3 were significantly lower than JLBC. The pollen germination rates of LNTL F_2 and its wild soybean had the same trend under four conditions, with the highest under mixed planting in farmland soil or comparable. That of LNTL F_3 and its wild soybean also had the same trend under four conditions, with the highest under pure planting in farmland soil and the lowest under mixed planting in wasteland soil. Under the same planting conditions, the pollen germination rate of LNTL F_2 and F_3 was 7.49–15.08% lower than that of LNTL (Figure 5).

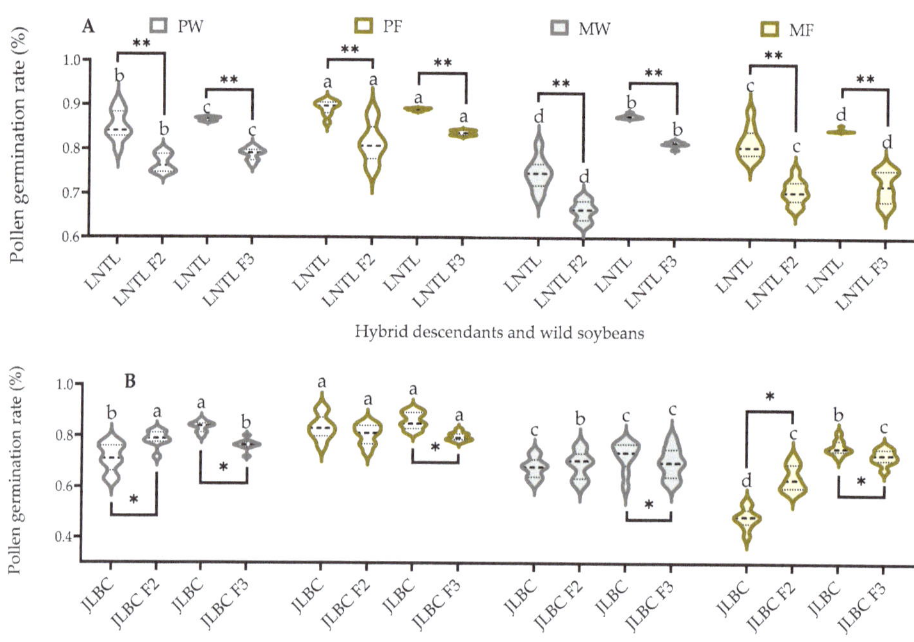

Figure 5. Vitro pollen germination rate of F_2, F_3, and wild soybeans ((**A**): LNTL; (**B**): JLBC) at 60 min under four planting condition. Note: PW: pure planting in wasteland soil; PF: pure planting in farmland soil; MW: mixed planting with weeds in wasteland soil; MF: mixed planting in farmland; * and ** indicates significant difference ($p < 0.05$) and extremely significant difference ($p < 0.01$) of the same trait between hybrid descendants and its wild soybean. Different lowercase letters indicate significant difference ($p < 0.05$) of hybrid descendants or wild soybeans among four planting conditions.

2.6. Pod and Filled Seed Number per Plant

Mean values for pod and filled seed number per plant of JLBC F_2 and JLBC were higher under pure planting than under mixed planting conditions, while mean values for JLBC F_3 and JLBC were higher in farmland soil than in wasteland soil. Pod and filled seed numbers per plant for JLBC F_2 and F_3 were 1.1–3.7 times higher than JLBC in all four conditions.

Mean values for pod and filled seed number per plant of LNTL F_2 and its wild soybean were significantly higher under mixed planting in farmland soil than in the other three planting conditions. In contrast, mean values for pod and filled seed number per plant of LNTL F_3 and LNTL were significantly higher under pure planting in farmland soil than in the other three conditions. Under the same planting conditions, the mean number of pods per plant of LNTL F_2 and F_3 was 8.46–24.28% higher than that of LNTL (Figure 6).

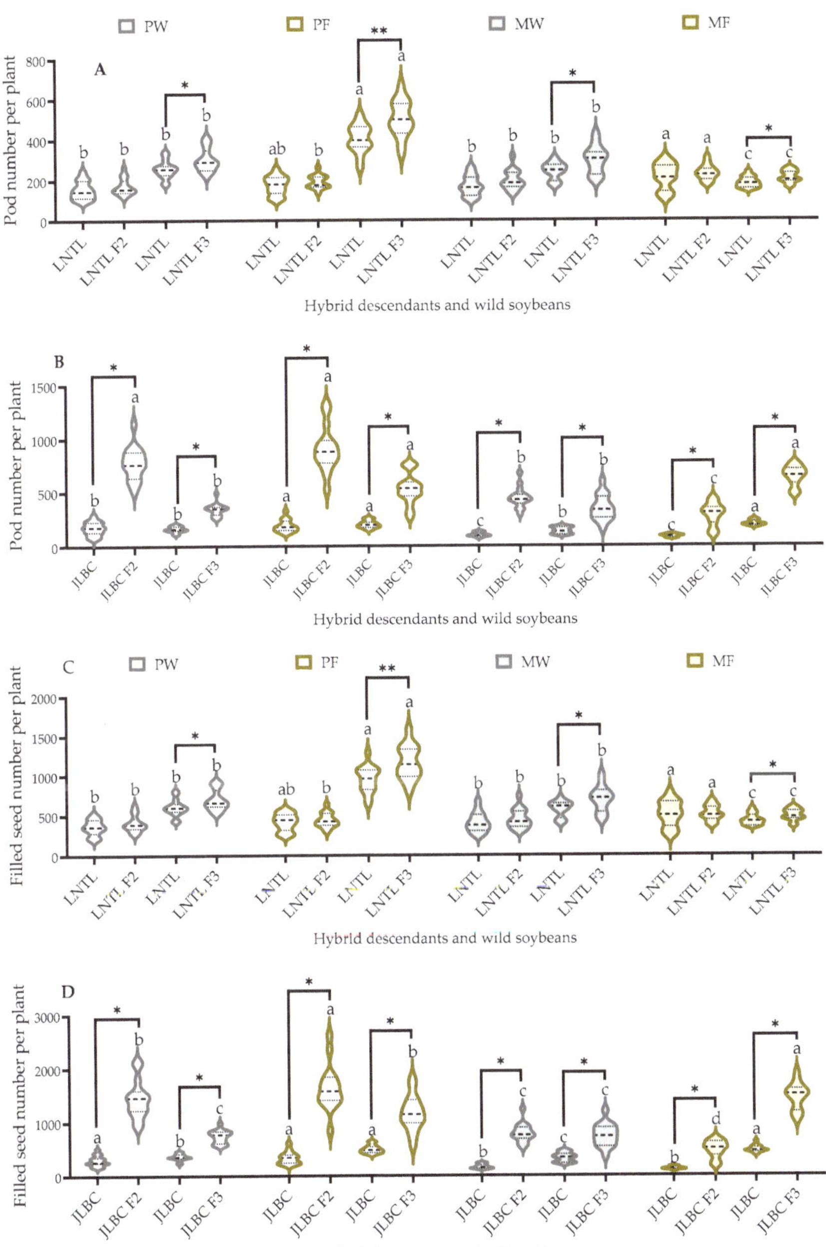

Figure 6. Pod number ((**A**): LNTL; (**B**): JLBC) and filled seed number ((**C**): LNTL; (**D**): JLBC) per plant of F_2, F_3 and wild soybeans under four planting condition. Note: PW: pure planting in wasteland soil; PF: pure planting in farmland soil; MW: mixed planting with weeds in wasteland soil; MF: mixed planting in farmland; * and ** indicates significant difference ($p < 0.05$) and extremely significant difference ($p < 0.01$) of the same trait between hybrid descendants and its wild soybean. Different lowercase letters indicate significant difference ($p < 0.05$) of hybrid descendants or wild soybeans among four planting conditions.

2.7. 100-Seed Weight

The mean values of 100-seed weight for self-pollinated seeds of JLBC F_2 and F_3 were significantly lower than JLBC. JLBC F_3 under mixed planting in farmland soil had significantly lower mean values than under other planting conditions; the values for JLBC F_2 and F_3 under other planting conditions were similar. Under the same planting condition, the mean 100-seed weight values for self-pollinated seeds of LNTL F_2 and F_3 were significantly higher than those of LNTL, with mean values 1.56–1.92 times greater than those of the wild parent. The mean 100-seed weight of LNTL F_3 under pure planting in farmland soil was significantly higher than that of the other three conditions, while others were similar (Figure 7).

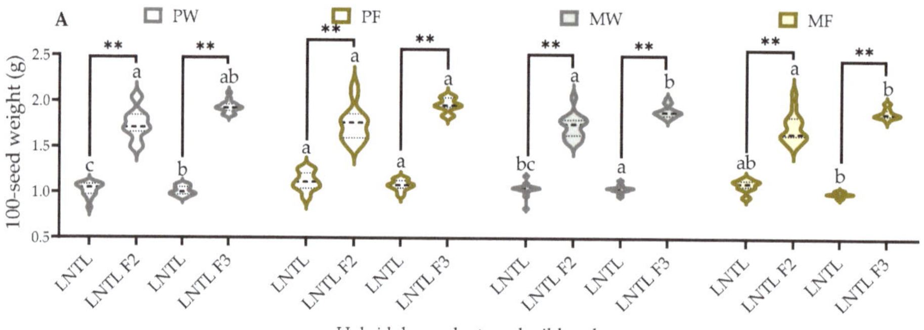

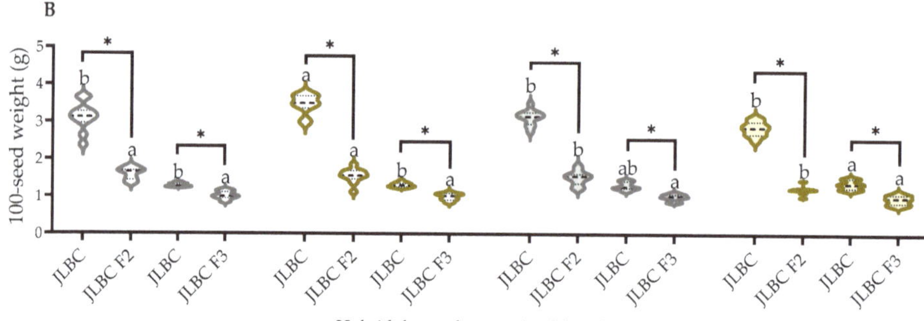

Figure 7. 100-seed weight of seeds of F_2, F_3, and wild soybean ((**A**): LNTL; (**B**): JLBC) under four planting conditions. Note: PW: pure planting in wasteland soil; PF: pure planting in farmland soil; MW: mixed planting with weeds in wasteland soil; MF: mixed planting in farmland; * and ** indicates significant difference ($p < 0.05$) and extremely significant difference ($p < 0.01$) of the same trait between hybrid descendants and its wild soybean. Different lowercase letters indicate significant difference ($p < 0.05$) of hybrid descendants or wild soybeans among four planting conditions.

2.8. Relative Composite Fitness

Taking wild soybean as the standard "1", the values of correspondingly F_2 and F_3 were valued as the relative composite fitness. The relative composite fitness of JLBC F_2 and F_3 was higher than that of JLBC under all four planting conditions, but not statistically significant. The relative composite fitness of JLBC F_2 among four planting conditions had no difference, while that of JLBC F_3 was higher under pure planting conditions or in farmland soil. The relative composite fitness of LNTL F_2 and F_3 was higher than that of LNTL under all four planting conditions, but the difference was not significant for F_2, while it was significant for F_3. There was no significant difference between LNTL F_2 and its wild parent among the four conditions, and both LNTL F_3 and LNTL had significantly higher fitness when pure planted in farmland soil than under the other three conditions, while there was no significant difference among the three conditions (Figure 8).

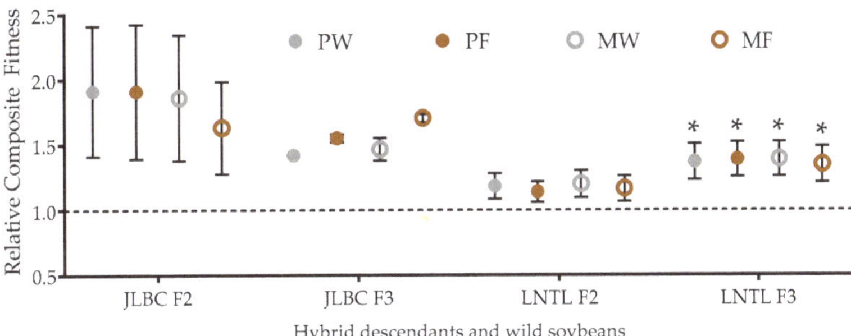

Figure 8. Comparison of composite fitness between wild soybeans and F_2, F_3. Note: PW: pure planting in wasteland soil; PF: pure planting in farmland soil; MW: mixed planting with weeds in wasteland soil; MF: mixed planting in farmland; The dashed line represents the composite fitness of wild soybean as 1, * indicates significant difference ($p < 0.05$) of the same trait between hybrid descendants and its wild soybean.

2.9. Hard Seededness and Germination Rate

Self-pollinated seeds of LNTL F_2 and F_3 were used to conduct this experiment. The hard seededness rate of LNTL F_2 seeds was 89.50%, which was extremely significantly lower than that of LNTL (98.50%), and there was no significant difference between LNTL F_3 seeds and LNTL seeds. After scarification, there was no significant difference in germination rate between F_2/F_3 seeds and wild soybean seeds (Table 1).

Table 1. Hard seededness rate and germination rate with scarification of LNTL F_2, F_3 seeds.

Material	Hard Seededness Rate (%)	Germination Rate with Seed Scarification (%)
LNTL	98.50 ± 0.96 **	94.44 ± 0.93
LNTL F_2	89.50 ± 0.96	93.86 ± 0.53
LNTL	100	98.00 ± 0.00
LNTL F_3	96.50 ± 0.02	94.50 ± 0.02

Note: ** indicates extremely significant difference ($p < 0.01$) between hybrids and its wild soybean.

2.10. Seed Coat Structure

Self-pollinated seeds of LNTL F_2 were used to conduct this experiment. There are obvious pits on the surface of the HR soybean seed coat, and the shape of the pits is irregular (Figure 9A,B). There are three main types of pits: Shallow long pits, deep round pits, shallow round pits, and some pits with a crack width of 0.1–0.3 μm. The surface of the HR soybean has almost no attachment, and the stratum corneum is directly exposed to the outside. Both LNTL and its LNTL F_2 seed coat surface are covered by a thick layer of sediment, similar to a bulge at the basin margin, and the entire seed coat surface is honeycomb-shaped (Figure 9E,I); there are no cracks on the surface of wild soybean or F_2 seed coat. At the hila of wild soybean and F_2, there is a middle dent and several multiple irregular cracks on both sides of the dent, with a width of 3–20 μm (Figure 9F,J). At the same time, no honeycomb-like sediment attachment was observed around the hila of LNTL and F_2 seeds.

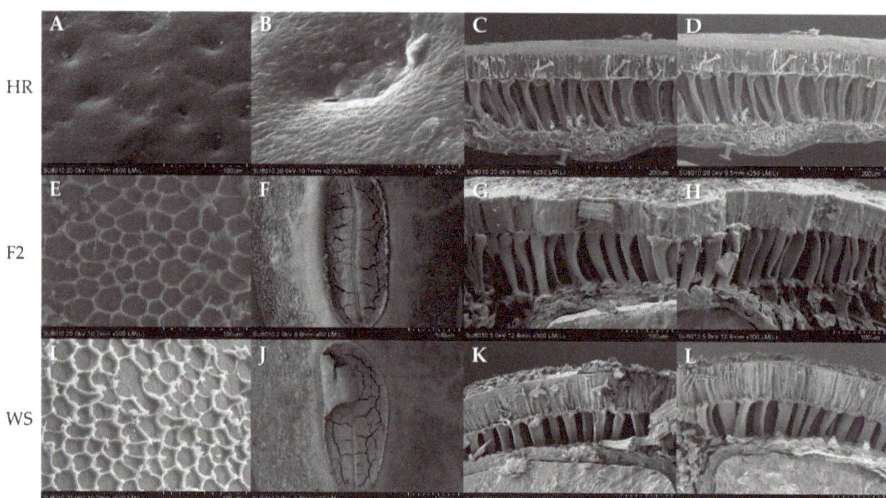

Figure 9. SEM of seed coat structure of transgenic soybean, LNTL wild soybean and F_2 seeds. Note: (**A**) Seed coat surface of TS ($\times$500); (**B**) depress and crack on seed coat surface of TS ($\times$2000); (**C,D**) seed coat layers of TS ($\times$250); (**E**) seed coat surface of LNTL F_2 ($\times$500); (**F**) hilum surface of LNTL F_2 ($\times$60); (**G,H**) seed coat layers of LNTL F_2 ($\times$300); (**I**) seed coat surface of LNTL F_2 ($\times$500); (**J**) hilum surface of LNTL ($\times$60); (**K,L**) seed coat surface of LNTL ($\times$300).

The seed coat structure of HR soybean, LNTL, and LNTL F_2 all contains four cell layers, followed by the palisade epidermis, hourglass cells, parenchyma, and aleurone layer; LNTL and F_2 seeds also have a stratum corneum over the seed coat. Among them, the aleurone layer has monolayer cells, which are not easy to recognize with SEM (Figure 9C,D,G,H,K,L).

The palisade epidermis of LNTL wild soybean was comparable to that of the F_2 seeds, and both were higher than that of HR soybean. The hard seededness rates of LNTL, F_2, and HR soybean seeds decreased from 98.50%, 89.50%, and 1.00%, respectively. However, the proportion of palisade epidermis thickness in the seed coat decreases with the decrease in hard seededness rate. The thickness of hourglass cells and their proportion increased with the decrease in hardness rate. The parenchyma layers of LNTL and F_2 seeds were significantly thinner than those of HR soybean (Figure 10).

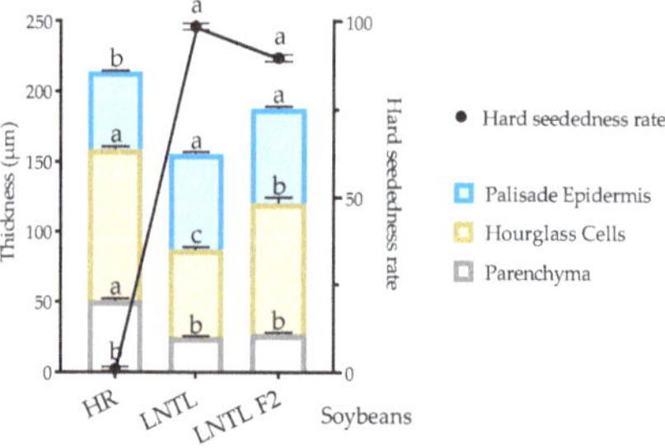

Figure 10. Relationship between hard seededness rate and thickness of seed coat layers of transgenic soybean, LNTL wild soybean and F_2 seeds. Note: Different lowercase letters indicate significant difference ($p < 0.05$) among hybrid descendants, wild soybean, and transgenic soybean.

2.11. Mineral Element in Seedcoat

Self-pollinated seeds of LNTL F_2 were used to conduct this experiment. The content of Ca in the seed coat palisade epidermis of LNTL F_2 seeds was significantly lower than that of LNTL; however, there was no significant difference for other mineral elements that were measured (Figure 11).

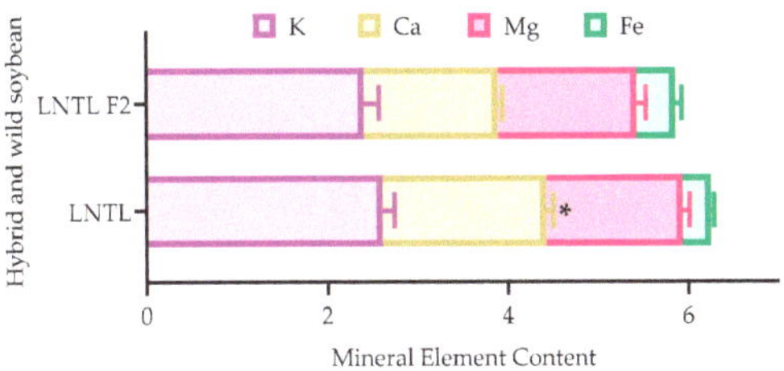

Figure 11. Main mineral element content of seed coat palisade epidermis of LNTL wild soybean and F_2 seeds. Note: * indicates significant difference ($p < 0.05$) between hybrid descendants and wild soybean.

2.12. Seed Vitality in Soil

For self-pollinated seeds of both JLBC F_3 and JLBC, under both 3 cm and 10 cm of soil, the trend of natural emergence rate increased with time, and the emergence rate with seed scarification hardly changed over time. The emergence rate of JLBC F_3 seeds was higher with seed scarification and lower without seed scarification than JLBC, respectively (Figure 12A).

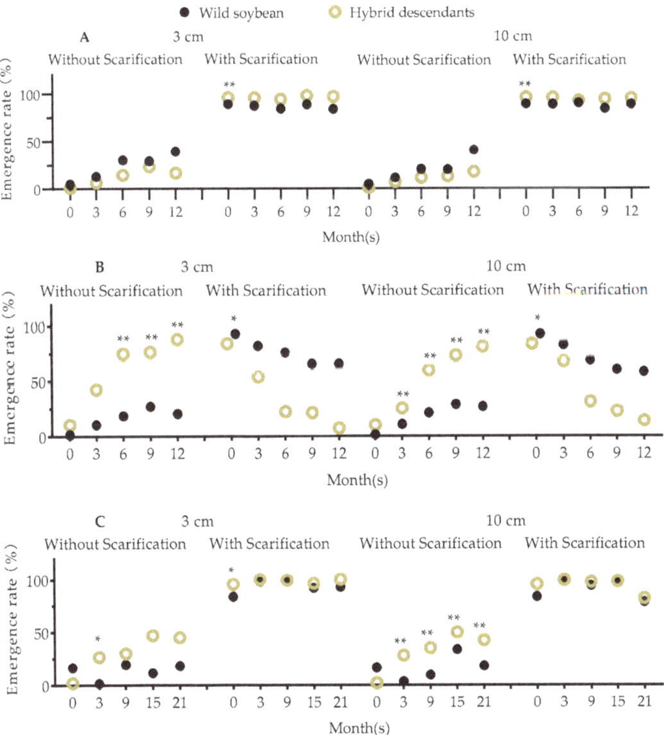

Figure 12. The emergence rate with or without scarification of seeds of JLBC F_2 (**A**), LNTL F_2 (**B**) and LNTL F_3 (**C**) after burying in 3 cm or 10 cm soil. Note: * and ** indicates significant difference ($p < 0.05$) and extremely significant difference ($p < 0.01$) between hybrid descendants and its wild soybean.

For self-pollinated seeds of both LNTL F_2 and LNTL, under both 3 cm and 10 cm soil, the trend of natural emergence rate increased with time, and the natural emergence rate of LNTL F_2 seeds after burying for 6 months was significantly higher than that of LNTL. After seed scarification, the trend of the natural emergence rate of both soybeans decreased with time, and the natural emergence rate of F_2 seeds after burying for 3 months was significantly higher than that of LNTL (Figure 12B).

For self-pollinated seeds of LNTL F_3 and LNTL, both under 3 cm and 10 cm of soil, the trend of natural emergence rate of both seeds increased with time and decreased after burying for 15 months. The natural emergence rate of LNTL F_3 seeds after burying was higher than that of LNTL but not significantly. The emergence rate with scarification of LNTL F_3 seeds was higher than that of LNTL but only significant at one time point (Figure 12C).

3. Discussion

3.1. Fitness of F_2, F_3 Compared with Parents

The F_2 and F_3 of this experiment were obtained by hybridizing HR soybeans (as the pollen donor) with wild soybeans (as the seed parent). The genetic difference between cultivated soybeans and wild soybeans derives from the domestication by humans of the wild species into the cultivated crop species [23–26]. In the domestication of legumes, selection favors enhanced aboveground traits, including greater seed size and palatability, reduced seed dormancy, and other desirable agronomic traits. The hybrids of cultivated soybean and wild soybean usually have a growth advantage over wild soybean [6,21]. In the context of crop improvement, hybridization of domestic soybeans (as the seed parent) and wild soybeans (as the pollen donor) can improve the resistance of hybrids and even promote the diversity of varieties [27,28]. However, if it is allowed to grow outside of cultivation, these same fitness advantages create potential ecological risks, particularly in regions of transgenic HR soybean cultivation; in this context, the advantages of hybrids do not bode well.

In our previous study, it was found that F_1 hybrids of HR soybean and wild soybeans, including LNTL and JLBC, had lower fitness than the wild soybean parents [14]. In this study, F_2 and F_3 of both LNTL and JLBC, regardless of soil conditions and whether there was weed competition, showed significantly elevated mean values relative to their wild parents for multiple fitness-related traits, including plant height, number of pods per plant, number of filled seeds per plant, filled seed weight per plant, aboveground dry biomass, and 100-seed weight. The mean composite fitness of LNTL F_3 under all four planting conditions was significantly higher than that of LNTL. As the generations increased, the adverse effects of hybridization are gradually eliminated through gene segregation and recombination [29,30]. The wild soybean LNTL and JLBC were collected at high latitudes, and in-field experiments were at lower latitudes. The fitness of wild soybeans was decreased due to the shorter photoperiod and other unsuitable environmental factors [31,32]. After receiving pollen of HR soybean adapted to the climate of the experimental location, F_2 and F_3 inherited adaptability to the local climate and environment, which ultimately led to the improvement of the survival competitiveness of the hybrid descendants.

It is worth noting that regardless of the planting conditions, LNTL F_2 and F_3 pollen viability was significantly higher than that of wild soybeans, while JLBC F_3 had lower pollen viability. Pollen activity reflects the quality of pollen, affects seed formation, and is an important indicator for valuing reproductive ability [33]. The probability of interspecific hybridization and the fertility of hybrid descendants depend largely on the homology of the genomes and the degree of homology, which determines the possibility of pairing and recombination between the chromosomes of the parents [34]. HR soybean and wild soybean both belong to *Glycine*, and they have the same chromosome number (2n = 40), but there are differences in chromosome behavior and the division cycle of meiosis [5,35]. Therefore, in meiosis, hybrid descendants would have abnormal chromosome behavior. And different populations of wild soybean have varying degrees of chromosomal abnormalities, which also reflect the diversity of the germplasm resources of wild soybeans.

3.2. Effects of Soil Nutrition and Competition on Fitness of F_2, F_3

As the substrate for crops, there are various elements and substances that affect the development and reproduction of plants in the soil [36,37]. Unlike other crops, legumes have the ability to symbiotically fix nitrogen with nitrogen-fixing bacteria [38,39]. Therefore, the growth of soybean is not only affected by soil nutrition, especially nitrogen in the soil [40]. It was proven that the nitrogen-fixing capacity of cultivated soybean and wild soybean and the interaction mode with rhizosphere microorganisms are different [17,41,42].

In this experiment, soils from farmland and wasteland were used to plant the hybrids. The results showed that when there was no weed competition, except for LNTL F_2, the fitness of JLBC F_2, JLBC F_3, LNTL F_3, and their wild soybean under the farmland soil was significantly higher than that of the wasteland soil. The nitrogen form in soil may partly explain this anomalous difference in LNTL F_2. Nitrate nitrogen and ammonium nitrogen, which are called available nitrogen, are effective forms of nitrogen nutrients in soil and can be directly absorbed and utilized by roots [43,44]. For JLBC F_3 and LNTL F_2, the content of available nitrogen was 10.71 mg/kg in wasteland soil and 23.59 mg/kg in farmland soil, both were not high enough for growth. At the same time, there was no significant difference in total nitrogen content between farmland soil and wasteland soil this year. This may explain the similar fitness of JLBC F_3 and LNTL F_2 and their wild soybeans in both wasteland and farmland soils. The restriction on the growth of JLBC F_3 and LNTL F_2 in wild soybean may be due to a lack of available phosphorus. Under limiting phosphorus, the uptake and utilization of nitrogen and other metabolic pathways will also be affected [45–48]. Therefore, when the nutrients were relatively abundant, the available nitrogen and available phosphorus, which were significantly different between farmland and wasteland soil, also had a significant impact on the growth of JLBC F_2, LNTL F_3, and wild soybean.

All hybrids have similar patterns in different soils to wild soybeans, suggesting that the utilization pattern of soil nutrients of hybrids and symbiotic nitrogen fixation are inherited from the seed producer, the wild soybean.

Weeds not only compete with crops for light [49,50] and nutrition in soil [51], but also change the environment and microorganisms of the rhizosphere through root exudates, which affects the growth of soybeans [52,53]. When there was weed competition, the number of pods and filled seeds per plant of JLBC F_3, LNTL F_2, and their wild soybean in farmland soil were significantly higher than those in wasteland soil, but the fitness of JLBC F_2, LNTL F_3 and their wild soybean was exactly opposite. This difference came from differences in the nutrient content of the soil used in the three-year trial. Weed dry biomass can reflect the nutrient level of the soil. It can also be seen that LNTL wild soybeans are less competitive with weeds than LNTL F_3. This increased competitiveness may come from the genes of the paternal HR soybeans [54]. Although the available nitrogen level in wasteland soil was relatively low, the nitrogen fixation ability of hybrids and wild soybeans could still maintain the nitrogen balance in the soil and the normal growth of plants. The number of pods and filled seeds per plant under LNTL F_3 mixed planting in wasteland soil was significantly higher than that of mixed planting in farmland soil. This phenomenon may also come from biological nitrification inhibition [55]. When there was weed competition, weeds, soybeans, soil nutrients, and the rhizosphere formed a complex interacting system [56–59]. The environment was changed to benefit the strong side, such as wild soybeans.

3.3. Seed Coat Structure and Seed Dormancy

Honeycomb epidermal attachments may be the first barrier to prevent the seed from absorbing water and expanding, and they are an important way for the seed to remain dormant. There is no obvious attachment on the surface of the seed coat of HR soybean, and the dormancy ability of crop seeds is almost completely lost. This attachment comes from the endocarp, known as bloom, and directly acts to change the gloss of the surface of the seed, reducing the chance that the seed will be found and eaten by animals [60,61]. Meanwhile, bloom has been proven to be related to seed oil content [62], and the change in soybean permeability in domestication was caused by human selection. The difference in bloom explained the difference in natural emergence rates between HR soybean and wild soybeans, but it still does not explain the difference between hybrid descendants and wild soybeans.

The emergence rate of seeds with scarification showed that there was no significant difference in seed viability between seeds of LNTL F_2, F_3, and wild soybean, which showed similar embryonic activity. The hard seededness of wild soybean ensures long-term seed dormancy. With time, buried seeds of all hybrids and wild soybeans were more likely to break dormancy, and embryonic activity decreased. Point mutations in *Gm*Hs1-1 cause the loss of hard seededness and this gene correlates with the content of calcium in the seed coat [63]. In the experiment, the calcium content in the seed coat of LNTL F_2 seeds was significantly lower than that of wild soybean, indicating that LNTL F_2 seeds partially inherited the soft seed coat of HR soybean, resulting in its hard seededness being weaker than wild soybean. However, some soybeans promote water absorption and break dormancy while maintaining the calcium content of the seed coat by cracking through the seed coat. This is the case with irregular cracks on the surface of HR soybeans seeds observed by SEM, but LNTL F_2 seeds did not have this character. The formation of such cracks may come from changes in the seed coat layers. The shape and number of hourglass cells are often thought to be strongly related to seed dormancy and viability [64,65]. Palisade epidermis and parenchyma of LNTL F_2 seeds were both similar to those of wild soybean, but the hourglass cells were significantly higher than those of wild soybean and lower than those of HR soybean. Hourglass cells are associated with the accumulation of various enzymes associated with water absorption and germination, such as catalase [66,67]. The difference in the hourglass cell layer could exactly explain the decline in hard seededness rate of LNTL F_2 seeds compared to wild soybean.

Therefore, the hybrid seeds of wild soybean and HR soybean reduced the hard seededness compared to wild soybeans through the thickening of hourglass cells and the reduction of calcium content in the palisade epidermis.

4. Materials and Methods

Herbicide-resistant transgenic soybeans T14R1251-70 were provided by the National Soybean Improvement Center of Nanjing Agricultural University. The HR soybean, containing one single-copy *cp4-epsps*, was obtained by Agrobacterium-mediated co-transformation of the receptor soybean NJR44-1, which is an elite line bred by the National Soybean Improvement Center of Nanjing Agricultural University. The HR soybean withstands 3600 g a.i. ha-1 41% glyphosate isopropylammonium AS (Roundup Ultra; Monsanto, St. Louis, MO, USA). Wild soybean populations were collected from Tieling, Liaoning Province, and Baicheng, Jilin Province. Crossed seeds were obtained by artificial hybridization of wild soybeans as the seed producer and HR soybeans as the pollen donor from 2016 to 2017 [22]. The crossed seeds were harvested from different seed producers individually and then stored at 4 °C until further use. Experiments were conducted in a greenhouse and net house at the Pailou Experimental Farm (32°20′ N, 118°37′ E), Nanjing Agricultural University, China, from 2018 to 2020.

4.1. Seed Treatment and Seeding

Scarify the seed coat of wild soybeans and hybrid descendants. Seeds were sown in a plastic cup with a hole at the bottom (a diameter of 7 cm and a height of 7.5 cm). The substrate for seeding was farmland soil and wasteland soil, as described in Table 2. Seedings were placed in a net chamber for normal water management, and all test materials were randomly placed in the net chamber and cultured under natural light and photoperiod, during which the temperature fluctuated between 20 and 38 °C.

Table 2. Soil physicochemical properties per year.

	Soils	Organic Matter g/kg	Total Nitrogen g/kg	Total Phosphorus g/kg	Total Potassium g/kg	Available Phosphorus mg/kg	Available Nitrogen mg/kg
JLBC F_2	Wasteland soil	2.79 ± 0.10	0.37 ± 0.01	0.56 ± 0.01	22.04 ± 0.46	22.39 ± 0.52	44.15 ± 0.2
	Farmland soil	38.51 ± 0.35 *	2.20 ± 0.03 *	1.76 ± 0.01 *	18.94 ± 0.19	47.81 ± 0.33 *	163.74 ± 0.54 *
JLBC F_3 and LNTL F_2	Wasteland soil	4.82 ± 0.22	0.27 ± 0.37	0.17 ± 0.11	9.79 ± 0.09	0.1 ± 0.03	10.71 ± 1.25
	Farmland soil	9.74 ± 0.81 *	0.37 ± 0.04	0.26 ± 0.12 *	10.07 ± 0.10	1.68 ± 0.31 *	23.59 ± 2.61 *
LNTL F_3	Wasteland soil	7.78 ± 0.40	0.72 ± 0.02	0.25 ± 0.01	20.94 ± 0.42	9.99 ± 0.86	51.91 ± 1.38
	Farmland soil	11.19 ± 1.50	1.06 ± 0.11	0.36 ± 0.07	21.10 ± 0.48	28.21 ± 1.32 *	145.41 ± 21.08 *

Note: * indicates significant difference between wasteland soil and farmland soil ($p < 0.05$).

4.2. Emergence Rate and Cotyledon, True Leaf Size

When the cotyledons of the plants are unearthed and completely green (about 2 weeks after sowing), the number of seedlings of soybean plants is counted. When the first compound leaf of the plant has formed and the leaves are wrinkled but not fully expanded, the longest and widest cotyledons and true leaves are determined using Vernier calipers; each single plant is a replicate, and 20 plants per material are randomly selected for measurement.

4.3. cp4-Epsps in Hybrids

After the first ternately compound leaf of the plant was unfolded, the *cp4-epsps* was detected by PCR with a specific primer (5′-GGCACAAGGGATACAAACC-3′; 5′-ACCGCCGAACATGAAGGAC-3′). Count the number of plants carrying resistance genes and plants without resistance genes, and use the chi-square test to verify whether the resistance of hybrid separation ratio results conform to Mendel's law of 3:1. The specific formula is as follows:

$$\chi^2 = \frac{[|b \times A_1 - a \times A_2| - (a+b)/2]^2}{a \times (A_1 + A_2)} \tag{1}$$

χ^2 represents the chi-square value, such as $\chi^2 < 3.84$, that is, $p > 0.05$, indicating that the inheritance law of resistance genes in hybrids conforms to Mendel's law of inheritance; A_1 indicates the number of plants carrying resistance genes; A_2 indicates the number of plants that do not carry resistance genes; F_2: $a = 3$, $b = 1$; F_3: $a = 5$, $b = 1$.

4.4. Planting Conditions

Wasteland soil and farmland soil were collected at the Pailou base. Take three copies of the soil and entrust Nanjing Zhongding Biological Company to test the physical and chemical properties of the soil (Table 2).

Four planting conditions were set: Pure planting in wasteland soil (PW), pure planting in farmland soil (PF), mixed planting with weeds in wasteland soil (MW), and mixed planting with weeds in farmland (MF). For the emergence rate test, 60 plants with consistent growth of HR soybean, LNTL, JLBC, and hybrid descendants were selected, and 15 plants were transplanted under four planting conditions. Under single planting conditions, a pod (23 cm in diameter and 25 cm in height) with bamboo was set for LNTL, JLBC, and hybrid descendants growth. When mixed planting with weeds, *Setariaviridis* (L.) *Beauv.* 0.5 g, *Digitariasanguinalis* (L.) Scop. 0.5 g, *Echinochloacolonu* (L.) Link. 0.5 g, and *Eleusine indica* (L.) Gaertn. 0.25 g were sown evenly in pots (52 cm diameter and 35 cm height).

4.5. Fitness Determination

Investigate fitness indicators during plant vegetative and reproductive periods. Emergence rate: Two weeks after sowing, count the number of all seedlings unearthed with green cotyledons; true leaf size: When the first compound leaf has been formed and the leaf is wrinkled but not fully expanded, the cotyledon length width and true leaf length width are measured by vernier calipers; plant height: At the third-ternately-compound stage, the length from the tip of the main stem to the cotyledon ring was measured; pollen vitality: Randomly collect flower buds on plants at full bloom period (flag petals are 1–2 mm higher than sepals), culture in vitro for 60 min, and count the number of germinated pollen under a microscope; aboveground dry biomass: After harvesting, the aboveground part of the plant is dried to a constant weight and weighed; number of pods per plant: After harvesting, the total number of pods per plant is counted and artificially threshed; number of filled seeds per plant: After harvesting, select the filled seeds from all single seeds (with regular shape, no depression, and no shrink), count the number, and weigh them; composite fitness: Taking wild soybean as the standard "1", the seedling emergence rate, cotyledon length × cotyledon width + true leaf length × true leaf width, plant height, aboveground dry biomass, pollen germination rate of 60 min, number of pods and filled seeds per plant, 100-seed weight to wild soybean were valued, and the composite relative fitness is the average of the values.

4.6. Seed Hard Seededness Rate and Scarified Emergence Rate

Fifty seeds were randomly selected from all the plants under the pure planting in soil with 4 repeats. The number of seeds that did not swell (seed size did not change) after 7 days of soaking in distilled water was counted. Hard seededness rate = number of unswollen seeds/total number of seeds×100%. After the hard seededness rate is determined, scarify the seed coat of the remaining hard wild soybean and hybrid seed without damaging the embryo. Incubate the scarified seeds at a constant temperature of 25 °C for 7 days; count the germinating seeds with a radicle length twice that of the seed length. Emergence rate (%) = total number of germinated seeds / total number of seeds × 100%.

4.7. Seed Coat Structure and Elemental Content

Select 3 filled seeds with a complete seed coat from plants purely planted in farmland soil. Cut the seeds along the seed ridge corresponding to the center point of the seed hilum to avoid damage to the embryo. Stick the cut seeds on the sample stage; use a Hitachi-1010 ion sputterer to spray gold on the surface; use a Hitachi-SU8010 scanning electron microscope for observation and photography; and use an SEM accelerating voltage of 20 kV. Photoshop (version 21.1.2; Adobe Systems Incorporated, San Jose, CA, USA) was used to measure the thickness of each structure. The elemental content of the palisade layer of the seed coat was determined with an X-ray spectrometer (HORIBA).

4.8. The Seed Vitality under Soil

Eighty seeds were randomly selected from each of the 15 plants purely planted in farmland soil, and they were packed into nylon mesh bags with a 0.2 mm pore size and buried in the research base of Nanjing Agricultural University in December of that year, 3 cm and 10 cm deep from the soil surface. The number of seeds that had emergence, the emergence rate after scarifying the seed coat, and the number of ungerminated seeds checked for rot and mildew were recorded.

4.9. Data Analysis

All data are statistically analyzed using SPSS (SPSS 22.0). Duncan's multiple range test in the univariate ANOVA test was used to analyze the differences in fitness indexes of the same material under four planting conditions, the thickness of different cell layers of transgenic soybean, wild soybean, and hybrid descendant seed coat, and the proportion of total thickness. The independent sample T test was used to analyze the differences in fitness indexes and composite fitness, in hard seededness rate and emergence rate after nicking hard seed coat, and in mineral element content between wild soybean and hybrid, and the data were plotted with Prism GraphPad.

5. Conclusions

The fitness of the F_2 and F_3 of herbicide-resistant transgenic soybean line T14R1251-70 and wild soybean LNTL and JLBC was significantly increased under farmland and wasteland soil conditions, as well as with or without weed competition, and the competitiveness was significantly enhanced. Self-pollinated seeds produced by hybrid descendants were similar to wild soybeans with a hard seed coat but had a lower hard seededness rate due to the seed coat structure. The decrease in hard seededness was due to the thicker hourglass cells and the lower Ca content in the seed coat. Hybrid descendants containing modified gene *cp4-epsps* can complete life histories and produce a large number of seeds, which can persist in the soil for a long time, germinate, and grow under suitable conditions. So, the hybrid descendants of herbicide-resistant transgenic soybean and wild soybean have potential ecological risks.

Author Contributions: Conceptualization, X.-L.S.; formal analysis, R.L., X.-Q.J. and J.-L.L.; funding acquisition, S.Q. and X.-L.S.; methodology, R.L., X.-Q.J. and J.-L.L.; project administration, S.Q. and X.-L.S.; validation, R.L. and X.-Q.J.; writing—original draft, R.L.; writing—review and editing, X.-L.S. and K.M.O. All authors will be informed about each step of manuscript processing including submission, revision, revision reminder, etc., via emails from our system or assigned Assistant Editor. All authors have read and agreed to the published version of the manuscript.

Funding: This research was funded by grants from the National Natural Science Foundation of China, grant number 32071656 and the National Special Transgenic Project of China, grant number 2016ZX08012005.

Data Availability Statement: Data is contained within the article. The data presented in this study are available in figures and tables.

Acknowledgments: Yu-qi Hu, Ze-wen Sheng and Jin-yue Liu participated in material planting.

Conflicts of Interest: The authors declare no conflict of interest.

References

1. ISAAA. *Global Status of Commercialized Biotech/GM Crops in 2018: Biotech Crops Continue to Help Meet the Challenges of Increased Population and Climate Change*; The International Service for the Acquisition of Agri-Biotech Applications (ISAAA): Ithaca, NY, USA, 2018, Volume 54.
2. ISAAA. GM Approval Database. Available online: https://www.isaaa.org/gmapprovaldatabase/citations/default.asp (accessed on 17 August 2022).
3. Wang, K.J.; Li, X.H. Interspecific gene flow and the origin of semi-wild soybean revealed by capturing the natural occurrence of introgression between wild and cultivated soybean populations. *Plant Breed.* **2011**, *130*, 117–127. [CrossRef]
4. Kim, D.Y.; Heo, J.H.; Pack, I.S.; Park, J.-H.; Um, M.S.; Kim, H.J.; Park, K.W.; Nam, K.-H.; Oh, S.D.; Kim, J.K.; et al. Natural hybridization between transgenic and wild soybean genotypes. *Plant Biotechnol. Rep.* **2021**, *15*, 299–308. [CrossRef]
5. Wang, K.J.; Li, X.H. Genetic diversity and gene flow dynamics revealed in the rare mixed populations of wild soybean (*Glycine soja*) and semi-wild type (Glycine gracilis) in China. *Genet Resour. Crop Ev.* **2013**, *60*, 2303–2318. [CrossRef]
6. Kitamoto, N.; Kaga, A.; Kuroda, Y.; Ohsawa, R. A model to predict the frequency of integration of fitness-related QTLs from cultivated to wild soybean. *Transgenic Res.* **2012**, *21*, 131–138. [CrossRef] [PubMed]
7. Mizuguti, A.; Yoshimura, Y.; Matsuo, K. Flowering phenologies and natural hybridization of genetically modified and wild soybeans under field conditions. *Weed Biol. Manag.* **2009**, *9*, 93–96. [CrossRef]
8. Nakayama, Y.; Yamaguchi, H. Natural hybridization in wild soybean (*Glycine max* ssp. *soja*) by pollen flow from cultivated soybean (*Glycine max* ssp. *max*) in a designed population. *Weed Biol. Manag.* **2002**, *2*, 25–30. [CrossRef]
9. Liu, B.; Xue, K.; Liu, L.-P.; Zhou, Y.; Han, J. Progress on the Gene Flow From Genetically Modified Soybeans to Wild Soybeans. *J. Ecol. Rural Environ.* **2020**, *36*, 833–841. [CrossRef]
10. Kan, G.-Z. Fitness of Hybrids between Wild Soybeans(*Glycine soja*) and the Glyphosateresistant Transgenic Soybean (*Glycine max*). *Soybean Sci.* **2015**, *34*, 177–184.
11. Kim, H.J.; Kim, D.Y.; Moon, Y.S.; Pack, I.S.; Park, K.W.; Chung, Y.S.; Kim, Y.J.; Nam, K.-H.; Kim, C.-G. Gene flow from herbicide resistant transgenic soybean to conventional soybean and wild soybean. *Appl. Biol. Chem.* **2019**, *62*, 54. [CrossRef]

12. Jenczewski, E.; Ronfort, J.; Chevre, A.-M. Crop-to-wild gene flow, introgression and possible fitness effects of transgenes. *Environ. Biosaf. Res.* **2003**, *2*, 9–24. [CrossRef]

13. Wang, K.J.; Li, X.H. Phylogenetic relationships, interspecific hybridization and origin of some rare characters of wild soybean in the subgenus *Glycine soja* in China. *Genet Resour. Crop Evol.* **2012**, *59*, 1673–1685. [CrossRef]

14. Liu, J.Y.; Sheng, Z.W.; Hu, Y.Q.; Liu, Q.; Qiang, S.; Song, X.L.; Liu, B. Fitness of F1 hybrids between 10 maternal wild soybean populations and transgenic soybean. *Transgenic Res.* **2021**, *30*, 105–119. [CrossRef] [PubMed]

15. Baskin, C.C. Breaking physical dormancy in seeds-focussing on the lens. *New Phytol.* **2003**, *158*, 229–232. [CrossRef]

16. Thiruppathi, D. Why so stubborn? MtKNOX4-regulated MtKCS12 manifests hardseededness. *Plant Physiol.* **2021**, *186*, 1367–1368. [CrossRef]

17. Liu, A.; Ku, Y.S.; Contador, C.A.; Lam, H.M. The Impacts of Domestication and Agricultural Practices on Legume Nutrient Acquisition Through Symbiosis With Rhizobia and Arbuscular Mycorrhizal Fungi. *Front. Genet.* **2020**, *11*, 583954. [CrossRef]

18. Liang, R.; Ji, X.; Sheng, Z.; Liu, J.; Qiang, S.; Song, X. Fitness and Rhizobacteria of F2, F3 Hybrids of Herbicide-Tolerant Transgenic Soybean and Wild Soybean. *Plants* **2022**, *11*, 3184. [CrossRef]

19. Guan, Z.J.; Zhang, P.F.; Wei, W.; Mi, X.C.; Kang, D.M.; Liu, B. Performance of hybrid progeny formed between genetically modified herbicide-tolerant soybean and its wild ancestor. *AoB Plants* **2015**, *7*, plv121. [CrossRef] [PubMed]

20. Yook, M.J.; Park, H.R.; Zhang, C.J.; Lim, S.H.; Jeong, S.C.; Chung, Y.S.; Kim, D.S. Environmental risk assessment of glufosinate-resistant soybean by pollen-mediated gene flow under field conditions in the region of the genetic origin. *Sci. Total Environ.* **2021**, *762*, 143073. [CrossRef] [PubMed]

21. Kuroda, Y.; Kaga, A.; Tomooka, N.; Yano, H.; Takada, Y.; Kato, S.; Vaughan, D. QTL affecting fitness of hybrids between wild and cultivated soybeans in experimental fields. *Ecol. Evol.* **2013**, *3*, 2150–2168. [CrossRef]

22. Hu, Y.-Q.; Sheng, Z.-W.; Liu, J.-Y.; Liu, Q.; Qiang, S.; Song, X.-L.; Liu, B. Sexual compatibility of transgenic soybean and different wild soybean populations. *J. Integr. Agr.* **2022**, *21*, 36–48. [CrossRef]

23. Wang, X.; Chen, L.; Ma, J. Genomic introgression through interspecific hybridization counteracts genetic bottleneck during soybean domestication. *Genome Biol.* **2019**, *20*, 22. [CrossRef]

24. Kim, M.Y.; Van, K.; Kang, Y.J.; Kim, K.H.; Lee, S.H. Tracing soybean domestication history: From nucleotide to genome. *Breed Sci.* **2012**, *61*, 445–452. [CrossRef]

25. Sedivy, E.J.; Wu, F.; Hanzawa, Y. Soybean domestication: The origin, genetic architecture and molecular bases. *New Phytol.* **2017**, *214*, 539–553. [CrossRef] [PubMed]

26. Kim, M.S.; Lozano, R.; Kim, J.H.; Bae, D.N.; Kim, S.T.; Park, J.H.; Choi, M.S.; Kim, J.; Ok, H.C.; Park, S.K.; et al. The patterns of deleterious mutations during the domestication of soybean. *Nat. Commun.* **2021**, *12*, 97. [CrossRef] [PubMed]

27. Hu, X.W.; Zhang, R.; Wu, Y.P.; Baskin, C.C. Seedling tolerance to cotyledon removal varies with seed size: A case of five legume species. *Ecol. Evol.* **2017**, *7*, 5948–5955. [CrossRef]

28. Xiong, Y.W.; Gong, Y.; Li, X.W.; Chen, P.; Ju, X.Y.; Zhang, C.M.; Yuan, B.; Lv, Z.P.; Xing, K.; Qin, S. Enhancement of growth and salt tolerance of tomato seedlings by a natural halotolerant actinobacterium *Glutamicibacter halophytocola* KLBMP 5180 isolated from a coastal halophyte. *Plant Soil* **2019**, *445*, 307–322. [CrossRef]

29. Clo, J.; Ronfort, J.; Gay, L. Fitness consequences of hybridization in a predominantly selfing species: Insights into the role of dominance and epistatic incompatibilities. *Heredity* **2021**, *127*, 393–400. [CrossRef]

30. Winn, A.A.; Elle, E.; Kalisz, S.; Cheptou, P.O.; Eckert, C.G.; Goodwillie, C.; Johnston, M.O.; Moeller, D.A.; Ree, R.H.; Sargent, R.D.; et al. Analysis of Inbreeding Depression in Mixed-Mating Plants Provides Evidence for Selective Interference and Stable Mixed Mating. *Evolution* **2011**, *65*, 3339–3359. [CrossRef]

31. Wang, Y.; Xu, C.; Sun, J.; Dong, L.; Li, M.; Liu, Y.; Wang, J.; Zhang, X.; Li, D.; Sun, J.; et al. GmRAV confers ecological adaptation through photoperiod control of flowering time and maturity in soybean. *Plant Physiol.* **2021**, *187*, 361–377. [CrossRef]

32. Singh, R.K.; Bhatia, V.S.; Bhat, K.V.; Mohapatra, T.; Singh, N.K.; Bansal, K.C.; Koundal, K.R. SSR and AFLP based genetic diversity of soybean germplasm differing in photoperiod sensitivity. *Genet. Mol. Biol.* **2010**, *33*, 319–324. [CrossRef]

33. Bai, Z.Y.; Ding, X.L.; Zhang, R.J.; Yang, Y.H.; Wei, B.G.; Yang, S.P.; Gai, J.Y. Transcriptome Analysis Reveals the Genes Related to Pollen Abortion in a Cytoplasmic Male-Sterile Soybean (*Glycine max* (L.) Merr.). *Int. J. Mol. Sci.* **2022**, *23*, 12227. [CrossRef] [PubMed]

34. Singh, R.J.; Hymowitz, T. The Genomic Relationships among 6 Wild Perennial Species of the Genus Glycine Subgenus Glycine Willd. *Theor. Appl. Genet.* **1985**, *71*, 221–230. [CrossRef] [PubMed]

35. Singh, R.J.; Nelson, R.L. Intersubgeneric hybridization between *Glycine max* and G-tomentella: Production of F-1, amphidiploid, BC1, BC2, BC3, and fertile soybean plants. *Theor. Appl. Genet.* **2015**, *128*, 1117–1136. [CrossRef] [PubMed]

36. Greenfield, L.M.; Hill, P.W.; Paterson, E.; Baggs, E.M.; Jones, D.L. Do plants use root-derived proteases to promote the uptake of soil organic nitrogen? *Plant Soil* **2020**, *456*, 355–367. [CrossRef]

37. Otlewska, A.; Migliore, M.; Dybka-Stepien, K.; Manfredini, A.; Struszczyk-Swita, K.; Napoli, R.; Bialkowska, A.; Canfora, L.; Pinzari, F. When Salt Meddles Between Plant, Soil, and Microorganisms. *Front. Plant Sci.* **2020**, *11*, 553087. [CrossRef]

38. Benezech, C.; Doudement, M.; Gourion, B. Legumes tolerance to rhizobia is not always observed and not always deserved. *Cell Microbiol* **2020**, *22*, e13124. [CrossRef]

39. Peix, A.; Ramírez-Bahena, M.H.; Velázquez, E.; Bedmar, E.J. Bacterial Associations with Legumes. *Crit. Rev. Plant Sci.* **2014**, *34*, 17–42. [CrossRef]

40. AbdElgawad, H.; Abuelsoud, W.; Madany, M.M.Y.; Selim, S.; Zinta, G.; Mousa, A.S.M.; Hozzein, W.N. Actinomycetes Enrich Soil Rhizosphere and Improve Seed Quality as well as Productivity of Legumes by Boosting Nitrogen Availability and Metabolism. *Biomolecules* **2020**, *10*, 1675. [CrossRef]

41. Liu, J.; Yu, X.; Qin, Q.; Dinkins, R.D.; Zhu, H. The Impacts of Domestication and Breeding on Nitrogen Fixation Symbiosis in Legumes. *Front. Genet.* **2020**, *11*, 00973. [CrossRef]

42. Zheng, Y.; Liang, J.; Zhao, D.L.; Meng, C.; Xu, Z.C.; Xie, Z.H.; Zhang, C.S. The Root Nodule Microbiome of Cultivated and Wild Halophytic Legumes Showed Similar Diversity but Distinct Community Structure in Yellow River Delta Saline Soils. *Microorganisms* **2020**, *8*, 207. [CrossRef]

43. Zhou, H.L.; Zhao, Q.; He, R.; Zhang, W.; Zhang, H.J.; Wang, H.Y.; Ao, X.; Yao, X.D.; Xie, F.T. Rapid Effect of Enriched Nitrogen on Soybean Nitrogen Uptake, Distribution, and Assimilation During Early Flowering Stage. *J. Soil Sci. Plant Nutr.* **2022**, *22*, 3798–3810. [CrossRef]

44. Oaks, A. Primary Nitrogen Assimilation in Higher-Plants and Its Regulation. *Can. J. Bot.* **1994**, *72*, 739–750. [CrossRef]

45. Li, R.; Chen, H.; Yang, Z.; Yuan, S. Research status of soybean symbiosis nitrogen fixation. *Oil Crop Sci.* **2020**, *5*, 6–10. [CrossRef]

46. Wang, Q.; Ma, M.; Jiang, X.; Guan, D.; Wei, D.; Cao, F.; Kang, Y.; Chu, C.; Wu, S.; Li, J. Influence of 37 Years of Nitrogen and Phosphorus Fertilization on Composition of Rhizosphere Arbuscular Mycorrhizal Fungi Communities in Black Soil of Northeast China. *Front. Microbiol.* **2020**, *11*, 539669. [CrossRef]

47. Zhu, Q.; Wang, H.; Shan, Y.Z.; Ma, H.Y.; Wang, H.Y.; Xie, F.T.; Ao, X. Physiological Response of Phosphorus-Efficient and Inefficient Soybean Genotypes under Phosphorus-Deficiency. *Russ. J. Plant Physl+* **2020**, *37*, 175–184. [CrossRef]

48. Ribet, J.; Drevon, J.J. Phosphorus Deficiency Increases the Acetylene-Induced Decline in Nitrogenase Activity in Soybean (Glycine-Max (L) Merr). *J. Exp. Bot.* **1995**, *46*, 1479–1486. [CrossRef]

49. Mckenzie-Gopsill, A.G.; Lee, E.; Lukens, L.; Swanton, C.J. Rapid and early changes in morphology and gene expression in soya bean seedlings emerging in the presence of neighbouring weeds. *Weed Res.* **2016**, *56*, 267–273. [CrossRef]

50. Wiles, L.J.; Wilkerson, G.G. Modeling Competition for Light between Soybean and Broadleaf Weeds. *Agr. Syst.* **1991**, *35*, 37–51. [CrossRef]

51. Chetan, F.; Rusu, T.; Chetan, C.; Urda, C.; Rezi, R.; Simon, A.; Bogdan, I. Influence of Soil Tillage Systems on the Yield and Weeds Infestation in the Soybean Crop. *Land* **2022**, *11*, 1708. [CrossRef]

52. Rockenbach, A.P.; Rizzardi, M.A. Competition at the soybean V6 stage affects root morphology and biochemical composition. *Plant Biol.* **2020**, *22*, 252–258. [CrossRef]

53. Wagner, A. Competition for nutrients increases invasion resistance during assembly of microbial communities. *Mol. Ecol.* **2022**, *31*, 4188–4203. [CrossRef] [PubMed]

54. Martinez-Romero, E.; Aguirre-Noyola, J.L.; Taco-Taype, N.; Martinez-Romero, J.; Zuniga-Davila, D. Plant microbiota modified by plant domestication. *Syst. Appl. Microbiol.* **2020**, *43*, 126106. [CrossRef] [PubMed]

55. Subbarao, G.V.; Rondon, M.; Ito, O.; Ishikawa, T.; Rao, I.M.; Nakahara, K.; Lascano, C.; Berry, W.L. Biological nitrification inhibition (BNI)—Is it a widespread phenomenon? *Plant Soil* **2007**, *294*, 5–18. [CrossRef]

56. Gong, T.; Xin, X.F. Phyllosphere microbiota: Community dynamics and its interaction with plant hosts. *J. Integr. Plant Biol.* **2020**, *63*, 297–304. [CrossRef]

57. Gutierrez, A.; Grillo, M.A. Effects of Domestication on Plant-Microbiome Interactions. *Plant Cell Physiol.* **2022**, *63*, 1654–1666. [CrossRef] [PubMed]

58. Pantigoso, H.A.; Newberger, D.; Vivanco, J.M. The rhizosphere microbiome: Plant-microbial interactions for resource acquisition. *J. Appl. Microbiol.* **2022**, *133*, 2864–2876. [CrossRef]

59. Xia, H.Y.; Wang, Z.G.; Zhao, J.H.; Sun, J.H.; Bao, X.G.; Christie, P.; Zhang, F.S.; Li, L. Contribution of interspecific interactions and phosphorus application to sustainable and productive intercropping systems. *Field Crop Res.* **2013**, *154*, 53–64. [CrossRef]

60. Gijzen, M.; Weng, C.; Kuflu, K.; Woodrow, L.; Yu, K.; Poysa, V. Soybean seed lustre phenotype and surface protein cosegregate and map to linkage group E. *Genome* **2003**, *46*, 659–664. [CrossRef]

61. Qutob, D.; Ma, F.; Peterson, C.A.; Bernards, M.A.; Gijzen, M. Structural and permeability properties of the soybean seed coat. *Botany* **2008**, *86*, 219–227. [CrossRef]

62. Zhang, D.; Sun, L.; Li, S.; Wang, W.; Ding, Y.; Swarm, S.A.; Li, L.; Wang, X.; Tang, X.; Zhang, Z.; et al. Elevation of soybean seed oil content through selection for seed coat shininess. *Nat. Plants* **2018**, *4*, 30–35. [CrossRef]

63. Sun, L.; Miao, Z.; Cai, C.; Zhang, D.; Zhao, M.; Wu, Y.; Zhang, X.; Swarm, S.A.; Zhou, L.; Zhang, Z.J.; et al. GmHs1-1, encoding a calcineurin-like protein, controls hard-seededness in soybean. *Nat. Genet* **2015**, *47*, 939–943. [CrossRef]

64. Satya Srii, V.; Nagarajappa, N.; Vasudevan, S.N. Is seed coat structure at fault for altered permeability and imbibition injury in artificially aged soybean seeds? *Crop Sci.* **2022**, *62*, 1573–1583. [CrossRef]

65. Kuchlan, M.K.; Dadlani, M.; Samuel, D.V.K. Seed Coat Properties and Longevity of Soybean Seeds. *J. New Seeds* **2010**, *11*, 239–249. [CrossRef]

66. Zablatzka, L.; Balarynova, J.; Klcova, B.; Kopecky, P.; Smykal, P. Anatomy and Histochemistry of Seed Coat Development of Wild (*Pisum sativum* subsp. elatius (M. Bieb.) Asch. et Graebn. and Domesticated Pea (*Pisum sativum* subsp. *sativum* L.). *Int. J. Mol. Sci* **2021**, *22*, 4602. [CrossRef] [PubMed]
67. Smykal, P.; Vernoud, V.; Blair, M.W.; Soukup, A.; Thompson, R.D. The role of the testa during development and in establishment of dormancy of the legume seed. *Front. Plant Sci.* **2014**, *5*, 351. [CrossRef]

Article

Topsoil and Vegetation Dynamics 14 Years after *Eucalyptus grandis* Removal in Eastern Cape Province of South Africa

Kuhle Mthethwa [1] and **Sheunesu Ruwanza** [1,2,*]

1 Department of Environmental Science, Rhodes University, Makhanda 6140, South Africa; kuhlemthethwa.h@gmail.com
2 Department of Environmental Science, Centre of Excellence for Invasion Biology, Rhodes University, Makhanda 6140, South Africa
* Correspondence: ruwanza@yahoo.com; Tel.: +27-46-603-7009

Abstract: A great deal of effort has been made to clear invasive alien plants in South Africa, yet it remains unclear if the clearing efforts are yielding positive soil and vegetation recovery trajectories. A few short-term studies have been conducted to monitor soil and vegetation recovery after alien plant removal in South Africa, but convincing, long-term monitoring studies are scarce yet needed. We investigated topsoil and vegetation recovery following *Eucalyptus grandis* removal 14 years ago by Working for Water in Makhanda, Eastern Cape province of South Africa. The detailed topsoil and vegetation surveys were conducted on forty 10 m × 10 m plots that were in paired cleared and natural sites. The results show no significant differences for the measured soil pH, total N, total C, K, Ca, and Na between the cleared and natural sites, an indication that the two sites are becoming similar. Similarly, the gravimetric soil moisture content shows no significant differences between the two sites, although monthly variations are observed. The topsoils in the cleared sites are hydrophobic as compared to those in the natural sites, which are wettable. We observed no significant vegetation diversity differences between the two sites, with native woody species, such as *Crassula pellucida* and *Helichrysum petiolare*, frequently occurring in the cleared sites. We recorded low reinvasion by *E. grandis* and other secondary invaders like *Acacia mearnsii* and *Rubus cuneifolius* in the cleared sites. Based on these results, we conclude that 14 years after *E. grandis* clearing, both topsoil and vegetation recovery are following a positive trajectory towards the natural sites. However, both reinvasion and secondary invasion have the potential to slow down soil and native vegetation recovery. Recommendations such as timeous follow up clearing and incorporating restoration monitoring in the WfW clearing programme are discussed.

Keywords: biological invasion; invasive alien plants; ecological restoration; plant–soil recovery; follow-up clearing

Citation: Mthethwa, K.; Ruwanza, S. Topsoil and Vegetation Dynamics 14 Years after *Eucalyptus grandis* Removal in Eastern Cape Province of South Africa. *Plants* **2023**, *12*, 3047. https://doi.org/10.3390/plants12173047

Academic Editors: Congyan Wang and Hongli Li

Received: 14 July 2023
Revised: 18 August 2023
Accepted: 22 August 2023
Published: 24 August 2023

1. Introduction

Invasion by invasive alien plants is a major threat to South Africa's socio-economic and ecological environment [1]. For example, O'Connor and van Wilgen [2] reported that an invasion of South Africa's rangelands by invasive alien plants, such as *Acacia*, *Eucalyptus*, *Opuntia*, and *Prosopis* species, can negatively affect livestock grazing and subsequently reduce livestock production by an estimated ZAR 340 million per year. In the Western and Eastern Cape provinces of South Africa, invasion by some of the above-mentioned alien plant species has been shown to negatively affect water resources and the country's natural vegetation [3]. From a social standpoint, invasion by invasive alien plants, such as *A. dealbata*, increases local people's vulnerability through a reduction in crop yields and grazing lands [4]. In urban areas, invasive alien plants contribute towards the homogenisation of city habitats, clogging water canals resulting in flooding, soil erosion, and disrupting ecosystem services, such as water infiltration [5,6]. These negative socio-economic and ecological effects caused by invasive alien plants need to be reversed, thus the need for restoration after alien plant control and management [7].

In South Africa, the control and management of invasive alien plants have been performed by the Working for Water (WfW) programme since its inception in 1995 [8]. The WfW programme is a poverty alleviation initiative that champions the control of invasive alien plants to protect and maximise water resources in the country [9]. The programme assumes that clearing invasive alien plants will result in the passive restoration of ecosystems to their original condition [9]. To date, an estimated ZAR 15 billion has been used to control invasive alien plants at a rate of approximately 200,000 condensed ha per year [10]. Generally, the programme is regarded as a success, although several challenges have been reported, e.g., having dual objectives, which negatively affect budget prioritisation, inefficiencies associated with having multiple country-wide projects, lack of restoration goals post alien plant clearing, and funding limitations [9,10]. Besides the above-mentioned challenges, little is known regarding how cleared areas recover post alien plant clearing by WfW [11]. Although few studies have been conducted in South Africa to assess ecosystem recovery after invasive alien plant removal by WfW, some studies have shown little vegetation recovery due to secondary invasion and a lack of native species soil seed banks [11,12], whereas others have reported a positive vegetation recovery trajectory [11,13,14]. Most of the above-mentioned studies were short-term monitoring studies that were conducted less than five years after the initial alien plant clearing, thus failing to give a clear picture of the restoration trajectory. Therefore, there is an urgent need to conduct long-term restoration studies after invasive alien plant clearing to understand recovery trajectories and develop effective restoration guidelines across varied contexts [11,15–17].

Long-term restoration monitoring after alien plant clearing is needed if the WfW clearing initiative is to yield positive ecosystem recovery outcomes. Although short-term restoration studies after alien plant clearing are important, they fail to investigate the recovery trajectories over time and justify the restoration funding since some goals might not have been achieved [18]. Therefore, this means that monitoring interventions cannot be implemented or tracked over time [18]. In contrast, long-term monitoring of invasive alien plant-cleared sites has the potential to assess the recovery trajectory over a long period as well as assess changes over varied environmental and climatic events. In addition, long-term monitoring can improve restoration effectiveness by implementing interventions that will also be monitored. In addition, it allows for restoration decisions to be made based on generated long-term data [19]. Therefore, long-term ecological restoration monitoring following alien plant removal is essential for investigating ecosystem recovery trajectories, thus providing important information that can be used to inform future restoration initiatives [20].

Most ecological restoration studies following alien plant removal have monitored the recovery of vegetation by measuring species abundance, composition, and diversity [11] but neglected soil monitoring. For example, Ruwanza et al. [11,13] conducted both short- and medium-term vegetation recovery monitoring following *E. camaldulensis* removal in the Western Cape province of South Africa and reported that the vegetation composition is dominated by grasses and herbs during the early stages of restoration but changes over time as shrubs and trees start to recruit in the cleared sites. Fill et al. [15] reported the dominance of native riparian shrubs following the removal of invasive alien plants along the riparian zones of the Rondegat River in South Africa; however, a high diversity of alien grasses was also reported. Very few restoration studies have monitored changes in the soil properties after alien plant removal [14,21], yet soil recovery has the potential to influence plant composition through soil–plant interactions [22,23]. Both Ndou and Ruwanza [14] and Kerr and Ruwanza [24] reported mixed results (both increases and decreases in soil nutrients depending on the measured property) in soil recovery following *Acacia* and *Eucalyptus* (respectively) removal in the Eastern Cape province of South Africa.

Nsikani et al. [25] reported that the soil nitrogen levels remain high in soils after *A. saligna* removal, an indication that soil legacy persistence has the potential to negatively affect vegetation recovery through promoting the growth of weedy secondary invaders. Methodologically, the bulk of the above-mentioned studies [14,24,25] on soil recovery after alien plant removal have assessed topsoil because it is the main source of soil nutrients and organic matter that is used by recruiting vegetation. Also, topsoil is assessed in restoration studies because it is a major repository of soil microbes that are known to influence the decomposition of plant debris, thus shaping both the above and below-ground vegetation recruitment trajectory. Although assessment of both the top and below-ground soil properties can yield more accurate results, the assessment of topsoil (which was performed in this study) can provide valuable information that can be used to assess the ecosystem recovery after alien plant removal. Our emphasis on topsoil measurements is centred on their role in influencing vegetation recovery after alien plant clearing, i.e., topsoil supply recruiting plants with valuable nutrients.

This study is motivated by the need for long-term monitoring of soil and vegetation recovery to gauge the effectiveness of alien plant clearing by WfW. To our knowledge, few long-term ecological restoration monitoring studies have been conducted in South Africa [14], yet billions of Rands have been invested in alien plant clearing. This paper presents the results of topsoil and vegetation monitoring 14 years after *E. grandis* removal by WfW. We used a comparative approach to assess physico-chemical properties in topsoil and native vegetation recovery following the initial *E. grandis* removal in 2008. Our results can provide important information that can be used for the adaptive management of alien plant-cleared areas.

2. Results

2.1. Effects of Alien Plant Clearing on Soil Properties

The topsoil (hereafter soil) from both the cleared and natural sites were sand (70% and 60%, respectively) and loam (30% and 40%, respectively) soils. Only soil P and Mg were significantly ($p < 0.01$) higher in the natural as compared to the cleared sites (Table 1). Soil P was almost twice higher in the natural as compared to the cleared sites. All other measured soil properties, namely, pH, total C, total N, K, Na, and Ca showed no significant ($p > 0.05$) differences between the cleared and natural sites (Table 1).

Table 1. Comparison of soil physical and chemical attributes between cleared and natural sites. Data are means ± SE and *t*-test results are shown.

	Cleared	Natural	*t*-Values	*p*-Values
Soil pH	4.43 ± 0.19	4.11 ± 0.04	1.66	0.115
	Total nutrient concentrations			
P Bray II (mg/kg)	3.66 ± 0.47	7.87 ± 1.27	3.11	0.006
C (%)	2.90 ± 0.25	3.51 ± 0.32	1.50	0.151
N (%)	0.23 ± 0.02	0.30 ± 0.04	1.50	0.150
	Exchangeable cations (%)			
K	6.22 ± 1.63	8.77 ± 2.50	0.85	0.404
Na	2.49 ± 0.19	2.80 ± 0.23	1.02	0.323
Ca	38.43 ± 1.20	30.28 ± 4.31	1.82	0.085
Mg	14.46 ± 1.65	20.19 ± 0.40	3.38	0.003

Gravimetric soil moisture content varied significantly between the cleared and natural sites ($p < 0.05$) but not across months ($p > 0.05$; Figure 1A). Significant differences in gravimetric soil moisture content were only visible in June (mean = 11.20 in cleared and 18.45 in natural) but not in May (mean = 19.22 in cleared and 18.20 in natural) and July (mean = 14.81 in cleared and 18.68 in natural) (Figure 1A). The month of June had the lowest gravimetric soil moisture content in the cleared sites (Figure 1A). There were no significant interactions ($p < 0.05$) between the sites and months for gravimetric soil moisture content

(Figure 1A). Soil penetration resistance levels showed no significant ($p > 0.05$) differences between cleared and natural sites (Figure 1B). In contrast, monthly comparisons for soil penetration resistance levels showed significant ($p < 0.05$) differences, with July recording the lowest soil penetration resistance levels compared to May and June (Figure 1B). The average soil penetration resistance level across all months was 3.08 in May, 3.48 in June, and 3.04 in July. There were no significant interactions ($p > 0.05$) between the sites and months for soil penetration resistance levels (Figure 1B).

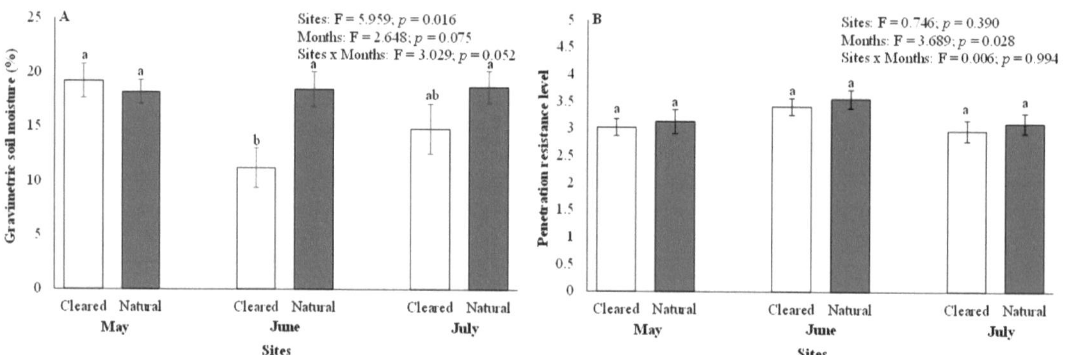

Figure 1. (**A**) Gravimetric soil moisture content, and (**B**) soil penetration resistance levels in cleared and natural sites. Bars represent means $\pm$ SE and the ANOVA results are shown. Bars with different letter superscripts are significantly different at $p < 0.05$.

For soil water repellency, most of the soil in the cleared sites were slightly repellent in May (55%) and June (45%); however, in July the bulk of the soils were wettable (60%) (Figure 2). In the cleared sites, strongly repellent soils were observed across all months, with higher percentages in June (15%) and July (25%) as compared to May (5%). Some of the soil in the cleared sites were severely repellent for all three months (10% in May and 5% in June and July, respectively) as compared to the natural sites, which reported no strongly repellent soils (Figure 2). The bulk of the soils in the natural sites were wettable across all months (May = 90%, June = 85%, and July = 95%) (Figure 2). The remainder of the soils in the natural site were slightly repellent (Figure 2). A chi-squared analysis of the WDPT categories showed significant differences between the cleared and natural sites for all three months (May: $\chi^2 = 15.23$, $p = 0.002$; June: $\chi^2 = 11.17$, $p = 0.011$; July: $\chi^2 = 7.91$, $p = 0.050$).

2.2. Effects of Alien Plant Clearing in Vegetation

Although all measured indices of diversity (species richness, Shannon–Wiener, Simpson's index of diversity, and evenness index) were high in the natural compared to the cleared sites, statistical comparisons showed no significant ($p > 0.05$) differences between the two sites (Table 2). Of all the 50 positively identified plant species, 27 were trees and shrubs, 12 were forbs, and 11 were graminoids and sedges (Table A1). Five plant species, namely, *Asparagus suaveolens*, *Crassula pellucida*, *Helichrysum petiolare*, *Centella asiatica*, and *Conyza bonariensis*, had a frequency occupancy of more than 50% in the cleared sites, while eight species, namely, *A. suaveolens*, *C. pellucida*, *H. cymosum*, *C. asiatica*, *Senecio macrocephalus*, *Agrostis lachnantha*, *Digitaria sanguinalis*, and *Pennisetum clandestinum*, had a frequency occupancy of more than 50% in the natural sites (Table A1). Of all the 27 identified trees and shrubs, 11 were present in both the cleared and natural sites. Half of the identified forbs were in both the cleared and natural sites, and only five graminoids were present in both sites. Two woody invasive alien plants, namely, *A. mearnsii* and *Rubus cuneifolius*, occurred in the cleared sites with a frequency occupancy of less than 25% (Table A1).

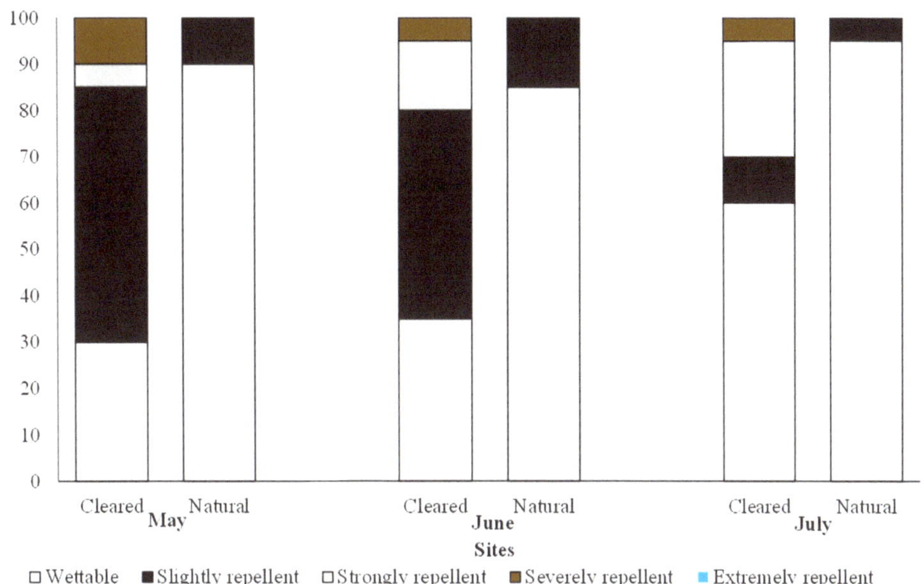

Figure 2. Distribution of the water repellency classes based on the water droplet penetration time method in the soil samples from cleared and natural sites. Chi-squared results are shown.

Table 2. Comparison of indices of diversity between the cleared and natural sites. Data are means $\pm$ SE and *t*-test results are shown.

	Cleared	Natural	*t*-Values	*p*-Values
Species richness	8.65 $\pm$ 0.51	10.25 $\pm$ 0.63	1.96	0.057
Shannon–Wiener	1.25 $\pm$ 0.06	1.40 $\pm$ 0.08	1.51	0.139
Simpsons index of diversity	0.72 $\pm$ 0.04	0.74 $\pm$ 0.02	0.69	0.495
Evenness index	0.59 $\pm$ 0.03	0.61 $\pm$ 0.02	0.68	0.679

3. Discussion

Fourteen years after the initial removal of *E. grandis* by WfW, our results show that both the soil physico-chemical properties and the vegetation diversity are recovering in the cleared sites. We observed no significant differences between the cleared and natural sites for most of the measured soil and vegetation variables, an indication that ecosystem recovery is taking place. These results were originally observed by Kerr and Ruwanza [24], who reported a positive vegetation recovery trajectory in one of the cleared sites. However, the same study noted varied clearing effects on the soil properties, both increased and decreased changes. Our results concur with the previous studies that have shown that soil and vegetation recovery tend to follow a positive restoration trajectory several years after the initial clearing [11,14]. Ndou and Ruwanza [14] reported that both soil and vegetation recovery was taking place on old *Acacia*-cleared sites (15 years) than on recently cleared sites (6 years). Similarly, Ruwanza et al. [11] assessed vegetation recovery seven years after *E. camaldulensis* removal along the Berg River and reported a positive vegetation recovery trajectory.

Our results on topsoil showed no significant differences between the cleared and natural sites for all measured soil properties except for P and Mg, an indication that soils in the cleared site resemble those in the natural sites. This contradicts the soil results by Kerr and Ruwanza [24], who observed varied soil nutrient changes. Several factors, including diminishing soil legacy effects after invasive alien plant removal, can explain our soil results [21,23,26]. It is well-documented that soil legacy effects caused by the invader can persist for several years after alien plant removal [21,23,26]; this depends on several factors, such as previous invasion extent, invasion by secondary invaders that add more soil nutrients, and external factors, e.g., grazing and fires, which influence soils [25]. Our results show a possibility that the soil legacy effect reported by Kerr and Ruwanza [24] could be diminishing and is no longer persistent in these cleared sites since the soils are now having similar properties to the ones in natural sites. It is known that the soil legacy effect can limit successful restoration post-alien plant removal [27]; however, in our case, the diminishing soil legacy effect could explain our soil results. Ndou and Ruwanza [14] reported that soil nutrients improve with increased clearing time, with the old cleared sites having similar soil nutrient levels to the natural sites, as compared to the recently cleared sites. In addition to diminishing soil legacy effects, the recovering native vegetation could also explain the observed topsoil nutrient results. It is possible that recruiting native species are using the excessive soil nutrients that were released by *E. grandis* before clearing. The above-mentioned speculation that recovering vegetation is playing a role in soil nutrient changes is plausible, given that similar trends have been observed in abandoned agricultural fields [28]. Studies in abandoned agricultural fields have reported that as woody species colonise grass-dominated abandoned fields, soil nutrients tend to decrease due to increased utilisation by recruiting plants [29]. It is not clear why the soil P and Mg were higher in the natural than in the cleared sites; however, organic matter content from native plant litter could explain this result. Some native species, such as *Maytenus acuminata* and *Pellaea mucronate*, were only present in the natural sites, hence the litter deposition from these species can influence the soil P through soluble P leaching from litter.

We did not observe variations between the cleared and natural sites on soil penetration resistance levels and gravimetric soil moisture content, except for monthly moisture differences, with June recording the lowest soil moisture content. The lack of soil compaction and moisture differences between the two sites could be because of the recruiting vegetation in the cleared sites. Recruiting plant phenological development and increased canopy cover in the cleared sites could have resulted in both soil compaction and moisture being the same as in the vegetated natural sites; however, the seasonal differences in moisture could be because of winter temperature and rainfall patterns. The soil moisture measurements were conducted in the austral winter when rainfall is low; thus, the soils are mostly dry and compact during that time. Several studies have shown that reductions in precipitation tend to lower soil moisture content to as much as 40% during dry months [30], and this reduction in soil moisture also results in increased soil compaction. It was anticipated that the above-mentioned observations in soil nutrients, moisture, and compaction in the cleared sites should have resulted in an improved soil water repellence, but that was not the case, as we recorded a greater percentage of repellent soils in the cleared rather than the natural sites. A palpable explanation is that external factors, such as livestock trampling, which is happening at a low-to-moderate scale in the cleared sites, could explain the reported soil repellence results. The impact of livestock trampling on soil is two-fold, (i) it can trigger soil compaction through decreased soil physical quality and hydraulic conductivity, and (ii) it can result in the detachment and shearing of topsoil layers [31]. However, the effects of livestock trampling on soil repellence remains unknown, with some studies suggesting a reduction in soil water repellence [31], whilst other studies claim an increase in soil water repellency due to increased soil compaction [32].

Our results on vegetation show that native species are recruiting in the cleared sites, an indication that passive native vegetation recovery is taking place. The few long-term studies that have been conducted in South Africa have shown successful native vegetation recovery several years after alien plant removal [11,14]. The above-mentioned studies reported that native species diversity, composition, and cover increase as years since clearing increase [11,14]. Several factors can explain the recruitment of native species in our cleared sites. Firstly, the cleared and natural sites are close to each other; therefore, the natural sites could act as seed suppliers to the cleared sites. The proximity of the cleared sites to the natural sites can assist with native species seed dispersal by animals, such as birds, or through the wind from natural patches to the cleared sites [33]. Secondly, the native recruiting species that were recorded by Kerr and Ruwanza [24] at the same cleared sites could have been established by now, thus acting as nurse plants that facilitate the recruitment of other plants [34,35]. Previous studies have reported that the availability of nurse plants in restoration sites facilitates germination, establishment, and growth of other plant species through (i) attracting birds and insects to disperse seeds underneath them, (ii) providing nutrient-rich microhabitats underneath their canopy that facilitate the germination and growth of other plants, and (iii) buffer recruiting native plants from the harsh environmental conditions [34,35]. Thirdly, it is possible that a native soil seed bank still exists at the cleared sites. Several studies have reported that a soil seed bank of native species can remain in the soil for several years and recruit after the invader has been cleared [36,37]. Fourthly, plant–soil positive interactions could favour the recruitment of native species in the cleared sites. The reported soil nutrient recovery in the cleared sites could benefit recruiting native species through nutrient availability. In turn, recruiting plant species could influence soil properties through litter deposition [38]. Lastly, although livestock grazing can trigger both positive and negative effects on native species recruitment on the cleared sites, it is possible that grazing is assisting with seed dispersal in the cleared sites. Grazing was observed to be more dominant in the cleared sites due to the accessibility by animals since the vegetation is still low and recruiting. Indeed, grazing livestock can disperse native seeds through endozoochory (seed dispersal via ingestion) or epizoochory (seed dispersal accidentally via attachment to animal body) [39,40]. This dispersal has the potential to influence the native species diversity and composition in the cleared sites.

Although Kerr and Ruwanza [24] noted the dominance of secondary invaders and reinvasion by *E. grandis* on our cleared sites, we recorded low abundances of secondary invaders. Even if the observed secondary invasion is diminishing, it still has the potential to slow down native vegetation recovery through competition for resources such as nutrients and water, which alternately hinder native vegetation recovery [11,23]. The observed reinvasion and secondary invasion speak to the challenge of effective follow-up clearing by WfW [11]. As per the South African WfW clearing guidelines, our cleared sites are outside the initial three-year follow-up clearing plan, implying that the property owner is responsible for managing follow-up clearing to remove recruiting *E. grandis* and secondary invaders. However, this could be challenging for the property owner due to a lack of funding, equipment, and human capital to effectively implement follow-up clearing.

4. Materials and Methods

4.1. Study Area

The study was conducted at a private farm (33°20′24.72″ S; 26°27′11.81″ E) that is approximately 8 km from Makhanda (previously known as Grahamstown; Figure 3) in the Eastern Cape province of South Africa. The farm is currently being used for small-scale livestock grazing. Vegetation in the study areas is dominated by grassy fynbos and small bushveld shrubs [41]. The soils in the area are sandy, acidic, and nutrient-poor, derived from quartzite formation [41]. Rain falls throughout the year with a bimodal distribution, peaking in October-November and February-March [41]. Mean annual rainfall is 545 mm and temperature averages 26 °C in austral summer and 6 °C in austral winter [41].

Within the farm, we identified two cleared and adjacent natural sites. One of the paired cleared and natural sites were surveyed by Kerr and Ruwanza [24], however, it was difficult to identify the exact surveyed plots used in the above-mentioned study since there were removed after the termination of their experiment in 2016. Kerr and Ruwanza [24] assessed similar measurements that were assessed in this study. Our sites were approximately 500 m apart, and the paired cleared and natural areas within each site were separated by farm roads. Clearing of *E. grandis* was performed in 2008 by WfW [24]. Clearing involved the felling of *E. grandis* trees using chainsaws and the spraying of herbicides on cleared stamps to avoid re-sprouting. Felled trees were stack burnt and follow-up treatments to remove re-sprouting alien plants and saplings were conducted on a 4–6 month interval for three years after the initial clearing [24]. After follow-up completion around 2011, the cleared sites were handed over to the property owner for maintenance. Two nearby natural sites acted as reference sites, and these were dominated by native species with a canopy cover of more than 80% [24].

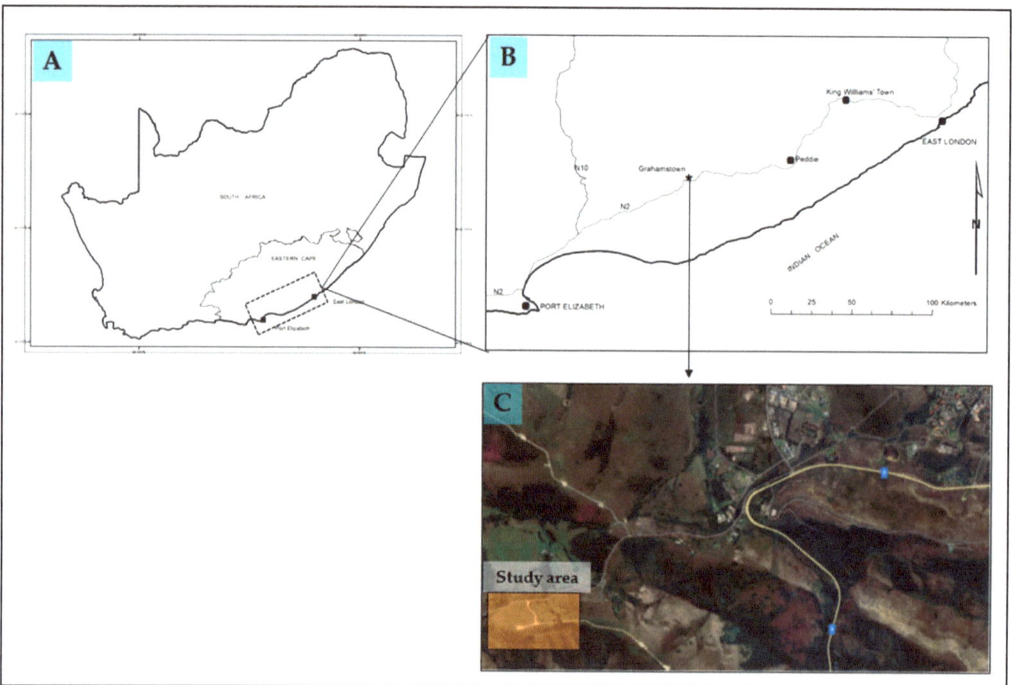

Figure 3. Map showing (**A**) location of study area in South Africa, (**B**) location of study area in the Eastern Cape province of South Africa, and (**C**) farm location (generated using Google Earth Pro Version 7.3 software). National Road (N2) is shown in yellow with highway name in blue.

4.2. Experimental Design and Data Collection

On each of the paired cleared and natural sites, soil and vegetation surveys were conducted on 10 m × 10 m plots with a buffer zone of 5 m. Each plot was replicated 10 times per site. The plots were marked with metal droppers to allow revisitation during repeated soil measurements. In total 40 plots were surveyed (10 plots per site × 4 sites (2 cleared and 2 natural)). Within each plot, soil cores measuring 10 cm in diameter and 10 cm in depth were collected at the centre of each plot for three months (May to July 2022). Soils were collected using a soil core after hand removal of stones and debris. Collected soils were packed in brown bags and immediately transported to Rhodes University laboratory for gravimetric soil moisture and water repellency measurements. Soil penetration resistance levels were conducted under field conditions, 30 cm from the plot centre where soils were collected. In June, an additional equal number of soils were collected for soil chemical analysis, which was assessed once due to financial limitations and the assumption that no soil chemical variations were expected within one winter season. Soil chemical analyses were conducted at a commercial laboratory, namely, Bemlab (Pty) Limited.

All collected soils were sieved using a 2 mm sieve upon arrival at the laboratory. To measure gravimetric soil moisture, sieved soils were weighed wet, oven-dried at 105 °C for three days and re-weighed to determine moisture content, which was expressed as a percentage [42]. The water droplet penetration time (WDPT) method was used to assess soil water repellency. Sieved soils were placed in Petri dishes, levelled, and air-dried for seven days under laboratory conditions where temperatures averaged 6 °C ($\pm$2 °C), similar to winter temperatures in Makhanda. After seven days, the WDPT test was conducted by placing five drops of distilled water on the soil surface using a hypodermic syringe and recording the time taken by each drop to penetrate the soil [24,43]. The average penetration time for the five drops was taken as the WDPT for each sample. The WDPT categories used were wettable (below 5 s), slightly repellent (5–60 s), strongly repellent (60–600 s), severely repellent (600–3600 s), and extremely repellent (above 3600 s) as described by Bisdom et al. [44] and Kerr and Ruwanza [24]. Soil penetration resistance levels (a measure of soil compaction) were performed using a pocket penetrometer (SOILTEST, Inc., Evanston, IL, USA). Measurements were taken in kg cm^{-2} as described by Leung and Meyer [45]. The penetrometer was pushed into the soil following the removal of debris and a metal ring was pushed to scale to record the penetration resistance measurement [45]. Soil pH, a measure of acidity and alkalinity of the soil was analysed in 1:5 soil-KCl extract as described by Rhoades [46]. Soil P was analysed using the Bray-II extract method as described by Bray and Krutz [47]. Soil total C was analysed using the modified Walkley-Black method as described by Chan et al. [48]. Soil total N was analysed by complete combustion using a Eurovector Euro EA Elemental Analyser (Euro EA; Eurovector, Milan, Italy). Exchangeable cations of K, Na, Ca, and Mg were extracted in a 1:10 ammonium acetate solution using the centrifuge procedure described by Thomas [49]. The soils were filtered and analysed using atomic absorption spectrometry (SP428, LECO Corporation, St. Joseph, MI, USA).

In June, detailed vegetation surveys were conducted in each plot. All identified trees and shrubs were counted in each plot, whereas forbs and graminoids were enumerated in a 1 × 1 m sub-plot positioned at the centre of the plot. Species were assigned to four growth forms based on morphology, namely, trees, shrubs, forbs (non-graminoid herbaceous plants), and graminoids [50]. All plant species were identified using local plant books such as Manning [51] and Manning and Goldblatt [52] as well as the PlantzAfrica online directory [53]. Species that could not be identified were taken to Selmar Schonland Herbarium at the Albany Museum in Makhanda for identification.

4.3. Data Analysis

All statistical analyses were performed using TIBCO STATISTICA version 14.0 software (TIBCO Software Inc., Palo Alto, California, USA) [54]. Normality tests were performed using the Kolmogorov–Smirnov test and data were normally distributed. The effect of clearing on gravimetric soil moisture and penetration resistance levels was analysed using repeated measures ANOVA since data were collected over three months. Where repeated ANOVAs were significantly different, Tukey's HSD test was used to determine differences between sites and across months at $p < 0.05$. Comparisons between cleared and natural sites for WDPT categories were performed using the Chi-squared test. Measured soil properties of pH, P, total N, total C, K, Ca, Mg, and Na were compared between cleared and natural sites using a *t*-test since data were collected once. Species richness, Shannon–Wiener diversity index, Simpson's index of diversity, and Evenness index were calculated per plot and compared between cleared and natural sites using a *t*-test since data were collected once.

5. Conclusions and Recommendations

In conclusion, we observed improved soil properties, vegetation diversity, and composition since the last monitoring assessment by Kerr and Ruwanza [24], evidence that ecosystem recovery on these cleared sites is following a positive restoration trajectory towards the natural sites. However, we observed evidence of reinvasion and secondary invasion in low abundance, and this is likely to slow down ecosystem recovery if not attended to. From a management standpoint, some interventions are needed if the current positive recovery trajectory is to be maintained. Although these interventions are yet to be tested and are not prescriptive as they aim to steer restoration conversations, we believe that these interventions need to be considered if clearing by WfW is to yield successful native vegetation recovery. Firstly, long-term restoration monitoring should be included in alien plant clearing and management plans, and such monitoring should be performed until restoration is completely achieved. Secondly, clearing managers should develop an effective and timeous follow-up clearing programme that monitors reinvasion and secondary invasion on cleared sites. Thirdly, there is a need to support landowners with resources to manage the cleared sites post the initial WfW follow-up period. Support to landowners could be in the form of financial resources to buy follow-up clearing chemicals, pay human capital, and information on how to manage the cleared sites. Lastly, future research on cleared sites should assess both topsoil and below-ground soil properties to provide accurate and detailed information on recovery after alien plant removal.

Author Contributions: Conceptualization, S.R. and K.M.; methodology, S.R. and K.M.; validation, S.R. and K.M.; investigation, K.M.; resources, S.R.; writing—original draft preparation, S.R.; writing—review and editing, S.R. and K.M.; supervision, S.R.; project administration, S.R.; funding acquisition, S.R. All authors have read and agreed to the published version of the manuscript.

Funding: This research was funded by National Research Foundation (NRF) South Africa, under the programme NRF Research Development Grant for Y-Rated Researchers (Grant number 137789). Additional funding for fieldwork was provided by DSI-NRF Centre of Excellence for Invasion Biology (CIB).

Data Availability Statement: The data that support our research findings are available from the corresponding author on request.

Acknowledgments: Thanks to the National Research Foundation (NRF: research grant UID number 137789) for funding to conduct fieldwork. Thanks to DSI-NRF Centre of Excellence for Invasion Biology (CIB) for additional funding. Also, thanks to the Department of Environmental Science at Rhodes University for providing equipment and transport for fieldwork. We also thank Uzziah Mutumbi and Ruwimbo Mutirwara who assisted with fieldwork. We are grateful to the farm owner, namely, Geoff Antrobus for permission to conduct this study at the farm.

Conflicts of Interest: The authors declare no conflict of interest.

Appendix A

Table A1. Fifty frequently occurring species in cleared and natural sites.

Species Name	Cleared	Natural
Acacia mearnsii	25	0
Anthospermum aethiopicum	5	0
Anthospermum spathulatum	30	15
Aspalathus subtingens	0	35
Asparagus suaveolens	50	80
Athanasia dentata	5	15
Chasmanthe aethiopica	0	5
Clutia daphnoides	0	5
Conyza scabrida	0	5
Crassula pellucida	50	60
Dovyalis rhamnoides	5	5
Eucalyptus grandis	30	0
Halleria lucida	5	15
Helichrysum cymosum	20	70
Helichrysum petiolare	65	45
Indigofera sp.	5	10
Maytenus acuminata	0	5
Metalasia muricata	5	20

Table A1. *Cont.*

Species Name	Cleared	Natural
Passerina rigida	0	15
Podocarpus latifolius	0	5
Protea neriifolia	0	10
Psychotria capensis	0	5
Pteronia incana	0	25
Searsia crenata	25	15
Rubus cuneifolius	5	0
Solanum linnaeanum	10	0
Zanthoxylum capense	0	5
Asplenium rutifolium	5	20
Centaurea benedicta	15	0
Centella asiatica	70	65
Conyza bonariensis	95	20
Erigeron canadensis	5	0
Erigeron sp.	10	10
Euphorbia epicyparissias	0	15
Pellaea mucronate	0	25
Pteridium aquilinum	0	10
Senecio macrocephalus	40	60
Thesium gnidiodes	45	40
Wahlenbergia procumbens	5	0
Agrostis lachnantha	30	55
Alloteropsis semialata	25	0
Aristida transvaalensis	5	0
Carex sp.	20	20
Cynodon dactylon	5	0
Cyperus albostriatus	20	0
Digitaria sanguinalis	25	65
Ehrharta erecta	35	0
Isolepis cernua	5	0
Pennisetum clandestinum	20	50
Sporobolus africanus	20	5

References

1. van Wilgen, B.W.; Measey, J.; Richardson, D.M.; Wilson, R.U.; Zengeya, T.A. Biological invasions in South Africa: An overview. In *Biological Invasions in South Africa*; van Wilgen, B.W., Measey, J., Richardson, D.M., Wilson, J.R., Zengeya, T.A., Eds.; Springer: Berlin/Heidelberg, Germany, 2020; pp. 3–32

2. O'Connor, T.; van Wilgen, B.W. The impact of invasive alien plants on rangelands in South Africa. In *Biological Invasions in South Africa*; van Wilgen, B.W., Measey, J., Richardson, D.M., Wilson, J.R., Zengeya, T.A., Eds.; Springer: Berlin/Heidelberg, Germany, 2020; pp. 457–486.

3. Le Maitre, D.C.; Blignaut, J.N.; Clulow, A.; Dzikiti, S.; Everson, C.G.; Örgens, A.H.M.; Gush, M.B. Impacts of plant invasions on terrestrial water flows in South Africa. In *Biological Invasions in South Africa*; van Wilgen, B.W., Measey, J., Richardson, D.M., Wilson, J.R., Zengeya, T.A., Eds.; Springer: Berlin/Heidelberg, Germany, 2020; pp. 429–456.

4. Ngorima, A.; Shackleton, C.M. Livelihood benefits and costs from an invasive alien tree (*Acacia dealbata*) to rural communities in the Eastern Cape, South Africa. *J. Environ. Manag.* **2019**, *229*, 158–165. [CrossRef] [PubMed]

5. van Wilgen, B.W.; Scott, D.F. Managing fires on the Cape Peninsula: Dealing with the inevitable. *J. Mediterr. Ecol.* **2001**, *2*, 197–208.

6. Gaertner, M.; Larson, B.M.H.; Irlich, U.M.; Holmes, P.M.; Stafford, L.; van Wilgen, B.W.; Richardson, D.M. Managing invasive species in cities: A framework from Cape Town, South Africa. *Landsc. Urban Plan.* **2016**, *151*, 1–9. [CrossRef]

7. Holmes, P.M.; Esler, K.J.; Gaertner, M.; Geerts, S.; Hall, S.A.; Nsikani, M.M.; Richardson, D.M.; Ruwanza, S. Biological invasions and ecological restoration in South Africa. In *Biological Invasions in South Africa*; van Wilgen, B.W., Measey, J., Richardson, D.M., Wilson, J.R., Zengeya, T.A., Eds.; Springer: Berlin/Heidelberg, Germany, 2020; pp. 665–700.

8. van Wilgen, B.W.; Le Maitre, D.C.; Cowling, R.M. Ecosystem services, efficiency, sustainability and equity: South Africa's Working for Water programme. *Trends Ecol. Evol.* **1998**, *13*, 379. [CrossRef]

9. van Wilgen, B.W.; Forsyth, G.G.; Le Maitre, D.C.; Wannenburg, A.; Kotze, J.D.F.; ven den Berg, E.; Henderson, L. An assessment of the effectiveness of a large, national-scale invasive alien plant control strategy in South Africa. *Biol. Conserv.* **2012**, *148*, 28–38. [CrossRef]

10. van Wilgen, B.W.; Wilson, J.R.; Wannenburgh, A.; Foxcroft, L.C. The extent and effectiveness of alien plant control projects in South Africa. In *Biological Invasions in South Africa*; van Wilgen, B.W., Measey, J., Richardson, D.M., Wilson, J.R., Zengeya, T.A. Eds.; Springer: Berlin/Heidelberg, Germany, 2020; pp. 597–628.

11. Ruwanza, S.; Gaertner, M.; Esler, K.J.; Richardson, D.M. Medium-term vegetation recovery after removal of invasive *Eucalyptus camaldulensis* stands along a South African river. *S. Afr. J. Bot.* **2018**, *119*, 63–68. [CrossRef]

12. Reinecke, M.K.; Pigot, A.; King, J.M. Spontaneous succession of riparian fynbos: Is unassisted recovery a viable restoration strategy? *S. Afr. J. Bot.* **2008**, *74*, 412–420. [CrossRef]

13. Ruwanza, S.; Gaertner, M.; Esler, K.J.; Richardson, D.M. The effectiveness of active and passive restoration on recovery of indigenous vegetation in riparian zones in the Western Cape, South Africa: A preliminary assessment. *S. Afr. J. Bot.* **2013**, *88*, 132–141. [CrossRef]

14. Ndou, E.; Ruwanza, S. Soil and vegetation recovery following alien tree clearing in the Eastern Cape Province of South Africa *Afr. J. Ecol.* **2016**, *54*, 460–470. [CrossRef]

15. Fill, J.M.; Kritzinger-Klopper, S.; Van Wilgen, B.W. Short-term vegetation recovery after alien plant clearing along the Rondegat River, South Africa. *Restor. Ecol.* **2018**, *26*, 434–438. [CrossRef]

16. Holmes, P.M.; Esler, K.J.; Richardson, D.M.; Witkowski, E.T.F. Guidelines for improved management of riparian zones invaded by alien plants in South Africa. *S. Afr. J. Bot.* **2008**, *74*, 538–552. [CrossRef]

17. Blanchard, R.; Holmes, P.M. Riparian vegetation recovery after invasive alien tree clearance in the fynbos biome. *S. Afr. J. Bot.* **2008**, *74*, 421–431. [CrossRef]

18. Hagen, D.; Evju, M. Using short-term monitoring data to achieve goals in a large-scale restoration. *Ecol. Soc.* **2013**, *18*, 29. [CrossRef]

19. Gerhart, V.J.; Waugh, W.J.; Glenn, E.P.; Pepper, I.L. Ecological restoration. In *Monitoring and Characterization of the Environment*; Artiola, J.F., Pepper, I.L., Brusseau, M.L., Eds.; Academic Press: San Diego, CA, USA, 2004; pp. 357–375.

20. Prach, K.; Marrs, R.; Pyšek, P.; van Diggelen, R. Manipulation of succession. In *Linking Restoration and Ecological Succession*; Walker, L.R., Walker, J., Hobbs, R.J., Eds.; Springer Science + Business Media: New York, NY, USA, 2007; pp. 121–149.

21. Rodríguez-Echeverría, S.; Afonso, C.; Correia, M.; Lorenzo, P.; Roiloa, S.R. The effect of soil legacy on competition and invasion by *Acacia dealbata* Link. *Plant Ecol.* **2013**, *214*, 1139–1146. [CrossRef]

22. Zhu, S.-C.; Zheng, H.-X.; Liu, W.-S.; Liu, C.; Guo, M.-N.; Huot, H.; Morel, J.L.; Qiu, R.-L.; Chao, Y.; Tang, Y.-T. Plant-soil feedbacks for the restoration of degraded mine lands: A Review. *Front. Microbiol.* **2022**, *12*, 751794. [CrossRef]

23. Nsikani, M.M.; Geerts, S.; Ruwanza, S.; Richardson, D.M. Secondary invasion and weedy native species dominance after clearing invasive alien plants in South Africa: Status quo and prognosis. *S. Afr. J. Bot.* **2020**, *132*, 338–345. [CrossRef]

24. Kerr, T.F.; Ruwanza, S. Does *Eucalyptus grandis* invasion and removal affect soils and vegetation in the Eastern Cape Province, South Africa? *Austral Ecol.* **2016**, *41*, 328–338. [CrossRef]

25. Nsikani, M.M.; Novoa, A.; van Wilgen, B.W.; Keet, J.; Gaertner, M. *Acacia saligna*'s soil legacy effects persist up to 10 years after clearing: Implications for ecological restoration. *Austral Ecol.* **2017**, *42*, 880–889. [CrossRef]

26. Tererai, F.; Gaertner, M.; Jacobs, S.M.; Richardson, D.M. *Eucalyptus camaldulensis* invasion in riparian zones reveals few significant effects on soil physico-chemical properties. *River Res. Appl.* **2015**, *31*, 590–601. [CrossRef]

27. Grove, S.; Parker, I.M.; Haubensak, K.A. Persistence of a soil legacy following removal of a nitrogen-fixing invader. *Biol. Invasions* **2015**, *17*, 2621–2631. [CrossRef]

28. Memiaghe, H.R. Old Field Restoration: Vegetation Response to Soil Changes and Restoration Efforts in Western Cape Lowlands. Master's Thesis, Stellenbosch University, Stellenbosch, South Africa, 2008.

29. Scott, A.J.; Morgan, J.W. Recovery of soil and vegetation in semi-arid Australian old fields. *J. Arid Environ.* **2012**, *76*, 61–71. [CrossRef]

30. Feng, H.; Liu, Y. Combined effects of precipitation and air temperature on soil moisture in different land covers in a humid basin. *J. Hydrol.* **2015**, *531*, 1129–1140. [CrossRef]

31. Stavi, I.; Barkai, D.; Knoll, Y.M.; Zaady, E. Livestock grazing impact on soil wettability and erosion risk in post-fire agricultural lands. *Sci. Total Environ.* **2016**, *573*, 1203–1208. [CrossRef] [PubMed]

32. Hallett, P.D.; Bachmann, J.; Czachor, H.; Urbanek, E.; Zhang, B. Hydrophobicity of soil. In *Encyclopedia of Agrophysics*; Gliński, J., Horabik, J., Lipiec, J., Eds.; Encyclopedia of Earth Sciences Series; Springer: Dordrecht, The Netherlands, 2011.

33. Galatowitsch, S.; Richardson, D.M. Riparian scrub recovery after clearing of invasive alien trees in headwater streams of the Western Cape, South Africa. *Biol. Conserv.* **2005**, *122*, 509–521. [CrossRef]

34. Zahawi, R.A.; Augspurger, C.K. Tropical forest restoration: Tree islands as recruitment foci in degraded lands of Honduras. *Ecol. Appl.* **2006**, *16*, 464–478. [CrossRef]

35. Ren, H.; Yang, L.; Liu, N. Nurse plant theory and its application in ecological restoration in lower subtropics of China. *Prog. Nat. Sci.* **2008**, *18*, 137–142. [CrossRef]

36. Fourie, S. Composition of the soil seed bank in alien-invaded grassy fynbos: Potential for recovery after clearing. *S. Afr. J. Bot.* **2008**, *74*, 445–453. [CrossRef]

37. Vosse, S.; Esler, K.J.; Richardson, D.M.; Holmes, P.M. Can riparian seed banks initiate restoration after alien plant invasion? Evidence from the Western Cape, South Africa. *S. Afr. J. Bot.* **2008**, *74*, 432–444. [CrossRef]

38. Bennett, J.A.; Klironomos, J. Mechanisms of plant–soil feedback: Interactions among biotic and abiotic drivers. *New Phytol.* **2019**, *222*, 91–96. [CrossRef]
39. Shiponeni, N.N.; Milton, S.J. Seed dispersal in the dung of large herbivores: Implications for restoration of Renosterveld shrubland old fields. *Biodiver. Conserv.* **2006**, *15*, 3161–3175. [CrossRef]
40. Baltzinger, C.; Karimi, S.; Shukla, U. Plants on the move: Hitch-hiking with ungulates distributes diaspores across landscapes. *Front. Ecol. Evol.* **2019**, *7*, 38. [CrossRef]
41. Mucina, L.; Rutherford, M.C. *The Vegetation of South Africa, Lesotho and Swaziland*; Strelitzia 19; South African National Biodiversity Institute: Pretoria, South Africa, 2006.
42. Black, C.A. *Methods of Soil Analysis: Part I Physical and Mineralogical Properties*; American Society of Agronomy: Madison, WI, USA, 1965.
43. Doerr, S.H.; Thomas, A.D. The role of soil moisture in controlling water repellency: New evidence from forest soils in Portugal. *J. Hydrol.* **2000**, *231–232*, 134–147. [CrossRef]
44. Bisdom, E.B.A.; Dekker, L.W.; Schoute, J.F.T. Water repellency of sieve fractions from sandy soils and relationships with organic material and soil structure. *Geoderma* **1993**, *56*, 105–118. [CrossRef]
45. Leung, Y.-F.; Meyer, K. Soil compaction as indicated by penetration resistance: A comparison of two types of penetrometers. In *Protecting Our Diverse Heritage: The Role of Parks, Protected Areas, and Cultural Sites, Proceedings of the George Wright Society/National Park Service Joint Conference, San Diego, CA, USA, 14–18 April 2003*; George Wright Society: Hancock, MI, USA, 2003; pp. 370–375.
46. Rhoades, J.D. Cation exchanger capacity. In *Methods of Soil Analysis*, 2nd ed.; Page, A.L., Miller, R.H., Keeney, D.R., Eds.; American Society of Agronomy: Madison, WI, USA, 1982; pp. 147–157.
47. Bray, R.H.; Krutz, L.T. Determination of total, organic and available forms of phosphorus in soils. *Soil Sci.* **1945**, *59*, 39–45. [CrossRef]
48. Chan, K.Y.; Bowman, A.; Oates, A. Oxidizible organic carbon fractions and soil quality changes in an Oxic Paleustalf under different pasture leys. *Soil Sci. Soc. Am. J.* **2001**, *166*, 61–67. [CrossRef]
49. Thomas, G.W. Exchangeable cations. In *Methods of Soil Analysis*, 2nd ed.; Page, A.L., Miller, R.H., Keeney, D.R., Eds.; American Society of Agronomy: Madison, WI, USA, 1982; pp. 159–162.
50. Goldblatt, P.; Manning, J. *Cape Plants. A Conspectus of the Cape Flora of South Africa*; Strelitzia 9; National Botanical Institute: Pretoria, South Africa, 2000.
51. Manning, J. *Field Guide to Fynbos*; Struik: Cape Town, South Africa, 2007.
52. Manning, J.C.; Goldblatt, P. *Plants of the Greater Cape Floristic Region 1: The Core Cape Flora*; Strelitzia 29; South African National Biodiversity Institute: Pretoria, South Africa, 2012.
53. South African National Biodiversity Institute (SANBI). *PlantZAfrica.com.* 2017. Available online: http://pza.sanbi.org/ (accessed on 10 October 2022).
54. TIBCO Software Inc. Statistica (Data Analysis Software System), Version 14.0. 2019. Available online: http://tibco.com (accessed on 10 November 2022).

Article

Heavy Metal Contamination Alters the Co-Decomposition of Leaves of the Invasive Tree *Rhus typhina* L. and the Native Tree *Koelreuteria paniculata* Laxm

Zhelun Xu [1,2,3,†], Shanshan Zhong [1,2,3,†], Youli Yu [1,2,3], Yue Li [1,2,3], Chuang Li [1,2,3], Zhongyi Xu [1,2,3], Jun Liu [4], Congyan Wang [1,2,3,5,*] and Daolin Du [1,2,3,5,*]

[1] School of Emergency Management, Jiangsu University, Zhenjiang 212013, China; xuzhelun163@163.com (Z.X.); zhongshanshan126@126.com (S.Z.); yuyoulireally@163.com (Y.Y.); liyueshine@163.com (Y.L.); lc2745144120@163.com (C.L.); llx1345631824@163.com (Z.X.)
[2] School of the Environment and Safety Engineering, Jiangsu University, Zhenjiang 212013, China
[3] Jiangsu Province Engineering Research Center of Green Technology and Contigency Management for Emerging Polluants, Jiangsu University, Zhenjiang 212013, China
[4] Zhenjiang Environmental Monitoring Center of Jiangsu Province, Zhenjiang 212009, China; liujunq8@163.com
[5] Jiangsu Collaborative Innovation Center of Technology and Material of Water Treatment, Suzhou University of Science and Technology, Suzhou 215009, China
[*] Correspondence: liuyuexue623@ujs.edu.cn (C.W.); ddl@ujs.edu.cn (D.D.)
[†] These authors contributed equally to this work.

Abstract: Invasive and native plants can coexist in the same habitat; however, the decomposition process may be altered by the mixing of invasive and native leaves. Heavy metal contamination may further alter the co-decomposition of both leaf types. This study evaluated the effects of two concentrations (35 mg·L^{-1} and 70 mg·L^{-1}) and three types (Pb, Cu, and combined Pb + Cu) of heavy metal contamination on the co-decomposition of leaves of the invasive tree *Rhus typhina* L. and the native tree *Koelreuteria paniculata* Laxm, as well as the mixed effect intensity of the co-decomposition of the mixed leaves. A polyethylene litterbag experiment was performed over six months. The decomposition coefficient of the two trees, mixed effect intensity of the co-decomposition, soil pH and enzymatic activities, soil bacterial alpha diversity, and soil bacterial community structure were determined. A high concentration of Pb and combined Pb + Cu significantly reduced the decomposition rate of *R. typhina* leaves. A high concentration of Pb or Cu significantly reduced the decomposition rate of the mixed leaves. In general, *R. typhina* leaves decomposed faster than *K. paniculata* leaves did. There were synergistic effects observed for the co-decomposition of the mixed leaves treated with combined Pb + Cu, regardless of concentration, but there were antagonistic effects observed for the co-decomposition of the mixed leaves treated with either Pb or Cu, regardless of concentration. A high concentration of Pb or Cu may increase antagonistic effects regarding the co-decomposition of mixed-leaf groups. Thus, heavy metal contamination can significantly affect the intensity of the mixed effect on the co-decomposition of heterogeneous groups of leaves.

Keywords: invasive plants; decomposition rate; soil bacterial diversity; soil bacterial community structure; the mixed effect intensity of the co-decomposition

check for updates

Citation: Xu, Z.; Zhong, S.; Yu, Y.; Li, Y.; Li, C.; Xu, Z.; Liu, J.; Wang, C.; Du, D. Heavy Metal Contamination Alters the Co-Decomposition of Leaves of the Invasive Tree *Rhus typhina* L. and the Native Tree *Koelreuteria paniculata* Laxm. *Plants* **2023**, *12*, 2523. https://doi.org/10.3390/plants12132523

Academic Editor: Sheng Qiang

Received: 3 June 2023
Revised: 25 June 2023
Accepted: 30 June 2023
Published: 1 July 2023

1. Introduction

Invasive plants can cause a loss of biodiversity, by altering the structures and functions of native communities [1–4]. Currently, more than 500 species of invasive plants have already invaded China. This is thought to be due mainly to the wide range of habitats and climates present in the region, as well as the increasing human activities in recent decades [5,6]. In particular, *Rhus typhina* L. has a significant impact on the structure and function of native ecosystems, and is currently considered one of the most impactful invasive tree species in China [7–10]. *Rhus typhina*, which originates from North America, was introduced to China as an ornamental and green species [5,11–13]. At present, research on invasive plants is mainly based on studies of herbaceous ones. As a result, it has become crucial to the field of invasion ecology to clarify the mechanisms whereby woody invasive plants such as *R. typhina* achieve successful colonization in new regions.

One of the important factors for the successful colonization of invasive plants is the potential for plant–soil interactions between invasive plants and soil microorganisms. This occurs mainly through the decomposition process [14–17], as invasive plants may benefit more from plant–soil interactions than native ones do [18–21]. Furthermore, invasive plants may produce more leaves or have their shed leaves degraded faster than native ones do, which may provide more nutrient substrate for soil micro-organisms (particularly decomposers) and subsequently improve their performance during the invasion process [17,22–24]. Therefore, it is crucial to elucidate the key mechanisms that invasive plant species use to achieve successful colonization, based on plant–soil interactions via the decomposition process.

Currently, most regions of China are threatened by heavy metal contamination, mainly due to the development of heavy industries [25–28]. Two types of metals, Cu and Pb, may represent key co-contaminants [25–28]. However, heavy metal contamination may alter the plant–soil interactions throughout the decomposition process, thereby affecting the invasion process of invasive plants [16,29–31]. Therefore, there is an urgent need to investigate the decomposition process under the co-pollution conditions involving these two metals, to elucidate the mechanisms behind the successful colonization of invasive plants, particularly woody ones. However, progress in this area is limited at present.

This study estimated the effects of two concentrations (35 mg·L^{-1} and 70 mg·L^{-1}) and three types (Pb, Cu, and combined Pb + Cu) of heavy metal contamination on the co-decomposition of leaves of the invasive *R. typhina* and the native *Koelreuteria paniculata* Laxm. tree species, as well as enzymatic activities, bacterial alpha diversity, and bacterial community structures in the surrounding soil. In many parts of China, both trees are used for ecological greenery and horticultural ornamentals. They share similar habitats, and the two trees can coexist in the same area [32]. More importantly, the regions where the two trees live have been affected by severe heavy metal contamination, including Pb and Cu co-pollution [25–28]. Pb and Cu carry both environmental and ecological risks. They can decrease plant growth, as well as enhance the growth competitiveness and the allelopathy of invasive plants [33–36]. They represent the two main types of metals found in excess concentrations on arable land sites throughout China [37,38]. Pb and Cu are two of the more widely polluting metals in China, and the approximate actual soil contamination values of Pb^{2+} and Cu^{2+} in Zhenjiang, South Jiangsu, have been found to be similar ($\approx$30–36 mg·L^{-1}) [25–27,33,39].

This study tested the following hypotheses: (I) the decomposition rate of *R. typhina* leaves may be higher than that of *K. paniculata* leaves; (II) synergistic effects may exist regarding the co-decomposition of the mixed leaves; (III) the presence of Pb, Cu, or both, may increase the synergistic effects related to the co-decomposition of the mixed leaves.

2. Results

2.1. Differences in the Decomposition Variables

The *k* values of *R. typhina* leaves treated with high concentrations of Pb and combined Pb + Cu were lower than those of *R. typhina* leaves treated with the distilled water control (Figure 1; $p < 0.05$). The *k* values of the mixed leaves treated with high concentrations of Pb and Cu were lower than those of the mixture treated with the control (Figure 1; $p < 0.05$).

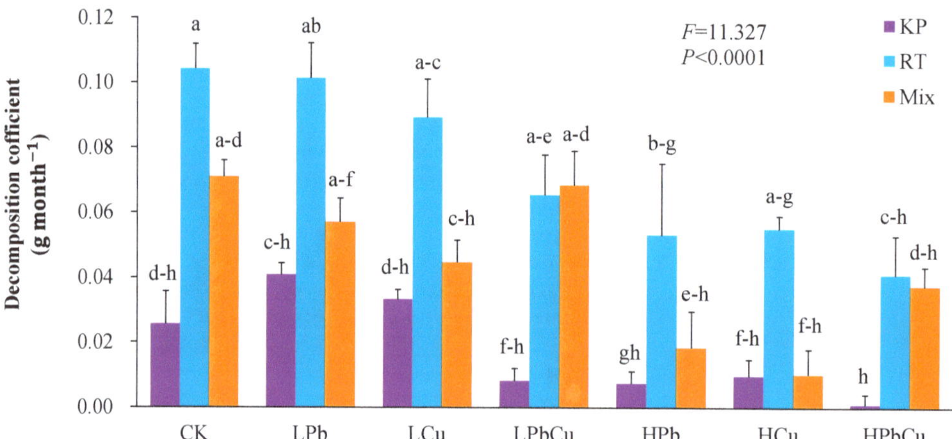

Figure 1. The decomposition coefficient for *Koelreuteria paniculata* Laxm (KP) and *Rhus typhina* L. (RT) leaves, and mixed leaves of this species (Mix). Bars (means and SE; *n* = 3) with different letters mean statistically significant differences (*p* < 0.05). Abbreviations: CK, control; LPb, a low concentration of Pb; LCu, a low concentration of Cu; LPbCu, a low concentration of combined Pb + Cu; HPb, a high concentration of Pb; HCu, a high concentration of Cu; HPbCu, a high concentration of combined Pb + Cu.

The *k* values of *R. typhina* leaves were higher than those of *K. paniculata* leaves, for all treatment types (Figure 1; *p* < 0.05).

The results of our three-way ANOVA analysis indicated that the type of heavy metal contamination, the type(s) of the leaves, and the interaction between the concentration of heavy metal contamination and the type(s) of the leaves significantly affected the *k* values (Table S1; *p* < 0.01).

The value of the observed *k* was higher than that of the expected *k* for the mixed leaves treated with a low concentration of combined Pb + Cu (Figure 2a; *p* < 0.05). However, the value of the observed *k* was lower than that of the expected *k* for the mixed leaves treated with high concentrations of either Pb or Cu alone (Figure 2a; *p* < 0.05).

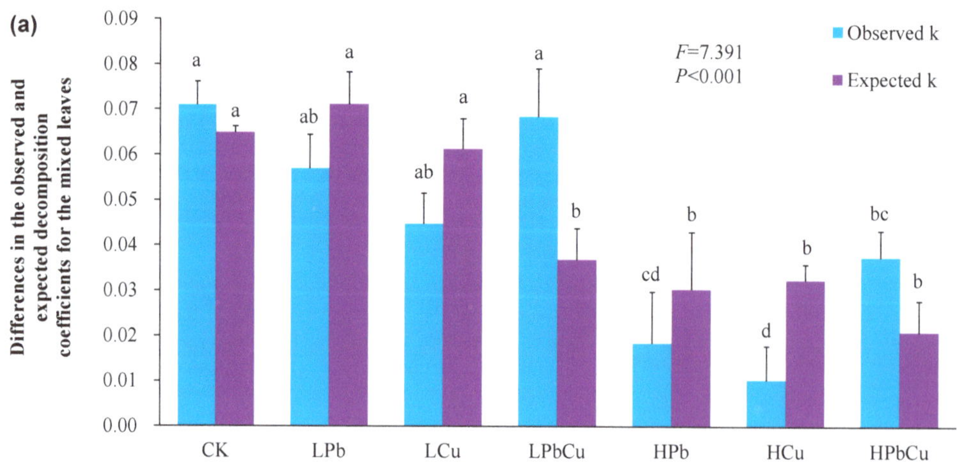

Figure 2. *Cont.*

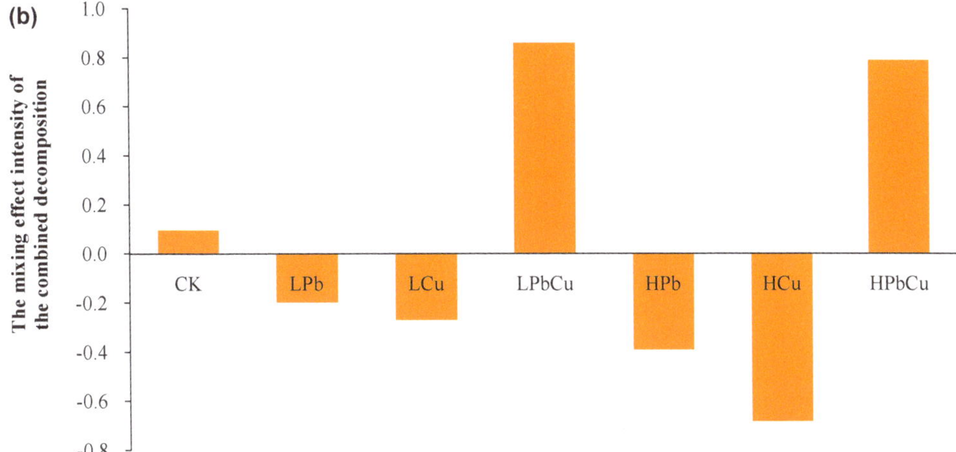

Figure 2. The observed (blue bars) and expected (purple bars) decomposition coefficients for the mixed *K. paniculata* and *R. typhina* leaves (**a**), and the mixing effect intensity of the co-decomposition (**b**). Bars (means and SE; $n = 3$) with different letters mean statistically significant differences ($p < 0.05$). Abbreviations have the same meanings as those presented in Figure 1.

The mixed effect intensity of the co-decomposition of leaf mixtures treated with the control and with combined Pb + Cu was higher than zero, regardless of concentration, but was lower than zero for those treated with either Pb or Cu alone, regardless of concentration (Figure 2b). The absolute value of the mixed effect intensities of co-decomposition under high concentrations of either Pb or Cu were significantly higher than those for bags treated with low concentrations of either Pb or Cu (Figure 2b).

2.2. Differences in Soil pH

Both heavy metal contamination and leaf type significantly increased soil pH compared to the control (Figure S1a; $p < 0.05$). The soil pHs of *K. paniculata* leaves treated with a low concentration of combined Pb + Cu, *K. paniculata* leaves treated with a high concentration of Pb, *K. paniculata* leaves treated with a high concentration of Cu, and *K. paniculata* leaves treated with a high concentration of combined Pb + Cu, were higher than that of *K. paniculata* leaves alone (Figure S1a; $p < 0.05$). Similarly, the soil pHs of *R. typhina* leaves treated with a high concentration of Pb and *R. typhina* leaves treated with a high concentration of Cu were higher than that of *R. typhina* leaves alone (Figure S1a; $p < 0.05$). The soil pH of the leaf mixture treated with a high concentration of Pb was higher than that of the leaf mixture control (Figure S1a; $p < 0.05$). Soil pHs under a high concentration of Pb, that of *K. paniculata* leaves treated with a high concentration of Pb, and that of mixed leaves treated with a high concentration of either Pb or Cu were higher than the soil pHs under a low concentration of Pb, *K. paniculata* leaves treated with a low concentration of Pb, and mixed leaves treated with a low concentration of either Pb or Cu, respectively (Figure S1a; $p < 0.05$). The soil pH of *K. paniculata* leaves treated with a low concentration of combined Pb + Cu was higher than that of *K. paniculata* leaves treated with a low concentration of Pb (Figure S1a; $p < 0.05$).

2.3. Differences in Soil Enzymatic Activities

Peroxidase activity levels of *K. paniculata* leaves treated with a low concentration of Pb, *R. typhina* leaves treated with a low concentration of either Pb or Cu, *K. paniculata* leaves treated with a low concentration of Cu, mixed leaves treated with a low concentration of combined Pb + Cu, mixed leaves treated with a high concentration of Pb, *R. typhina* leaves treated with a high concentration of Pb, and *K. paniculata* leaves treated with a high concentration of combined Pb + Cu were lower than those of the control (Figure S1b; $p < 0.05$). Peroxidase activity levels of *K. paniculata* leaves treated with a low concentration of Pb and *K. paniculata* leaves treated with a high concentration of combined Pb + Cu were lower than those of *K. paniculata* leaves in other conditions (Figure S1b; $p < 0.05$). The peroxidase activity level of the mixed leaves treated with a low concentration of Cu was higher than that of the mixed leaves (Figure S1b; $p < 0.05$).

Sucrase activity levels of *K. paniculata* leaves treated with a low concentration of Pb, *K. paniculata* leaves treated with a low concentration of combined Pb + Cu, and *K. paniculata* leaves treated with a high concentration of Pb were lower than those of *K. paniculata* leaves in all other conditions (Figure S1c; $p < 0.05$). The sucrase activity in the bag of *K. paniculata* leaves was higher than that that of the *R. typhina* and mixed leaf bags under the control condition (Figure S1c; $p < 0.05$).

Protease activity levels of the mixed leaves treated with a high concentration of Pb and the mixed leaves treated with a high concentration of combined Pb + Cu were lower than those of the mixed leaves treated with a high concentration of Cu (Figure S1d; $p < 0.05$).

Urease activity under a low concentration of Pb and in the mixed leaves treated with a low concentration of Pb was higher than that of the control. Urease activity levels under a high concentration of Pb and in the mixed leaves treated with a high concentration of combined Pb + Cu were lower than that of the control. The effects of Pb or Cu on urease activity were mostly concentration-dependent (Figure S1e; $p < 0.05$). The effects of Pb or Cu on urease activity were mostly concentration-dependent (Figure S1e; $p < 0.05$). Urease activity levels of *R. typhina* leaves treated with a low concentration of Pb and *R. typhina* leaves treated with a low concentration of Cu were higher than that of *R. typhina* leaves treated with a low concentration of combined Pb + Cu (Figure S1e; $p < 0.05$). Urease activity in the mixed leaves treated with a low concentration of Pb was higher than that of the mixed leaves treated with a low concentration of combined Pb + Cu (Figure S1e; $p < 0.05$). Urease activity levels under high concentrations of Cu and combined Pb + Cu were higher than that under a high concentration of Pb (Figure S1e; $p < 0.05$). Urease activity of *R. typhina* leaves treated with a high concentration of Cu was higher than that of *R. typhina* leaves treated with a high concentration of Pb and of *R. typhina* leaves treated with a high concentration of combined Pb + Cu (Figure S1e; $p < 0.05$). Urease activity of the mixed leaves treated with a high concentration of Pb was higher than that of the mixed leaves treated with a high concentration of combined Pb + Cu (Figure S1e; $p < 0.05$). Urease activity of *K. paniculata* leaves treated with a high concentration of heavy metal contamination was higher than that of *K. paniculata* leaves under other conditions (Figure S1e; $p < 0.05$). The urease activity of *R. typhina* leaves treated with a low concentration of combined Pb + Cu and *R. typhina* leaves treated with a high concentration of heavy metal contamination was higher than that of *R. typhina* leaves under other conditions (Figure S1e; $p < 0.05$). Urease activity levels of the mixed leaves treated with a low concentration of combined Pb + Cu and of the mixed leaves treated with a high concentration of heavy metal contamination were higher than that of the mixed leaves (Figure S1e; $p < 0.05$). The urease activity of *R. typhina* leaves treated with a high concentration of Cu was higher than that of *R. typhina* leaves treated with high concentrations of Pb, as well as a high concentration of combined Pb + Cu (Figure S1e; $p < 0.05$).

Phosphatase activity under a high concentration of Pb was higher than that under a high concentration of combined Pb + Cu (Figure S1f; $p < 0.05$). The phosphatase activity of *K. paniculata* leaves treated with a high concentration of Pb was lower than that of *K. paniculata* leaves under other conditions (Figure S1f; $p < 0.05$).

2.4. Differences in Soil Bacterial Alpha Diversity

The phylogenetic diversity indexes under a high concentration of Cu and a high concentration of combined Pb + Cu, regardless of leaf type, were higher than that of the control (Figure S2a; $p < 0.05$). The impacts of Cu and combined Pb + Cu on the phylogenetic diversity index were concentration-dependent (Figure S2a; $p < 0.05$). The phylogenetic diversity indexes under a high concentration of Cu and a high concentration of combined Pb + Cu were higher than those under a high concentration of Pb, regardless of leaf type (Figure S2a; $p < 0.05$).

The Sobs indexes of *R. typhina* leaves treated with a high concentration of Cu and a high concentration of combined Pb + Cu, and of mixed leaves treated with a high concentration of combined Pb + Cu were higher than those under the control condition (Figure S2b; $p < 0.05$). The Sobs indexes under a high concentration of Cu and for *R. typhina* leaves treated with a high concentration of Cu were higher than those under a low concentration of Cu and for *R. typhina* leaves treated with a low concentration of Cu (Figure S2b; $p < 0.05$). The Sobs index of *K. paniculata* leaves treated with a low concentration of combined Pb + Cu was higher than that of *K. paniculata* leaves treated with a low concentration of Pb (Figure S2b; $p < 0.05$). The Sobs indexes of the mixed leaves treated with a high concentration of Cu and the mixed leaves treated with a high concentration of combined Pb + Cu were higher than those of the mixed leaves treated with a high concentration of Pb (Figure S2b; $p < 0.05$).

Shannon's diversity indexes of the mixed leaves treated with a low concentration of Pb and a high concentration of Cu were higher than those of the control (Figure S2c; $p < 0.05$).

Simpson's dominance index under a high concentration of Cu and a high concentration of combined Pb + Cu was higher than that under the control condition (Figure S2d; $p < 0.05$). The Simpson's dominance index under a low concentration of combined Pb + Cu was higher than that under a high concentration of combined Pb + Cu (Figure S2d; $p < 0.05$). The Simpson's dominance index according to the type of heavy metal contamination with a high concentration was highest for a high concentration of Pb, followed by a high concentration of Cu, and finally a high concentration of combined Pb + Cu (Figure S2d; $p < 0.05$).

Pielou's evenness indexes of *R. typhina* leaves treated with a high concentration of Pb, the mixed leaves treated with a high concentration of Pb, the mixed leaves treated with a high concentration of Cu, the mixed leaves treated with a high concentration of combined Pb + Cu, and *R. typhina* leaves treated with a high concentration of combined Pb + Cu were higher than that of the control (Figure S2e; $p < 0.05$). The Pielou's evenness index of *R. typhina* leaves treated with a low concentration of combined Pb + Cu was higher than that of *R. typhina* leaves treated with a high concentration of combined Pb + Cu (Figure S2e; $p < 0.05$).

ACE's richness index for the mixed leaves treated with a high concentration of combined Pb + Cu was higher than that of leaves under the control condition (Figure S2f; $p < 0.05$). ACE's richness index for *K. paniculata* leaves treated with a low concentration of combined Pb + Cu was higher than that for *K. paniculata* leaves treated with a low concentration of Pb (Figure S2f; $p < 0.05$). The ACE's richness index for the mixed leaves treated with a low concentration of combined Pb + Cu was higher than that for the mixed leaves treated with a low concentration of Pb (Figure S2f; $p < 0.05$). ACE's richness index under a high concentration of combined Pb + Cu was higher than that under a low concentration of Pb, for the mixed leaves (Figure S2f; $p < 0.05$). ACE's richness index for *K. paniculata* leaves treated with a high concentration of combined Pb + Cu was higher than that for *K. paniculata* leaves (Figure S2f; $p < 0.05$).

Chao1's richness index for *K. paniculata* leaves treated with a low concentration of combined Pb + Cu was higher than that for *K. paniculata* leaves treated with a low concentration of Pb (Figure S2g; $p < 0.05$). Chao1's richness index for *K. paniculata* leaves treated with a low concentration of Cu was higher than that for *K. paniculata* leaves treated with a low concentration of Pb (Figure S2g; $p < 0.05$).

The results of our three-way ANOVA analysis indicated that the concentration of heavy metal contamination significantly affected soil pH, urease activity, the phylogenetic diversity index, Sobs index, Shannon's diversity index, Pielou's evenness index, ACE's richness index, and Chao1's richness index (Table S1; $p < 0.01$). The type of heavy metal contamination significantly affected soil pH, sucrase activity, protease activity, urease activity, the phylogenetic diversity index, Sobs index, Simpson's dominance index, Pielou's evenness index, ACE's richness index, and Chao1's richness index (Table S1; $p < 0.01$). Leaf type had a significant effect on soil pH (Table S1; $p < 0.01$). The interaction of the concentration of heavy metal contamination and the type of heavy metal contamination significantly affected soil pH, peroxidase activity, protease activity, urease activity, acid phosphatase activity, and the phylogenetic diversity index (Table S1; $p < 0.01$). The interaction of the concentration of heavy metal contamination and leaf type significantly affected protease activity, urease activity, and acid phosphatase activity (Table S1; $p < 0.01$). The interaction of the type of heavy metal contamination and leaf type significantly affected urease activity, acid phosphatase activity, and Pielou's evenness index (Table S1; $p < 0.01$). The interaction of the three factors significantly affected protease activity, urease activity, acid phosphatase activity, and Simpson's dominance index (Table S1; $p < 0.01$).

2.5. Differences in Soil Bacterial Community Structure

The mean value of Good's coverage indexes for soil bacterial communities across all samples was ≈98.47%. There were significant differences in soil bacterial beta diversity based on weighted UniFrac distances between different treatments (Figures S3–S6). The influence intensity of the concentration of heavy metal contamination and the type of heavy metal contamination on soil bacterial community structure was significantly higher than the influence intensity of leaf type (Figures S3–S6).

Under the perspective comparison among the effects of heavy metal contamination, Georgenia, *Bogoriellaceae*, *Microbacteriaceae*, *Enteractinococcus*, *Nocardioides*, *Streptomyces_thermocarboxydus*, *Streptomyces*, *Streptomycetaceae*, *Streptomycetales*, MWH_CFBk5, Flavobacterium, *Galbibacter_marinus*, *Galbibacter*, *Muricauda*, *Pricia*, *Salinimicrobium*, *Flavobacteriaceae*, *Flavobacteriales*, *Parapedobacter*, *Pedobacter*, *Sphingobacteriaceae*, *Sphingobacteriales*, *Bacteroidia*, *Micavibrionales*, *Devosiaceae*, *Rhizobiales*, *Alphaproteobacteria*, *Marinobacter_sp*, *Alcanivorax*, *Alcanivoracaceae*, *Oceanospirillales*, *Lysobacter_defluvii*, *Lysobacter*, and *Xanthomonadaceae* were the most altered taxa of soil bacterial taxa under the control (Figure 3A); *Actinomarinales*, *Lamia*, *Lamiaceae*, *Gordonia*, *Ornithinicoccus*, *Ornithinimicrobium*, *Intrasporangiaceae*, JG30 KF CM45, *Thermomicrobiaceae*, *Thermomicrobiales*, *Chloroflexia*, *Bacillus*, *Bacillaceae*, *Bacillales*, *Bacilli*, S0134_terrestrial_group, and *Hyphomicrobiaceae* were the most altered taxa of soil bacterial taxa under Pb treatment (Figure 3A); *Rhodothermia* and *Sphingopyxis* were the most changed taxa of soil bacterial taxa under Cu treatment (Figure 3A). *Thermocrispum*, *Pseudonocardiaceae*, *Pseudonocardiales*, *Micropepsaceae*, *Micropepsales*, *Altererythrobacter*, *Sphingomonadaceae*, *Sphingomonadales*, *Burkholderiaceae*, *Betaproteobacteriales*, *Chujaibacter*, *Rhodanobacter*, and *Rhodanobacteraceae* were the most greatly changed taxa of soil bacterial taxa under combined Pb + Cu treatment (Figure 3A).

Under the perspective comparison of the impacts of the leaves of the two trees, *Persicitalea* and *Spirosomaceae* were the most altered taxa of soil bacterial taxa under the control condition (Figure 3B); Bacillus and *Isosphaera* were the most altered taxa of soil bacterial taxa under *R. typhina* leaves (Figure 3B).

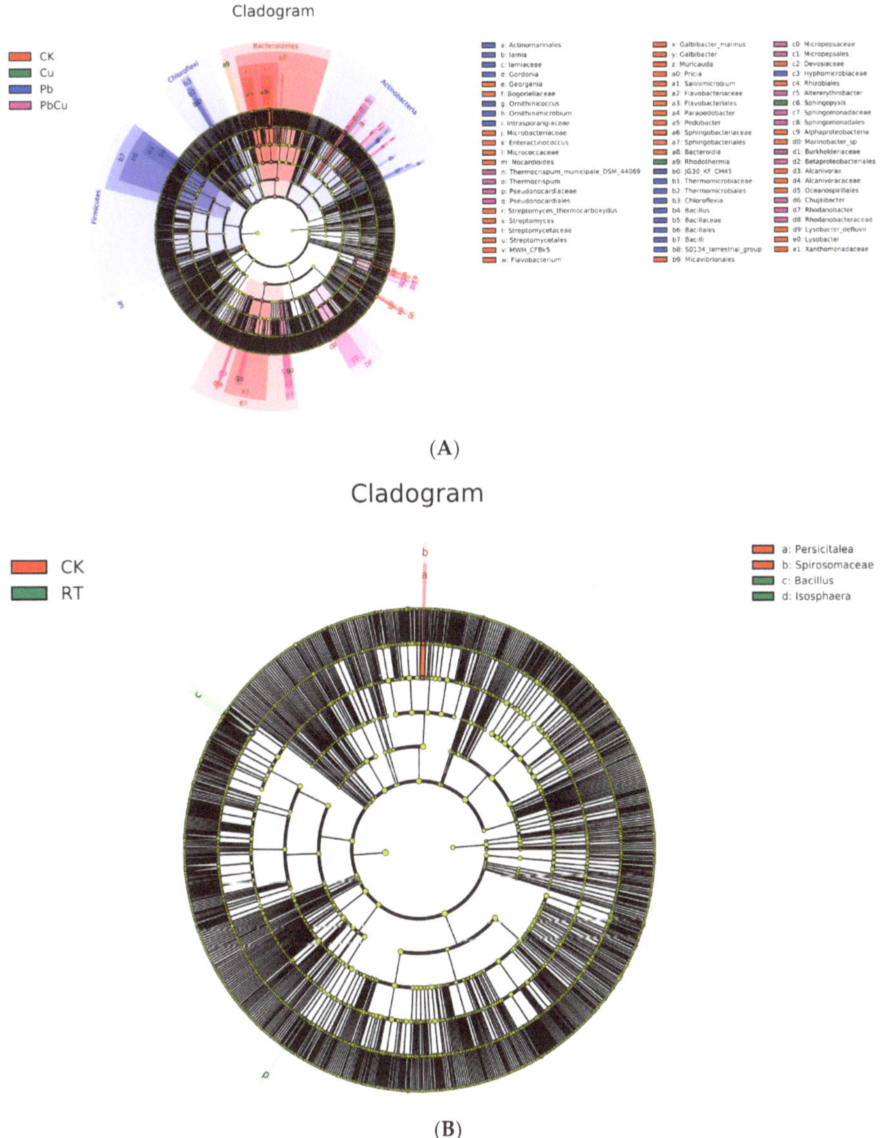

Figure 3. The LEfSe method identifies the significantly different abundant taxa of soil bacteria (subgraph (**A**), the type of heavy metal contamination; Subgraph (**B**), the type of the leaves). The taxa with significantly different abundances among treatments are signified by colored dots, and from the center outward, they mean the kingdom, phylum, class, order, family, genus, and species levels, respectively. The colored shadows mean trends of the significantly differed taxa. Only taxa meeting an LDA significance threshold of >2 are displayed. Abbreviations have the same meanings as those presented in Figure 1.

2.6. Contribution Intensity of Soil pH and Enzymatic Activities, and Soil Bacterial Alpha Diversity on K

The absolute value of the direct path coefficient of the Sobs, Shannon's diversity, Pielou's evenness, and ACE's richness indexes of soil bacteria was obviously larger than that of other factors (Figure 4).

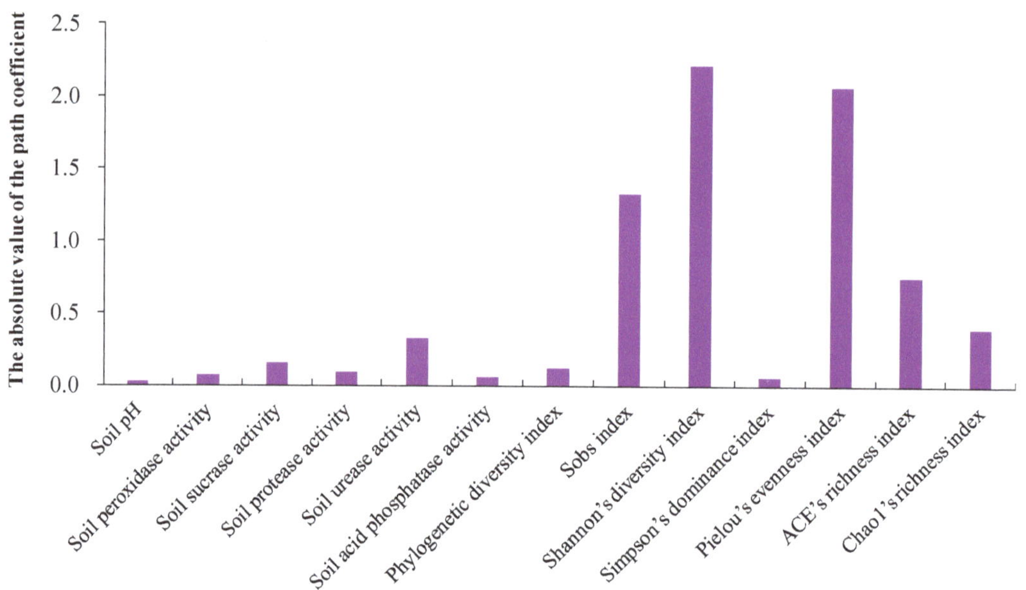

Figure 4. The influence intensity of soil variables and soil bacterial alpha diversity on the decomposition coefficient using the path analysis based on the absolute value of the path coefficient.

3. Materials and Methods

3.1. Experimental Design

Leaves from *R. typhina* and *K. paniculata* were collected from natural sources in Zhenjiang, southern Jiangsu, China (32.205–32.216° N; 119.518–119.527° E), over the first 10 days of October 2021. Zhenjiang has a humid, northern subtropical monsoon climate. The annual mean temperature in Zhenjiang was ≈17.1 °C in 2022. The monthly mean temperature reaches a maximum of ≈28.1 °C in July and decreases to a minimum of ≈3.7 °C in January. The annual precipitation was ≈1164.1 mm in 2022, and the monthly mean precipitation reaches a maximum of ≈432.1 mm in July before dropping to a minimum of ≈2.7 mm in December. Zhenjiang received ≈1909.0 h of sunlight in 2022, and its mean monthly sunlight reaches a maximum of ≈208.2 h in December before dropping to a minimum of ≈125.9 h in August [40]. The soil type in which these two trees grow is mainly yellow soil [41]. Leaves from each tree were collected from three plant communities separated by >100 m. Leaf samples from 10 individuals of the same species were randomly collected and mixed thoroughly with other samples from the same species/community. From each individual tree, ≈50 fully expanded and intact leaves from sun-exposed parts of the plant were randomly selected, to minimize the effects of sunlight on the leaf compounds. Leaf samples were then air-dried to standardize their weights.

The decomposition process of the two trees was mimicked using a polyethylene litter bag experiment in an artificial greenhouse at Jiangsu University (located at 32.206° N; 119.512° E), under the condition of natural light, from 15 October 2021 to 15 April 2022 (experimental period: ≈6 months). The air-dried leaves of the two trees were placed in polyethylene litterbags (size: 10 × 15 cm; mesh size ≈ 0.425 mm). Leaves from the two trees were arranged in one of the following three ways per bag: 6 g of *R. typhina* leaves, 6 g of *K. paniculata* leaves, or 6 g of an equal mixture of both leaf types. The polyethylene litterbags were buried in the flower pots (upper diameter: ≈25 cm; lower diameter: ≈13 cm) which were filled with the garden soil, with one polyethylene litterbag per flowerpot. Garden soil was chosen as the culture substrate in order to maximize the possibility of an invasion history recruited by invasive plants or a pollution history mediated by metals. The garden soil was not disinfected, so as not to disturb the presence of micro-organisms (particularly the decomposers).

The polyethylene litterbags were treated with the following six types of heavy metal contaminants: a low concentration of Pb, a low concentration of Cu, a low concentration of combined Pb + Cu, a high concentration of Pb, a high concentration of Cu, and a high concentration of combined Pb + Cu, with distilled water serving as a negative control in a seventh bag. The Pb and Cu solutions were formulated using lead acetate trihydrate (purity: ≥99.0%) and copper sulfate pentahydrate (purity: ≥99.0%), respectively. The low concentration for both the independent and combined Pb^{2+} + Cu^{2+} solutions was set to 35 mg·L^{-1}, to mimic the approximate actual soil contamination values of Pb^{2+} and Cu^{2+} in Zhenjiang, South Jiangsu. The high concentration for both the independent and combined Pb^{2+} and Cu^{2+} solutions was set to 70 mg·L^{-1}, which greatly exceeded the common contamination level in Zhenjiang, South Jiangsu, by a large margin, in order to simulate possible future scenarios where fields are more heavily contaminated [25–27,33,39].

The polyethylene litterbag experiment comprised three factors: the concentration of heavy metal contamination, the type of heavy metal contamination, and the type of leaves. Each of these factors had two or three levels: two concentrations (35 mg·L^{-1} or 70 mg·L^{-1}) of heavy metal contamination, three types (Pb, Cu, and combined Pb + Cu) of heavy metal contamination, and three types (*R. typhina* leaves, *K. paniculata* leaves, and the equally mixed) of the leaves of the two trees. Each treatment combination was carried out in triplicate.

After six months, all of the polyethylene litterbags were collected. Leaf samples from the two tree species were lightly scarified to remove the residual soil particles and thoroughly air-dried to standardize their weights so that decomposition variables could be more easily estimated. Soil samples were taken from within 1 cm of the polyethylene litterbags and passed through a 2 mm sieve to assess soil pH, enzymatic activities, bacterial alpha diversity, and soil bacterial community structure.

3.2. Determination of Decomposition Variables

The decomposition coefficient, which was used to evaluate the decomposition rate, was estimated using the following equation [42]:

$$X_t = X_o * e^{-kt} \tag{1}$$

where k is the decomposition coefficient, and X_O and X_t denote the dry weights of the leaves at the beginning of the experiment and at time t, respectively. The dry weights of the leaves were measured using an electronic balance with an accuracy of 0.001 g.

The expected k value for the equal mixture of leaves from the two trees was evaluated as follows [43,44]:

$$\text{Expected } k = \frac{x+y}{2} \tag{2}$$

where x and y denote the observed k values of the two trees.

The mixed effect intensity of the co-decomposition of the mixed leaves was assessed as follows [43,44]:

$$\text{The intensity of non} - \text{additive effects} = \frac{O}{E} - 1 \tag{3}$$

where O and E denote the observed and expected k of the mixture of the two leaf types, respectively. Thus, intensity values greater than zero correspond to synergistic co-decomposition effects, whereas intensity values less than zero indicate antagonistic co-decomposition effects. The stronger the response, the greater the deviation from zero.

3.3. Determination of Soil pH and Enzymatic Activities

Soil pH was determined in situ using a digital soil acidity meter (ZD Instrument Co., Ltd., Taizhou, China).

The activities of five soil enzymes closely related to soil nutrient cycling were estimated, including (1) peroxidase (E.C. 1.11.1.1) activity—analyzed via the pyrogallol method using a colorimetric assay at 430 nm; (2) sucrase (E.C. 3.2.1.26) activity—measured Via the 3,5-dinitrosalicylic acid method with a spectrophotometer at 508 nm; (3) protease (E.C. 3.4.11.4) activity—measured using the tyrosine method with colorimetric assay at 700 nm; (4) urease (E.C. 3.5.1.5) activity—estimated via the sodium phenolate-sodium hypochlorite method with a spectrophotometer at 578 nm; and (5) acid phosphatase (E.C. 3.1.3.2) activity—estimated via the disodium phenyl phosphate method with colorimetric assay at 660 nm [45–47].

3.4. Determination of Soil Bacterial Communities

Soil bacterial communities were assessed via high-throughput sequencing using the Illumina PE250 instrument at GENE DENOVO Co., Ltd. (Guangzhou, China). The V3–V4 region of bacterial 16S rRNA genes was amplified using the universal bacterial primers 341F/806R (forward primer: 5′-CCT AYG GGR BGC ASC AG-3′; reverse primer: 5′-GGA CTA CNN GGG TAT CTA AT-3′) [48,49]. The remaining methods for determining soil bacterial communities were the same ones used in our previous related studies [15,39].

3.5. Statistical Analysis

Differences in the values of decomposition variables, soil pHs, soil enzymatic activity levels, and soil bacterial alpha diversity levels between the different bags were assessed using a one-way analysis of variance (ANOVA) with Tukey's test. Three-way ANOVA was used to evaluate the effects of the concentration of heavy metal contamination, the type of heavy metal contamination, and the type of leaves, as well as their interactions with k values, soil pHs, soil enzymatic activities, and soil bacterial alpha diversities. The intensities of the contributions of soil pH, enzyme activities, and bacterial alpha diversity levels to k were evaluated using path analysis. $p \leq 0.05$ was considered a statistically significant difference. IBM SPSS Statistics 26.0 (IBM Corp., Armonk, NY, USA) was used for all statistical analyses.

4. Discussion

The decomposition process is essential for nutrient cycling [14–17]. A high concentration of Pb and combined Pb + Cu significantly reduced the decomposition rate of *R. typhina* leaves (Figure 1). A high concentration of either Pb or Cu also significantly reduced the decomposition rate of mixed *R. typhina* and *K. paniculata* leaves. Thus, the nutrient cycling rates of *R. typhina* leaves and the mixed leaves may have been suppressed by high concentrations of Pb or Cu. This finding may be due to the increased energy cost of metabolism and the decreased resource utilization efficiency of soil microbial degraders under conditions with high concentrations of metals [16,50–52]. However, Pb or Cu did not significantly affect the rate of decomposition of *K. paniculata* leaves (Figure 1). Thus, Pb or Cu may be detrimental to the invasion of *R. typhina*, via a reduced nutrient cycling rate compared to that of *K. paniculata* alone. This may be a good thing in terms of slowing down the invasions of invasive plants that may pose threats to local ecological structures and functions—particularly with regard to biodiversity.

Consistent with the first hypothesis of our study, the decomposition rate of *R. typhina* leaves was greater than that of *K. paniculata* leaves, regardless of the addition of Pb or Cu (Figure 1). Thus, the nutrient cycling rate of *R. typhina* may be higher than that of *K. paniculate*, and is not affected by either Pb or Cu. In general, the decomposition and nutrient cycling rates of invasive plants are typically higher than those of native plants [17,53–55]. This may be due to the higher levels of easily degradable compounds and lower proportions of recalcitrant materials that are difficult to degrade in *R. typhina* leaves, compared to those of *K. paniculata*. Another reason why *R. typhina* leaves may have degraded faster than *K. paniculata* leaves are in this study is likely that the altered soil bacterial community structure of the leaves of the two trees, in particular *R. typhina* leaves, can trigger the emergences of certain dominant soil bacterial communities (including Bacillus and Isosphaera) (Figure 3B). Therefore, one of the main factors underlying the success of invasive species may be the faster rate of nutrient cycling mediated by a higher decomposition rate compared to that of native plants.

In any given environment, a single plant species rarely occurs alone. Usually, two or more plant species occur together (including both invasive and native plants), meaning their leaves can also coexist and decompose together [14,15,17,55]. The mixed effect intensity of the co-decomposition treated with the control, and the combined Pb + Cu conditions (regardless of concentration) was positive, but was negative when treated with Pb or Cu (also regardless of concentration). Thus, there were synergistic effects for the co-decomposition of the mixed leaves treated under the control and combined Pb + Cu conditions, regardless of concentration, but there were antagonistic effects for the co-decomposition of the mixed leaves treated with Pb or Cu, regardless of concentration. Thus, the decomposition of invasive plants can increase the decomposition of native plants [14,24,56,57] treated under the control and combined Pb + Cu conditions, regardless of concentration, but the opposite is true under Pb + Cu conditions, regardless of concentration. Thus, the type of heavy metal contamination is one of the key factors that significantly affects the intensity of the mixed effect of the co-decomposition of mixed leaves. This result is not fully consistent with the second hypothesis of our study.

The absolute value of the mixed effect intensity of co-decomposition under conditions of high concentrations of Pb or Cu was markedly higher than that under conditions of low concentrations of Pb or Cu; however, there was a similarity between the absolute value of the mixed effect intensity of co-decomposition under a low concentration of combined Pb + Cu and a high concentration of combined Pb + Cu (Figure 2b). Thus, a high concentration of Pb or Cu can intensify the antagonistic effects on the co-decomposition of the mixed leaves, compared to those under a low concentration of either Pb or Cu. However, the concentration of combined Pb + Cu did not alter the antagonistic effects on the co-decomposition of the mixed leaves. Hence, the concentration of Pb or Cu is one of the key factors that significantly affects the antagonistic effects on co-decomposition of mixed-leaf samples. This result did not fully confirm the third hypothesis of this study.

A high concentration of Pb or Cu may exert a stronger inhibitory effect on the decomposition rate and nutrient cycling rate of the co-decomposition of mixed leaves, via intensified antagonistic effects on co-decomposition—but combined Pb + Cu, regardless of concentration, may exert a positive effect on the decomposition rate and nutrient cycling rate of the co-decomposition mixed leaves, via induced synergistic effects on the co-decomposition. The main reason for the differences observed in the intensity of the mixed effect of the co-decomposition of the mixed leaves under different types of heavy metal contamination may be due to the lower diversity of microbial decomposer species in the soil under high concentrations of Pb or Cu, as well as increased species diversity in terms of soil microbial degraders under combined Pb + Cu conditions, especially at higher concentrations (Figure S2). Furthermore, different types of heavy metal contamination can cause the emergence of different dominant soil bacterial communities (Figure 3A) and changes in soil bacterial beta diversity (Figures S3–S6), as was observed in this study. The concentration and type of heavy metal contamination significantly affected the phylogenetic diversity, species number, species diversity, species evenness, and species richness of surrounding soil bacteria (Table S1). The intensity of the influence of soil bacterial alpha diversity (especially in terms of species number, diversity, and richness) on the decomposition rate was significantly greater than that of other factors, based on the results of the path analysis (Figure S2). Previous studies have also shown that soil bacterial alpha diversity is positively correlated with the decomposition rates of plant species [58–60].

Invasive plants can improve soil enzymatic activities through the effects of higher nutrient levels on metabolic processes [61–64]. It is therefore expected that soil enzymatic activities may be enhanced following the decomposition of *R. typhina* leaves. However, in contrast to previous studies [61–64], the decomposition of *R. typhina* leaves decreased soil sucrase activity compared to that of *K. paniculata* leaves under the control condition (Figure S1). Thus, the decomposition of *R. typhina* leaves can reduce sucrase hydrolysis capacity. The reduced soil sucrase activity with regard to the decomposition of *R. typhina* leaves may be due to the reduced levels of available nutrients in the soil subsystem and higher microbial metabolic rates. Some invasive plants (including *R. typhina*) may reduce soil enzymatic activities [14,65,66].

5. Conclusions

A high concentration of Pb and combined Pb + Cu significantly decreased the decomposition rate of *R. typhina* leaves. A high concentration of either Pb or Cu alone significantly decreased the decomposition rate of a mixture of *R. typhina* and *K. paniculata* leaves. However, neither Pb nor Cu had any significant effect on the decomposition rate of *K. paniculata* leaves. Therefore, Pb or Cu may be detrimental to the invasiveness of *R. typhina*, in that they may reduce the rate of nutrient cycling compared to that of *K. paniculata*. *Rhus typhina* leaves were degraded faster than *K. paniculata* leaves were. Synergistic effects were found with regard to the co-decomposition of mixed leaves treated under the control, and combined Pb + Cu conditions in this study, regardless of concentration, but there were antagonistic effects observed on the co-decomposition of mixed leaves treated with either Pb or Cu alone, regardless of concentration. The type of heavy metal contamination is one of the main factors that significantly affects the intensity of the mixed effect of the co-decomposition of the mixed leaves. A high concentration of Pb or Cu can intensify the antagonistic effects on the co-decomposition of the mixed leaves, compared to low concentrations of either Pb or Cu. The concentration of combined Pb + Cu did not alter the antagonistic effects on the co-decomposition of the mixed leaves. Thus, the concentration of Pb or Cu is one of the crucial factors that significantly affects the antagonistic effects on the co-decomposition of the mixed leaves.

Supplementary Materials: The following supporting information can be downloaded at. https://www.mdpi.com/article/10.3390/plants12132523/s1. Table S1: Three-way ANOVA showing the effect of the main factors: the concentration of heavy metal contamination, the type of heavy metal contamination, and the type of the leaves, and their interactions on the decomposition coefficient (k), soil pH, soil enzymatic activities, and the alpha diversity of soil bacteria; Figure S1: Soil pH and soil enzymatic activities; Figure S2: Alpha diversity of soil bacteria; Figure S3: The heatmap of beta diversity estimates of soil bacteria at phylum level based on weighted UniFrac distances; Figure S4: PCA of soil bacteria based on weighted UniFrac distance; Figure S5: PCoA of soil bacteria based on weighted UniFrac distance; Figure S6: NMDS of soil bacteria based on weighted UniFrac distance.

Author Contributions: Z.X. (Zhelun Xu): data curation; investigation; writing—review and editing. S.Z.: data curation; investigation; methodology; writing—review and editing. Y.Y.: data curation; investigation; writing—review and editing. Y.L.: data curation; investigation; writing—review and editing. C.L.: data curation; investigation; writing—review and editing; Z.X. (Zhongyi Xu): data curation; writing—review and editing. J.L.: Data curation; Investigation; Writing—review and editing. C.W.: conceptualization; formal analysis; funding acquisition; project administration; supervision; writing—original draft. D.D.: funding acquisition; project administration; writing—review and editing. All authors have read and agreed to the published version of the manuscript.

Funding: This study was supported by Special Research Project of School of Emergency Management, Jiangsu University (grant No.: KY-C-01), National Natural Science Foundation of China (grant No.: 32071521), Carbon Peak and Carbon Neutrality Technology Innovation Foundation of Jiangsu Province (grant No.: BK20220030), Jiangsu Collaborative Innovation Center of Technology and Material of Water Treatment (no grant number), and Student Scientific Research Project, Jiangsu University (grant No.: 22A091).

Data Availability Statement: The datasets generated during and/or analyzed during the current study are available from the corresponding author on reasonable request.

Conflicts of Interest: The authors declare no conflict of interest.

References

1. Henry, A.L.; Gonzalez, E.; Bourgeois, B.; Sher, A.A. Invasive tree cover covaries with environmental factors to explain the functional composition of riparian plant communities. *Oecologia* **2021**, *196*, 1139–1152. [CrossRef]
2. Wang, C.Y.; Cheng, H.Y.; Wang, S.; Wei, M.; Du, D.L. Plant community and the influence of plant taxonomic diversity on community stability and invasibility: A case study based on *Solidago canadensis* L. *Sci. Total Environ.* **2021**, *768*, 144518. [CrossRef] [PubMed]
3. Sapsford, S.J.; Wakelin, A.; Peltzer, D.A.; Dickie, I.A. Pine invasion drives loss of soil fungal diversity. *Biol. Invasions* **2022**, *24*, 401–414. [CrossRef]
4. Pysek, P.; Jarosik, V.; Hulme, P.E.; Pergl, J.; Hejda, M.; Schaffner, U.; Vila, M. A global assessment of invasive plant impacts on resident species, communities and ecosystems: The interaction of impact measures, invading species' traits and environment. *Glob. Chang. Biol.* **2012**, *18*, 1725–1737. [CrossRef]
5. Yan, X.L.; Liu, Q.R.; Shou, H.Y.; Zeng, X.F.; Zhang, Y.; Chen, L.; Liu, Y.; Ma, H.Y.; Qi, S.Y.; Ma, J.S. The categorization and analysis on the geographic distribution patterns of Chinese alien invasive plants. *Biodivers. Sci.* **2014**, *22*, 667–676. [CrossRef]
6. Wang, C.Y.; Liu, J.; Xiao, H.G.; Zhou, J.W.; Du, D.L. Floristic characteristics of alien invasive seed plant species in China. *An. Acad. Bras. Ciências* **2016**, *88*, 1791–1797. [CrossRef]
7. Hou, Y.P.; Liu, L.; Chu, H.; Ma, S.J.; Zhao, D.; Liang, R.R. Effects of exotic plant *Rhus typhina* invasion on soil properties in different forest types. *Acta Ecol. Sin.* **2015**, *35*, 5324–5330. [CrossRef]
8. Huang, Q.Q.; Xu, H.; Fan, Z.W.; Hou, Y.P. Effects of *Rhus typhina* invasion into young *Pinus thunbergii* forests on soil chemical properties. *Ecol. Environ. Sci.* **2013**, *22*, 1119–1123. [CrossRef]
9. Wei, M.; Wang, S.; Wu, B.D.; Jiang, K.; Zhou, J.W.; Wang, C.Y. Variability of leaf functional traits of invasive tree *Rhus typhina* L. in North China. *J. Cent. South Univ.* **2020**, *27*, 155–163. [CrossRef]
10. Xu, Z.W.; Guo, X.; Caplan, J.S.; Li, M.Y.; Guo, W.H. Novel plant-soil feedbacks drive adaption of invasive plants to soil legacies of native plants under nitrogen deposition. *Plant Soil* **2021**, *467*, 47–65. [CrossRef]
11. Wang, G.M.; Jiang, G.M.; Yu, S.L.; Li, Y.H.; Liu, H. Invasion possibility and potential effects of *Rhus typhina* on beijing municipality *J. Integr. Plant Biol.* **2008**, *50*, 522–530. [CrossRef] [PubMed]
12. Zhang, Z.J.; Jiang, C.D.; Zhang, J.Z.; Zhang, H.J.; Shi, L. Ecophysiological evaluation of the potential invasiveness of *Rhus typhina* in its non-native habitats. *Tree Physiol.* **2009**, *29*, 1307–1316. [CrossRef] [PubMed]
13. Wang, C.Y.; Zhou, J.W.; Jiang, K.; Liu, J. Differences in leaf functional traits and allelopathic effects on seed germination and growth of *Lactuca sativa* between red and green leaves of *Rhus typhina*. *S. Afr. J. Bot.* **2017**, *111*, 17–22. [CrossRef]
14. Yu, Y.L.; Cheng, H.Y.; Wang, C.Y.; Du, D.L. Heavy drought reduces the decomposition rate of the mixed litters of two composite invasive alien plants. *J. Plant Ecol.* **2023**, *16*, rtac047. [CrossRef]
15. Zhong, S.S.; Xu, Z.L.; Yu, Y.L.; Cheng, H.Y.; Wang, S.; Wei, M.; Du, D.L.; Wang, C.Y. Acid deposition at higher acidity weakens the antagonistic responses during the co-decomposition of two Asteraceae invasive plants. *Ecotoxicol. Environ. Saf.* **2022**, *243*, 114012. [CrossRef] [PubMed]
16. Wang, C.Y.; Wei, M.; Wang, S.; Wu, B.D.; Du, D.L. Cadmium influences the litter decomposition of *Solidago canadensis* L. and soil N-fixing bacterial communities. *Chemosphere* **2020**, *246*, 125717. [CrossRef]
17. Dekanova, V.; Svitkova, I.; Novikmec, M.; Svitok, M. Litter breakdown of invasive alien plant species in a pond environment: Rapid decomposition of *Solidago canadensis* may alter resource dynamics. *Limnologica* **2021**, *90*, 125911. [CrossRef]
18. Aldorfová, A.; Münzbergová, Z. Conditions of plant cultivation affect the differences in intraspecific plant-soil feedback between invasive and native dominants. *Flora* **2019**, *261*, 151492. [CrossRef]
19. Stefanowicz, A.M.; Kapusta, P.; Stanek, M.; Frac, M.; Oszust, K.; Woch, M.W.; Zubek, S. Invasive plant *Reynoutria japonica* produces large amounts of phenolic compounds and reduces the biomass but not activity of soil microbial communities. *Sci. Total Environ.* **2021**, *767*, 145439. [CrossRef]
20. Czortek, P.; Krolak, E.; Borkowska, L.; Bielecka, A. Impacts of soil properties and functional diversity on the performance of invasive plant species *Solidago canadensis* L. on post-agricultural wastelands. *Sci. Total Environ.* **2020**, *729*, 139077. [CrossRef]
21. Yang, Q.; Carrillo, J.; Jin, H.y.; Shang, L.; Hovick, S.M.; Nijjer, S.; Gabler, C.A.; Li, B.; Siemann, E. Plant-soil biota interactions of an invasive species in its native and introduced ranges: Implications for invasion success. *Soil Biol. Biochem.* **2013**, *65*, 78–85. [CrossRef]

22. Ehrenfeld, J.G. Effects of exotic plant invasions on soil nutrient cycling processes. *Ecosystems* **2003**, *6*, 503–523. [CrossRef]

23. Prescott, C.E.; Zukswert, J.M. Invasive plant species and litter decomposition: Time to challenge assumptions. *New Phytol.* **2016**, *209*, 5–7. [CrossRef]

24. Hu, X.; Arif, M.; Ding, D.D.; Li, J.J.; He, X.R.; Li, C.X. Invasive plants and species richness impact litter decomposition in Riparian Zones. *Front. Plant Sci.* **2022**, *13*, 955656. [CrossRef] [PubMed]

25. Han, Z.X.; Wan, D.J.; Tian, H.X.; He, W.X.; Wang, Z.Q.; Liu, Q. Pollution assessment of heavy metals in soils and plants around a molybdenum mine in Central China. *Pol. J. Environ. Stud.* **2019**, *28*, 123–133. [CrossRef]

26. Chen, H.Y.; Teng, Y.G.; Lu, S.J.; Wang, Y.Y.; Wang, J.S. Contamination features and health risk of soil heavy metals in China. *Sci. Total Environ.* **2015**, *512–513*, 143–153. [CrossRef]

27. Cheng, M.M.; Wu, L.H.; Huang, Y.J.; Luo, Y.M.; Christie, P. Total concentrations of heavy metals and occurrence of antibiotics in sewage sludges from cities throughout China. *J. Soil. Sediment* **2014**, *14*, 1123–1135. [CrossRef]

28. Wang, H.T.; Wang, Q.J.; Cui, Z.Q.; Sun, Z.; Sun, C. Analysis of correlation among heavy metals elements and background value of heavy metal elements in soil of Nantong suburb. *J. Anhui Agric. Sci.* **2011**, *39*, 14062–14064. [CrossRef]

29. Wang, C.Y.; Jiang, K.; Zhou, J.W.; Liu, J.; Wu, B.D. Responses of soil N-fixing bacterial communities to redroot pigweed (*Amaranthus retroflexus* L.) invasion under Cu and Cd heavy metal soil pollution. *Agric. Ecosyst. Environ.* **2018**, *267*, 15–22. [CrossRef]

30. Wu, Y.M.; Leng, Z.R.; Li, J.; Jia, H.; Yan, C.L.; Hong, H.L.; Wang, Q.; Lu, Y.Y.; Du, D.L. Increased fluctuation of sulfur alleviates cadmium toxicity and exacerbates the expansion of *Spartina alterniflora* in coastal wetlands. *Environ. Pollut.* **2022**, *292*, 118399. [CrossRef]

31. Mesa-Marin, J.; Redondo-Gomez, S.; Rodriguez-Llorente, I.D.; Pajuelo, E.; Mateos-Naranjo, E. Microbial strategies in non-target invasive *Spartina densiflora* for heavy metal clean up in polluted saltmarshes. *Estuar. Coast. Shelf Sci.* **2020**, *238*, 106730. [CrossRef]

32. Xu, Z.L.; Zhong, S.S.; Yu, Y.L.; Wang, Y.Y.; Cheng, H.Y.; Du, D.L.; Wang, C.Y. *Rhus typhina* L. triggered greater allelopathic effects than *Koelreuteria paniculata* Laxm under ammonium fertilization. *Sci. Hortic.* **2023**, *309*, 111703. [CrossRef]

33. Jiang, K.; Wu, B.D.; Wang, C.Y.; Ran, Q. Ecotoxicological effects of metals with different concentrations and types on the morphological and physiological performance of wheat. *Ecotoxicol. Environ. Saf.* **2019**, *167*, 345–353. [CrossRef] [PubMed]

34. Wang, C.Y.; Wu, B.D.; Jiang, K.; Zhou, J.W. Effects of different types of heavy metal pollution on functional traits of invasive redroot pigweed and native red amaranth. *Int. J. Environ. Res.* **2018**, *12*, 419–427. [CrossRef]

35. Wei, M.; Wang, S.; Wu, B.D.; Cheng, H.Y.; Wang, C.Y. Heavy metal pollution improves allelopathic effects of Canada goldenrod on lettuce germination. *Plant Biol.* **2020**, *22*, 832–838. [CrossRef]

36. Lu, Y.J.; Wang, Y.F.; Wu, B.D.; Wang, S.; Wei, M.; Du, D.L.; Wang, C.Y. Allelopathy of three Compositae invasive alien species on indigenous *Lactuca sativa* L. enhanced under Cu and Pb pollution. *Sci. Hortic.* **2020**, *267*, 109323. [CrossRef]

37. Shang, E.; Xu, E.Q.; Zhang, H.Q.; Huang, C.H. Spatial-temporal trends and pollution source analysis for heavy metal contamination of cultivated soils in five major grain producing regions of China. *Environ. Sci.* **2018**, *39*, 4670–4683.

38. Chen, N.C.; Zheng, Y.J.; He, X.F.; Li, X.F.; Zhang, X.X. Analysis of the Report on the national general survey of soil contamination. *J. Agro-Environ. Sci.* **2017**, *36*, 1689–1692.

39. Wang, C.Y.; Wu, B.D.; Jiang, K.; Wei, M.; Wang, S. Effects of different concentrations and types of Cu and Pb on soil N-fixing bacterial communities in the wheat rhizosphere. *Appl. Soil Ecol.* **2019**, *144*, 51–59. [CrossRef]

40. Statistics, Z.B.o. *Zhenjiang Statistical Yearbook 2022*; China Statistics Press: Beijing, China, 2022.

41. Zhang, W.L.; Xu, A.G.; Zhang, R.L.; Ji, H.J. Review of soil classification and revision of China soil classification system. *Sci. Agric. Sin.* **2014**, *47*, 3214–3230.

42. Olson, J.S. Energy storage and the balance of producers and decomposers in ecological systems. *Ecology* **1963**, *44*, 322–331. [CrossRef]

43. Hoorens, B.; Aerts, R.; Stroetenga, M. Does initial litter chemistry explain litter mixture effects on decomposition. *Oecologia* **2003**, *137*, 578–586. [CrossRef] [PubMed]

44. Jones, G.L.; Scullion, J.; Worgan, H.; Gwynn-Jones, D. Litter of the invasive shrub *Rhododendron ponticum* (Ericaceae) modifies the decomposition rate of native UK woodland litter. *Ecol. Indic.* **2019**, *107*, 105597. [CrossRef]

45. Zhang, J.E. *Experimental Methods and Techniques Commonly Used in Ecology*; Chemical Industry Press: Beijing, China, 2006.

46. Perucci, P.; Casucci, C.; Dumontet, S. An improved method to evaluate the *o*-diphenol oxidase activity of soil. *Soil Biol. Biochem.* **2000**, *32*, 1927–1933. [CrossRef]

47. Guan, S.Y. *Soil Enzyme and Its Research Methods*; Agricultural Press: Beijing, China, 1986.

48. Klindworth, A.; Pruesse, E.; Schweer, T.; Peplies, J.; Quast, C.; Horn, M.; Glöckner, F.O. Evaluation of general 16S ribosomal RNA gene PCR primers for classical and next-generation sequencing-based diversity studies. *Nucleic Acids Res.* **2013**, *41*, e1. [CrossRef] [PubMed]

49. Wear, E.K.; Wilbanks, E.G.; Nelson, C.E.; Carlson, C.A. Primer selection impacts specific population abundances but not community dynamics in a monthly time-series 16S rRNA gene amplicon analysis of coastal marine bacterioplankton. *Environ. Microbiol.* **2018**, *20*, 2709–2726. [CrossRef]

50. Nwuche, C.O.; Ugoji, E.O. Effects of heavy metal pollution on the soil microbial activity. *Int. J. Environ. Sci. Technol.* **2008**, *5*, 409–414. [CrossRef]

51. Zhang, X.Y.; Gu, P.X.; Liu, X.Y.; Huang, X.; Wang, J.Y.; Zhang, S.Y.; Ji, J.H. Effect of crop straw biochars on the remediation of Cd-contaminated farmland soil by hyperaccumulator *Bidens pilosa* L. *Ecotoxicol. Environ. Saf.* **2021**, *219*, 112332. [CrossRef]
52. Zhang, C.; Nie, S.; Liang, J.; Zeng, G.M.; Wu, H.P.; Hua, S.S.; Liu, J.Y.; Yuan, Y.J.; Xiao, H.B.; Deng, L.J.; et al. Effects of heavy metals and soil physicochemical properties on wetland soil microbial biomass and bacterial community structure. *Sci. Total Environ.* **2016**, *557–558*, 785–790. [CrossRef]
53. Liao, C.Z.; Luo, Y.Q.; Jiang, L.F.; Zhou, X.H.; Wu, X.W.; Fang, C.M.; Chen, J.K.; Li, B. Invasion of *Spartina alterniflora* enhanced ecosystem carbon and nitrogen stocks in the Yangtze Estuary, China. *Ecosystems* **2007**, *10*, 1351–1361. [CrossRef]
54. Liao, C.Z.; Peng, R.H.; Luo, Y.Q.; Zhou, X.H.; Wu, X.W.; Fang, C.M.; Chen, J.K.; Li, B. Altered ecosystem carbon and nitrogen cycles by plant invasion: A meta-analysis. *New Phytol.* **2008**, *177*, 706–714. [CrossRef]
55. Maan, I.; Kaur, A.; Sharma, A.; Singh, H.P.; Batish, D.R.; Kohli, R.K.; Arora, N.K. Variations in leaf litter decomposition explain invasion success of *Broussonetia papyrifera* over confamilial non-invasive *Morus alba* in urban habitats. *Urban For. Urban Green.* **2022**, *67*, 127408. [CrossRef]
56. Mitchell, J.D.; Lockaby, B.G.; Brantley, E.F. Influence of Chinese Privet (*Ligustrum sinense*) on decomposition and nutrient availability in riparian forests. *Invasive Plant Sci. Manag.* **2011**, *4*, 437–447. [CrossRef]
57. Schuster, M.J.; Dukes, J.S. Non-additive effects of invasive tree litter shift seasonal N release: A potential invasion feedback. *Oikos* **2014**, *123*, 1101–1111. [CrossRef]
58. Hättenschwiler, S.; Tiunov, A.V.; Scheu, S. Biodiversity and litter decomposition in terrestrial ecosystems. *Annu. Rev. Ecol. Evol. Syst.* **2005**, *36*, 191–218. [CrossRef]
59. Hättenschwiler, S.; Fromin, N.; Barantal, S. Functional diversity of terrestrial microbial decomposers and their substrates. *Comptes Rendus Biol.* **2011**, *334*, 393–402. [CrossRef]
60. Brunel, C.; Gros, R.; Ziarelli, F.; Farnet Da Silva, A.M. Additive or non-additive effect of mixing oak in pine stands on soil properties depends on the tree species in Mediterranean forests. *Sci. Total Environ.* **2017**, *590–591*, 676–685. [CrossRef]
61. Yu, H.X.; Le Roux, J.J.; Jiang, Z.Y.; Sun, F.; Peng, C.L.; Li, W.H. Soil nitrogen dynamics and competition during plant invasion: Insights from *Mikania micrantha* invasions in China. *New Phytol.* **2021**, *229*, 3440–3452. [CrossRef]
62. Torres, N.; Herrera, I.; Fajardo, L.; Bustamante, R.O. Meta-analysis of the impact of plant invasions on soil microbial communities. *BMC Ecol. Evol.* **2021**, *21*, 172. [CrossRef]
63. Allison, S.D.; Nielsen, C.; Hughes, R.F. Elevated enzyme activities in soils under the invasive nitrogen-fixing tree *Falcataria moluccana*. *Soil Biol. Biochem.* **2006**, *38*, 1537–1544. [CrossRef]
64. Yu, Y.L.; Cheng, H.Y.; Wang, S.; Wei, M.; Wang, C.Y.; Du, D.L. Drought may be beneficial to the competitive advantage of *Amaranthus spinosus*. *J. Plant Ecol.* **2022**, *15*, 494–508. [CrossRef]
65. Murugan, R.; Beggi, F.; Prabakaran, N.; Maqsood, S.; Joergensen, R.G. Changes in plant community and soil ecological indicators in response to *Prosopis juliflora* and *Acacia mearnsii* invasion and removal in two biodiversity hotspots in Southern India. *Soil Ecol. Lett.* **2020**, *2*, 61–72. [CrossRef]
66. Stefanowicz, A.M.; Stanek, M.; Nobis, M.; Zubek, S. Species-specific effects of plant invasions on activity, biomass, and composition of soil microbial communities. *Biol. Fertil. Soils* **2016**, *52*, 841–852. [CrossRef]

Review

Ecology, Biology, Environmental Impacts, and Management of an Agro-Environmental Weed *Ageratum conyzoides*

Amarpreet Kaur [1], Shalinder Kaur [1,*], Harminder Pal Singh [2], Avishek Datta [3], Bhagirath Singh Chauhan [4], Hayat Ullah [3], Ravinder Kumar Kohli [5] and Daizy Rani Batish [1]

[1] Department of Botany, Panjab University, Chandigarh 160014, India; amanhayer411@gmail.com (A.K.); daizybatish@yahoo.com (D.R.B.)
[2] Department of Environment Studies, Panjab University, Chandigarh 160014, India; hpsingh_01@yahoo.com
[3] Department of Food, Agriculture and Bioresources, School of Environment and Resource Development, Asian Institute of Technology, Klong Luang, Pathumthani 12120, Thailand; datta@ait.ac.th (A.D.); hayatbotanist204@gmail.com (H.U.)
[4] The Centre for Crop Science, The University of Queensland, Gatton, QLD 4343, Australia; b.chauhan@uq.edu.au
[5] Amity University, Mohali 140306, Punjab, India; rkkohli45@yahoo.com
* Correspondence: shalinder@pu.ac.in; Tel.: +91-172-2534095

Abstract: *Ageratum conyzoides* L. (Billy goat weed; Asteraceae) is an annual herbaceous plant of American origin with a pantropical distribution. The plant has unique biological attributes and a raft of miscellaneous chemical compounds that render it a pharmacologically important herb. Despite its high medicinal value, the constant spread of the weed is noticeable and alarming. In many countries, the weed has severely invaded the natural, urban, and agroecosystems, thus presenting management challenges to natural resource professionals and farmers. Its interference with agricultural crops, grassland forbs, forest ground flora, and its ability to replace native plant species are of serious concern. Therefore, it is pertinent to monitor its continuous spread, its entry into new geographic regions, the extent of its impact, and the associated evolutionary changes. While management strategies should be improvised to control its spread and reduce its adverse impacts, the possible utilization of this noxious weed for pharmacological and agronomic purposes should also be explored. The objective of this review is to provide a detailed account of the global distribution, biological activities, ecological and environmental impacts, and strategies for the management of the agro-environmental weed *A. conyzoides*.

Keywords: agricultural weed; Asteraceae; billy goat weed; ecological impacts; environmental weed; invasive weed; weed management

Citation: Kaur, A.; Kaur, S.; Singh, H.P.; Datta, A.; Chauhan, B.S.; Ullah, H.; Kohli, R.K.; Batish, D.R. Ecology, Biology, Environmental Impacts, and Management of an Agro-Environmental Weed *Ageratum conyzoides*. *Plants* **2023**, *12*, 2329. https://doi.org/10.3390/plants12122329

Academic Editors: Congyan Wang and Hongli Li

Received: 4 May 2023
Revised: 3 June 2023
Accepted: 9 June 2023
Published: 15 June 2023

1. Introduction

Ageratum conyzoides L. (Family Asteraceae) is an aromatic annual herb native to Central and South America [1]. The genus "*Ageratum*" refers to the Greek term "*ageras*", signifying the seemingly non-ageing quality of this species (referring to its long lifespan), and the species epithet "*konyz*" refers to the Greek name of the plant species, *Inula helenium*, which the weed resembles [2]. The common name, "goat weed" or "Billy goat weed", is derived from an Australian male goat due to a close resemblance in odor [3]. It has two subspecies: "*latifolium*", found within the American continent, and "*conyzoides*", with a distribution throughout the tropical and subtropical regions of the world [2].

The plant was initially distributed across different continents owing to its ornamental value, but it has now naturalized and spread in nearly all types of ecosystems, colonizing aggressively, and presenting management issues to environmentalists, ecologists, conservation managers, and agronomists [4]. Apart from its invasive abilities, the plant is well known for its strong phytochemical composition, unique biological attributes, and versatile applications in agriculture and industry. While previous reviews regarding *A. conyzoides* primarily emphasized its potential as a medicinal [5] or industrial [6] crop, this holistic review aims to provide a succinct overview of its global distribution, biological activities, ecological and environmental impacts, and management strategies; thereby encompassing all facets of the plant's behavior. The review serves as an updated introduction to *A. conyzoides*, while highlighting the fields where most of the current studies are focused and identifying the research areas requiring further attention.

2. Global Distribution

The occurrence of *A. conyzoides* as a troublesome weed came to attention between 1960 and 1980 [7]. Many reports suggest its widespread distribution in various countries in Asia, Africa, Australia, and America (Figure 1). In Asia, its invasion has been reported in India, Malaysia, the Philippines, Malawi, Cambodia, Thailand, Vietnam, Bangladesh, Pakistan, Sri Lanka, Indonesia, China, and Japan [4,8–14]. In Africa, it is widely distributed in South Africa, East Africa, Zimbabwe, Mauritius, Angola, Ethiopia, Kenya, Liberia, Tanzania, Uganda, the Democratic Republic of the Congo, Egypt, Ghana, and Nigeria [9,11,15–20]. The herb is also found in Australia [21], Fiji, New Caledonia, the Cook Islands, the Solomon Islands, Vanuatu [8], the Mariana Islands, the Hawaiian Islands, the Virgin Islands, the Galapagos Islands [14], the Federated States of Micronesia [22], and Palau [23]. *A. conyzoides* has considerably enhanced its distribution range in the last few decades, and research utilizing ecological niche models suggests that the weed has the potential to perform better under climate change scenarios and to expand into uninvaded regions in the future [1,24,25]. A detailed account of its distribution on different continents and island ecosystems around the world is provided in Figure 1.

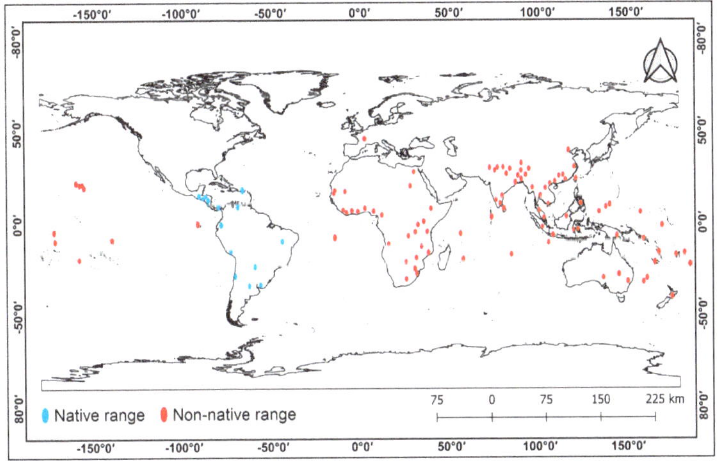

Figure 1. Distribution of *Ageratum conyzoides* in native and non-native regions across the globe (Source: Vélez-Gavilán [7]).

3. Ecology

Ageratum conyzoides grows erect up to 1 m with profuse branching, and is characterized by oval, pubescent leaves with toothed margins and a reddish stem [1] (Figure 2). White to mauve-colored hermaphroditic disc florets are arranged closely in the inflorescence, i.e., the capitulum [1]. The variations in flower color are associated with different chemotypes of the plant [26]. The seeds bear an aristate pappus that facilitates anemochory in the plant [13]. Vegetative reproduction occurs through stolons [4] (Figure 2). In the northern plains of India, *A. conyzoides* appears twice annually, with a short (June–July) and a prolonged (October–March) life cycle [27] (Figure 3). *A. conyzoides* has attained the status of a noxious biological pollutant, especially in croplands, by virtue of many ecological and biological traits that help in its successful invasion in alien environments.

Figure 2. Different growth stages of *Ageratum conyzoides*.

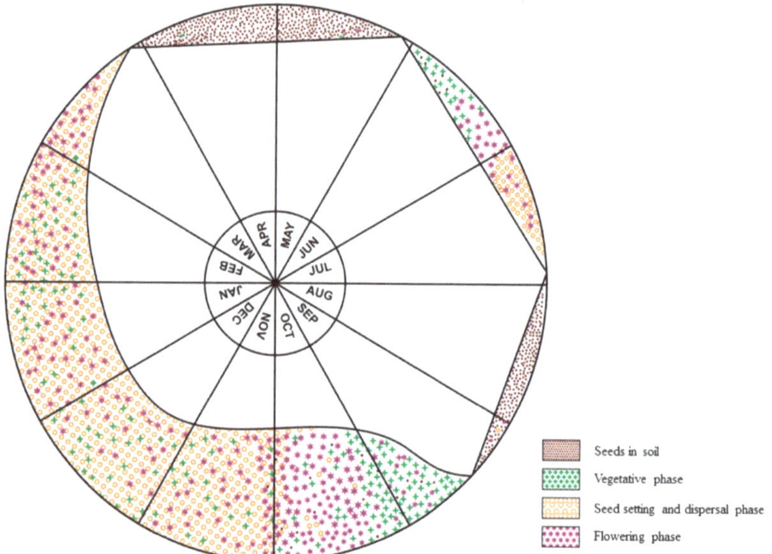

Figure 3. Life cycle pattern of *Ageratum conyzoides* in Chandigarh, India (Source: Arora [27]).

3.1. Environmental Suitability and Adaptability

Ageratum conyzoides is highly adaptable to different temperatures, moisture conditions, soil textures, and altitudinal ranges. Its growth is best suited to temperatures ranging from 20–25 °C, but it also survives well at 15–30 °C [27]. This explains its prevalence at higher altitudes (i.e., temperate climates) as well as on the plains (i.e., tropical climates) [28]. Its growth is not affected by soil fertility status, and the weed acclimatizes well to high light intensity and severe salinity stresses [29]. High phenotypic plasticity allows the plant to settle in novel surroundings via suitable biomass allocation [30].

3.2. Ecological Range

Within its natural geographic range, *A. conyzoides* is only considered to be an agricultural weed, but in invaded areas, its thickets can be spotted in agricultural lands, grasslands, wastelands, natural forests, wetlands, plantations, vegetable gardens, pastures, orchards, tea plantations, alongside water channels, disturbed sites, sites of fresh landslides, and roadsides [4,31] (Figure 4). The plant is an early colonizer of abandoned fields or shifting cultivation sites, and sometimes it dominates as a pioneer community [32]. It can easily colonize available gaps in widely spaced annual crops or plantations [2,33].

Figure 4. The spread of *Ageratum conyzoides* in various habitats.

3.3. Reproductive and Regenerative Potential

Fast growth, a short life cycle, early reproductive maturity, prolific seed production, and vegetative reproduction enable *A. conyzoides* to establish itself in an alien environment [2]. The large production of small, lightweight seeds with a wide range of dispersal enables its fast spread and colonization [13]. A study conducted in China revealed that *A. conyzoides* dispersed at a minimum speed of 2.4 km year^{-1}, mainly through human or wind-mediated dispersal [13]. A single plant is reported to produce 40,000–95,000 seeds with a germination rate of 50% [34]. Additionally, the weed may also proliferate quickly through vegetative reproduction by stolon production [4].

3.4. Allelopathy

Aqueous extracts, volatile oils, and the rhizospheric soil of *A. conyzoides* are known to possess allelochemicals that interfere with the growth and development of associated plants [35,36]. The germination, plumule, and radicle length of several crop species were reported to be affected by *A. conyzoides* [36–39]. In addition, the large amounts of weed residue left at the infestation sites interfered with the growth of the succeeding crops [40]. Allelochemicals are released from the plant by leachates from foliage and plant litter by rain, volatilization through aerial parts as root exudates, or the decomposition of organic matter. The mechanism behind the phytotoxic effect of these compounds has also been examined. It was found to alter the actions of the respiratory enzymes [41] and the growth hormone gibberellic acid, which is involved in seed germination [42]. However, further investigations are required to accurately identify its mode of action.

In addition to the above-mentioned characteristics, several other factors such as morphological and phenological adaptations, a short juvenile phase, long flowering and fruiting periods, the absence of natural enemies (pests, pathogens, and herbivores), the resistance to native predators due to the release of a wide array of secondary metabolites, and unpalatability due to the highly phytotoxic nature of the plant contribute to its unchecked prevalence in the invaded ranges [2,4]. In general, wide adaptability in invasive species has been reported to give rise to intraspecific variations that maximize their fitness under heterogeneous environmental conditions [43,44]. The presence of different chemotypes, ecotypes, and biotypes in *A. conyzoides* [5,26] indicates its ability to expand extensively across diverse geographical regimes by overcoming its physiological limitations and environmental barriers.

4. Biological Activity

The weed has been used for a wide range of biological activities since antiquity, as it possesses a raft of miscellaneous chemical compounds. The natural product chemistry and biological properties of *A. conyzoides* have been extensively investigated. Some of its prominent biological activities and potential applications are discussed further.

4.1. Natural Product Chemistry

A wide variety of secondary metabolites have been identified in the aqueous extracts and volatile oils obtained from various parts of *A. conyzoides* [3]. The essential oil contains phytochemicals such as phenols, phenolic esters, and coumarins, whereas the other parts of the plant contain terpenoids, steroids, chromenes, pyrrolizidine alkaloids, and flavonoids [5,45]. As per Dores et al. [46], the weed contains the maximum phenolic compounds in its roots (23 mg mL^{-1}), followed by flowers (19 mg mL^{-1}), and leaves (15 mg mL^{-1}), while the maximum content of flavonoids is observed in leaves (5.7 µg mL^{-1}), followed by flowers (5.4 µg mL^{-1}) and roots (4.8 µg mL^{-1}). The essential oil of *A. conyzoides* constitutes nearly 200 chemical compounds, of which precocene I and II, their derivatives, monoterpenes, and sesquiterpenes are the major constituents, accounting for 77% of the oil [35,47].

4.2. Pharmacological Properties

Ageratum conyzoides has ethnobotanical importance due to its use in traditional medicinal practices [5]. Leaves of the plant are commonly applied to heal burns and wounds across the world and are used in ayurvedic medicines prepared for fever, earache, cold, headache, rheumatism, diabetes, infertility, blood clotting, diarrhea, ear infections, etc. [3,48]. According to the Bodo community of Assam, the root extracts of *A. conyzoides* help fight malaria [49]. A recent study has shown that the leaves of *A. conyzoides* possess alpha-amylase inhibitory potential, which is beneficial in treating type II diabetes as well as its secondary complications, by lowering postprandial hyperglycemia [50,51]. Antitumor and co-chemotherapeutic effects of *A. conyzoides* have also been reported [52,53]. The correlation between in silico and classical pharmaceutical investigations implies that the antidiabetic and anticancerous activities of the weed stem mainly from chromenes (precocene I, precocene II, and VMDC) and sterols (betasterol and stigmasterol), respectively [45]. *A. conyzoides* has also proven to be a potential adjuvant agent in the treatment of polycystic ovary syndrome [54]. Extracts of the weed also exhibited antibacterial activity by inhibiting 40.4% of the 464 strains of drug-resistant gram-positive and gram-negative bacteria [55]. Crude extracts of the plant severely affected the flagella and ventral discs of *Giardia duodenalis* (the causal organism of giardiasis in humans), thus impeding their ability to attach to the surface of mucosal cells [56]. Phyto-formulations based on *A. conyzoides* possess acaricidal potential against acaricide-resistant ticks, infesting cattle, and buffalo [57]. It is suggested that the compounds isolated from *A. conyzoides* can combat numerous pathogenic strains that cause infections in humans and livestock [55].

4.3. Insecticidal Properties

The essential oil obtained from *A. conyzoides* can kill insects by modifying their digestive systems. Precocenes present in the essential oil possess antijuvenile hormonal activity, causing precocious metamorphosis in insects [3]. The oil may also induce abnormalities at the phenotypic or genotypic level in the larvae of *Aedes*, *Anopheles*, and *Culex* spp. [58]. The weed has also shown promising results against plant and animal pests such as *Rhipicephalus microplus*, *Phytophthora megakarya*, *Diaphania hyalinata*, *Tribolium castaneum*, *Helicoverpa armigera*, *Plutella xylostella*, etc. [11,59–63] indicating the plant's potential to be utilized for controlling insect pests.

4.4. Fungicidal Properties

The essential oil of *A. conyzoides* can serve as a substitute for synthetic fungicides due to its strong fungicidal and aflatoxin-inhibitory potential [64]. Flavones released by the weed showed results equivalent to those of the commercial fungicide carbengin by reducing fungal infections in citrus plantations [65]. Essential oils and extracts of the weed showed very strong antifungal activity against *Drechslera* sp., *Puccinia arachidis*, *Botryodiplodia theobromae*, *Fusarium verticillioides*, *Alternaria cucumeria*, *Curvularia lunata*, *Pyricularia oryzae*, *Rhizoctonia solani*, *Aspergillus flavus*, etc. [66–70], thereby providing an effective and eco-friendly solution for the management of fungal pathogens.

4.5. Herbicidal Properties

Ageratum conyzoides has recently been recognized as a novel agrochemical tool for weed control. Plant extracts hindered the growth of *Digitaria sanguinalis*, *Lactuca sativa*, *Amaranthus caudatus*, *Amaranthus spinosus*, *Echinochloa crus-galli*, *Monochoria vaginalis*, and *Aeschynomene indica* [71–73]. When intercropped in citrus orchards, it significantly inhibited weeds such as *Cyperus difformis*, *Bidens pilosa*, and *Digitaria sanguinalis* [74]. Application of the dried leaves of *A. conyzoides* killed 75% of the weeds present in the rice field and increased the grain yield by 14% compared to a commercial herbicide [72]. In another study, the extracts of *A. conyzoides* showed promising results in controlling weeds while at the same time conserving and increasing the soil microflora [75], thereby indicating its beneficial use as a bioherbicide.

A. conyzoides show promise as a valuable biosource for developing effective formulations for clinical, industrial, and agricultural uses. However, further scientific research is necessary to assess the chronic toxicological reactions, potential side effects, safe dosage levels, and long-term interactions and feedback [69]. A recent study indicated that the removal of pyrrolizidine alkaloids plays a significant role in determining the toxicity of *A. conyzoides* extracts [76]. This finding underscores the importance of considering and managing the presence of these compounds in formulations and products derived from the plant. Furthermore, there is a need for advanced molecular technologies, including RNAi, CRISPR/Cas9, multi-omics approaches, etc., which may aid in deciphering the action mechanisms and enhance these formulations [77]. It is not only vital for enhancing the economic value of *A. conyzoides* but also for developing sustainable and safe strategies for biomedical, environmental, and agricultural applications.

5. Ecological Impacts

Ageratum conyzoides damages ecosystems both economically and ecologically, either directly competing with the native plants for resources, and/or indirectly by altering ecosystem processes and ecological functioning such as soil nutrient cycling, pollination, etc. Nevertheless, there is a scarcity of research in this domain, specifically in terms of studies that offer empirical data to accurately assess the magnitude of the harm inflicted. The following section focuses on the primary ecosystems impacted by the invasion of *A. conyzoides*.

5.1. Agricultural Ecosystem

Ageratum conyzoides is a devastating agricultural weed that affects nearly 36 crop species and is found in 46 countries [9]. It affects staple food crops as well as commercially important cash crops [31,78,79]. The establishment of permanent seed banks in the fields of the lower Shivalik range of the Himalayas due to heavy infestations of *A. conyzoides* had left them useless and abandoned [77,80]. Apart from being an agricultural weed, it is also known to host pests and pathogens of various crops. Different begomovirus-satellite complexes have been identified in *A. conyzoides* [81]. Reports show that *Ageratum enation*, a virus capable of infecting several important food crops, is hosted by this weed [82]. *A. conyzoides* also hosts tomato yellow leaf curl, cotton leaf curl, and okra enation leaf curl viruses [83,84] and, therefore, may act as an alternate source of infection in okra and tomato crops. The weed is a natural host for various aphid species that act as vectors for carrying papaya ringspot virus type P, a causal organism of papaya ringspot [85]. The Capsicum chlorosis virus was also reported to be hosted by *A. conyzoides* in the eastern regions of Queensland, Australia [21]. Infestations of the plant result in heavy monetary losses, particularly in the case of farmers with small land holdings [6].

5.2. Forest Ecosystem

Due to its shade-tolerant nature, *A. conyzoides* can maintain dense populations under tree canopies in forests [33,79]. For example, the understory of the tree plantations of *Acacia catechu*, *Eucalyptus* spp., *Pinus* forests, and mixed forests in the lower Himalayas of Himachal Pradesh (India) has been observed to be occupied by *A. conyzoides* [80]. It has been reported that *A. conyzoides* is among several exotic species that are posing a serious threat to the dense interior forests of Gandhamardan Hill Range, Odisha, India [86], the protected forests of Tripura [87], and a forest range in the village of Changki, Nagaland, India [88].

5.3. Grasslands and Rangelands

The dominance and ecological impact of *A. conyzoides* in grassland ecosystems have been observed by several researchers [89,90]. The fast-spreading stolons of *A. conyzoides* greatly enhance the weed's potential to cover large areas of grasslands and rangelands, thus destroying native grasses and forbs and causing fodder shortages for livestock [4,90]. Furthermore, it reduces the carrying capacity of pastures and may lead to the disappearance of threatened and endemic species [4].

5.4. Soil Ecology

Ageratum conyzoides affects the soil chemistry, nutrient composition, and soil microbiota, thereby altering the environment of the invaded habitat [91]. A study reported a significant reduction in soil nitrogen and phosphorus in rice fields due to weed infestation [78], whereas other reports suggested that weed residues have enriched the soil nutrient content [31,92], despite their negative effects on associated crop species [31]. The weed modifies the soil environment through root exudation by mobilizing or chelating nutrients and, in turn, disturbing the natural soil composition [36].

5.5. Biodiversity

The ability of the weed to occupy available niches has reduced the availability of habitats for the local flora, which affects the biodiversity of the invaded areas [79]. *A. conyzoides* can easily outcompete medicinal-rich plants [4]. The weed has affected the biodiversity components of the invaded localities in the lower Himalayas by replacing native grasses and economically important herbs and creating homogenous stands [93].

5.6. Humans and Livestock

As a medicinal plant, *A. conyzoides* has limited uses due to its toxicity. *Ageratum conyzoides* at 500 and 1000 mg kg^{-1} can stimulate hematological disorders and may also affect the liver and kidneys in humans [94]. It can also cause dermatitis, nausea, bronchitis, and asthma. A study showed that it is one of the most common pollen allergens affecting patients with allergic rhinitis [95]. If livestock feeds on the plant due to a scarcity of fodder or immature taste buds, it can cause shivering, a very high fever, the production of bitter milk, anorexia, diarrhea, ulceration, or even lethal toxicity under extreme conditions [96].

The unchecked spread of *A. conyzoides* may have serious ecological and economic implications in different ecosystems, the magnitude of which is still not clearly known. Therefore, it becomes imperative to develop and implement appropriate management strategies to restrict its spread and mitigate its impact in an efficient, cost-effective, and eco-friendly manner.

6. Control and Management

Various methods have been proposed and practiced by agronomists and natural resource professionals for weed control. In this section, the most commonly and successfully used methods, in addition to their pros and cons, are considered.

6.1. Physical Methods

Physical control methods include uprooting, burning, or cutting (using blades, shrub masters, etc.), depending on the intensity of spread, the size of the area infected, and the stage of the weed that is being removed. However, this cannot be a feasible option if the area involved is large and is only a short-term solution if the plant has already set seeds [97]. Furthermore, the dangers of health problems, soil contamination by burning, and the likelihood of its re-emergence are also present [98].

6.2. Cultural Methods

In agricultural systems, the density of *A. conyzoides* can be reduced by using conservation tillage systems, and the residues left can also be used as mulch to enhance soil fertility [99]. Since the weed cannot germinate under anaerobic conditions, flooding the field for a short time can help control its infestations [100]. The density of *A. conyzoides* is best controlled by passing a wheel hoe at regular intervals along with hand weeding, or by using the stale seedbed technique along with inter-cultivation [101].

6.3. Chemical Methods

Both pre-emergence (oxadiazon, atrazine, oxyfluorfen, diuron, methazole, simazine, etc.) and post-emergence (glyphosate, 2, 4-D, etc.) herbicides are used to control infestations of *A. conyzoides*. In agroecosystems, the selection of the herbicide usually depends on the host crop species. Unlike grassy weeds, *A. conyzoides* appears late in maize fields and, therefore, is better controlled by a post-emergence spray of atrazine [102]. However, these herbicides pose environmental dangers; cannot control the regeneration of plants from root stumps, runners, suckers, stolons, or the seed bank present in the soil; and may result in the production of resistant species [97,98].

6.4. Biological Methods

The use of natural plant products, e.g., parthenin extracted from *Parthenium hysterophorus* and volatile essential oils from *Callistemon viminalis*, has been found to be effective in controlling *A. conyzoides* [103–105]. Monoterpenes, such as cineole and citronellol, which are found in members of the citrus family, also have potential for the management of *A. conyzoides* [106]. As a cover crop and mulch, both *Chromolaena odorata* and *Mikania micrantha* showed allelopathic properties against the growth of *A. conyzoides* [107]. Layering the soil with residues of *Dicranopteris linearis* inhibited the growth of *A. conyzoides* seedlings [108]. Even endophytic actinomycetes isolated from different plants are rich sources of herbicidal metabolites and can be employed against the weed [109].

Pathogens, insects, and nematodes have been introduced from the native ranges of *A. conyzoides* to serve as natural enemies of the weed; none, however, have proven effective. These were found to be polyphagous and, thus, had the strong potential to become pests of many other useful plants. Therefore, these are generally not recommended as a management strategy for the weed.

6.5. Field and Crop Management

Knowledge of the patterns of weed emergence can prove advantageous in planning their management. Herbicidal applications can be more beneficial if an understanding of the plant developmental cycle is achieved [110]. After harvesting, instead of being left fallow, crop fields can be used to grow legume crops or other useful plants to occupy the empty niche. This also applies to the areas where the weed is removed to prevent reinfestation. Palisade grass (*Urochloa brizantha*) cropping at regular intervals and sorghum intercropped with congo grass (*Brachiaria ruziziensis*) have also been reported to decrease the seed bank of *A. conyzoides* [111]. Growing sweet potato (*Ipomoea batatas*) varieties in fields dominated by *A. conyzoides* has been shown to reduce their growth, biomass, and yield traits [112]. Furthermore, vermicomposting is another viable, eco-friendly, economical, and proven solution for the effective and on-site management of *A. conyzoides* [79]. The compost of *A. conyzoides* prepared using a rotary drum composter was also found to be nutrient-rich and non-toxic and can be used effectively as a soil conditioner [91].

Though our understanding of the ecology and management possibilities of *A. conyzoides* has improved considerably, challenges remain to control its spread. We recommend a six-step management guide provided by Batish et al. [97] to manage *A. conyzoides*. It begins with targeting an infested area, followed by (a) the compilation of all the necessary information on the weed, (b) understanding weed biology, mode of spread, and invasive characteristics, (c) estimating its monetary, ecological, and socio-economic impacts, (d) creating awareness among local people, (e) taking appropriate control measures, and (f) devising preventive measures to avoid the re-emergence of the invasive weed.

Integrated weed management is the best approach to monitoring and regulating any weed, including *A. conyzoides*. Individual strategies involving physical, chemical, and biological methods have failed to provide long-term control, but an integrated approach utilizing all these techniques may prove to be relatively successful [91,113]. A suitable set of control measures (taking into consideration the extent and intensity of infestation) should be selected and employed with the involvement of government and non-governmental organizations, researchers, conservation managers, agriculturists, and local people. The least infested areas should be targeted first, followed by those with dense infestations. The waste accumulated may even be utilized for biogas or compost production [79,91,114]. With the help of the public, the further re-emergence and spread of the weed should be checked. Consistent follow-up work with the participation of both higher authorities and local communities is essential for the sustainable management of *A. conyzoides*. Furthermore, promoting its utilization at both commercial and non-commercial scales can offer economically viable and beneficial options for its management.

7. Conclusions and Way Forward

Ageratum conyzoides exhibits diverse biological characteristics that contribute to its significance both in medical and socio-economic contexts while also amplifying its invasive tendencies. This review drew attention to its expanding global distribution, the biological characteristics enabling its invasive success, the resulting ecological impacts of infestation, and the management options considered so far. In addition, the economic value of the species, along with traditional and modern applications, was highlighted. This discussion also sought to identify lesser-explored aspects and knowledge gaps in the ongoing research to suggest potential areas for future research. While the pharmacological and industrial applications of the plant have received considerable attention, the consequent impacts of its spread remain relatively unexplored. It is important to cover any possible lacunae in our understanding of its invasive behavior to strengthen its management at different levels. Additionally, the relationship between its toxicity and bioactivity, which is crucial for validating its medicinal properties, has not been adequately addressed. Given the invasive nature of *A. conyzoides* and its diverse biological activities, there is considerable anticipation for its potential as a botanical drug or pesticide. Consequently, there is a need to enhance *A. conyzoides*-based products using advanced tools and technologies to fully harness their therapeutic and pesticidal properties.

Author Contributions: Conceptualization, D.R.B., S.K., B.S.C. and R.K.K.; Data curation, A.K. and S.K.; Funding acquisition, D.R.B. and H.P.S.; Supervision, D.R.B., H.P.S. and R.K.K.; Writing—original draft, A.K.; Writing—review and editing, S.K., B.S.C., H.P.S., A.D. and H.U. All authors have read and agreed to the published version of the manuscript.

Funding: This research received no external funding.

Institutional Review Board Statement: Not applicable.

Informed Consent Statement: Not applicable.

Data Availability Statement: Not applicable.

Conflicts of Interest: The authors declare that they have no conflict of interest.

References

1. Ray, D.; Behera, M.D.; Jacob, J. Comparing invasiveness of native and non-native species under changing climate in north-east India: Ecological niche modelling with plant types differing in biogeographic origin. *Environ. Monit. Assess.* **2019**, *191*, 793. [CrossRef]
2. Kaur, S.; Batish, D.R.; Kohli, R.K.; Singh, H.P. Ageratum conyzoides: An alien invasive weed in India. In *Invasive Alien Plants: An Ecological Appraisal for the Indian Subcontinent*; Bhatt, J.R., Singh, J.S., Singh, S.P., Tripathi, R.S., Kohli, R.K., Eds.; CABI: Wallingford, UK, 2012; pp. 57–76.
3. Okunade, A.L. *Ageratum conyzoides* L. (Asteraceae). *Fitoterapia* **2002**, *73*, 507–510. [CrossRef]
4. Kohli, R.K.; Batish, D.R.; Singh, H.P.; Dogra, K.S. Status, invasiveness and environmental threats of three tropical American invasive weeds (*Parthenium hysterophorus* L., *Ageratum conyzoides* L., *Lantana camara* L.) in India. *Biol. Invasions* **2006**, *8*, 1501–1510. [CrossRef]
5. Yadav, N.; Ganie, S.A.; Singh, B.; Chhillar, A.K.; Yadav, S.S. Phytochemical constituents and ethnopharmacological properties of *Ageratum conyzoides* L. *Phytother. Res.* **2019**, *33*, 2163–2178. [CrossRef]
6. Rioba, N.B.; Stevenson, P.C. Ageratum conyzoides L. for the management of pests and diseases by small holder farmers. *Ind. Crops Prod.* **2017**, *110*, 22–29. [CrossRef]
7. Vélez-Gavilán, J. Ageratum conyzoides (billy goat weed). In *CABI Datasheet*; CABI: Wallingford, UK, 2016. [CrossRef]
8. Waterhouse, D.F. *The Major Invertebrate Pests and Weeds of Agriculture and Plantation Forestry in the Southern and Western Pacific*; Monograph No. 44; The Australian Centre for International Agricultural Research (ACIAR): Canberra, Australia, 1997.
9. Holm, L.; Pancho, J.V.; Herberger, J.P.; Plucknett, D.L. *A Geographical Atlas of World Weeds*; John Wiley and Sons: New York, NY, USA, 1979.
10. Akter, A.; Zuberi, M.I. Invasive alien species in northern Bangladesh: Identification, inventory and impacts. *Int. J. Biodivers. Conserv.* **2009**, *1*, 129–134.
11. Jaya; Singh, P.; Prakash, B.; Dubey, N.K. Insecticidal activity of *Ageratum conyzoides* L., *Coleus aromaticus* Benth. and *Hyptis suaveolens* (L.) Poit essential oils as fumigant against storage grain insect *Tribolium castaneum* Herbst. *J. Food Sci. Technol.* **2014**, *51*, 2210–2215. [CrossRef]
12. Kudo, Y.; Mutaqien, Z.; Simbolon, H.; Suzuki, E. Spread of invasive plants along trails in two national parks in West Java, Indonesia. *Tropics* **2014**, *23*, 99–110. [CrossRef]
13. Horvitz, N.; Wang, R.; Wan, F.H.; Nathan, R. Pervasive human-mediated large-scale invasion: Analysis of spread patterns and their underlying mechanisms in 17 of China's worst invasive plants. *J. Ecol.* **2017**, *105*, 85–94. [CrossRef]
14. PIER. Pacific Island Ecosystems at Risk. *Ageratum conyzoides*. 2023. Available online: http://www.hear.org/pier/species/ageratum_conyzoides.htm (accessed on 10 January 2023).
15. Foxcroft, L.C.; Richardson, D.M.; Wilson, J.R.U. Ornamental plants as invasive aliens, problems and solutions in Kruger National Park, South Africa. *Environ. Manag.* **2008**, *41*, 32–51. [CrossRef]
16. Hyde, M.A.; Wursten, B. Flora of Zimbabwe: Species Information *Ageratum conyzoides*. 2010. Available online: https://www.zimbabweflora.co.zw/speciesdata/species.php?species_id=158650 (accessed on 10 January 2023).

17. Amonum, J.I.; Ikyaagba, E.T.; Jimoh, S.O. Composition and diversity of understory plants in the tropical rain forest of Cross River national park (CRNP), Nigeria. *J. Res. For. Wildl. Environ.* **2016**, *8*, 11–20.

18. Kawooya, R.; Wamani, S.; Magambo, S.; Nalugo, R. Weed flora of cassava in west Nile zones of Uganda. *Afr. Crop Sci. J.* **2016**, *24*, 145–148. [CrossRef]

19. Quee, D.D.; Kanneh, S.M.; Yila, K.M.; Nabay, O.; Kamanda, P.J. Weed species diversity in cassava (*Manihot esculenta* Crantz) monoculture in Ashanti region of Ghana. *J. Exp. Biol. Agric. Sci.* **2016**, *4*, 499–504. [CrossRef]

20. Rejmánek, M.; Huntley, B.J.; Le Roux, J.J.; Richardson, D.M. A rapid survey of the invasive plant species in western Angola. *Afr. J. Ecol.* **2017**, *55*, 56–69. [CrossRef]

21. Sharman, M.; Thomas, J.E.; Tree, D.; Persley, D.M. Natural host range and thrips transmission of capsicum chlorosis virus in Australia. *Australas. Plant Pathol.* **2020**, *49*, 45–51. [CrossRef]

22. Donnegan, J.A.; Butler, S.L.; Kuegler, O.; Hiserote, B.A. *Federated States of Micronesia's forest resources, 2006*; Resource Bulletin PNW-RB-262; USDA Forest Service, Pacific Northwest Research Station: Portland, OR, USA, 2011.

23. Space, J.C.; Lorence, D.H.; LaRosa, A.M. *Report to the Republic of Palau: 2008 Update on Invasive Plant Species*; USDA, Pacific Southwest Research Station, Institute of Pacific Islands Forestry: Hilo, HI, USA, 2009. Available online: http://www.hear.org/pier/reports/p2008report.htm (accessed on 10 January 2023).

24. Thapa, S.; Chitale, V.; Rijal, S.J.; Bisht, N.; Shrestha, B.B. Understanding the dynamics in distribution of invasive alien plant species under predicted climate change in Western Himalaya. *PLoS ONE* **2018**, *13*, e0195752. [CrossRef]

25. Thiney, U.; Banterng, P.; Gonkhamdee, S.; Katawatin, R. Distributions of alien invasive weeds under climate change scenarios in mountainous Bhutan. *Agronomy* **2019**, *9*, 442. [CrossRef]

26. do Rosário, C.J.R.M.; Lima, A.S.; de JSMendonça, C.; Soares, I.S.; Júnior, E.B.A.; Gomes, M.N.; Costa-Junior, L.M.; Maia, J.G.S.; da Rocha, C.Q. Essential oil *Ageratum conyzoides* chemotypes and anti-tick activities. *Vet. Parasitol.* **2023**, *319*, 109942. [CrossRef]

27. Arora, V. Comparative assessment of eco-physiological functions between *Ageratum conyzoides* L. and *Parthenium hysterophorus* L. Ph.D. Thesis, Panjab University, Chandigarh, India, 1999.

28. Kosaka, Y.; Saikia, B.; Mingki, T.; Tag, H.; Riba, T.; Ando, K. Roadside distribution patterns of invasive alien plants along an altitudinal gradient in Arunachal Himalaya, India. *Mt. Res. Dev.* **2010**, *30*, 252–258. [CrossRef]

29. Sun, P.; Mantri, N.; Möller, M.; Shen, J.; Shen, Z.; Jiang, B.; Chen, C.; Miao, Q.; Lu, H. Influence of light and salt on the growth of alien invasive tropical weed *Ageratum conyzoides*. *Aust. J. Crop Sci.* **2012**, *6*, 739–748.

30. Chaudhary, N.; Narayan, R. The advancing dominance of *Ageratum conyzoides* L. and *Lantana camara* L. in dry tropical peri-urban vegetation in India. *Int. Res. J. Environ. Sci.* **2013**, *2*, 88–95.

31. Batish, D.R.; Kaur, S.; Singh, H.P.; Kohli, R.K. Role ofroot-mediated interactions in phytotoxic interference of *Ageratum conyzoides* with rice (*Oryza sativa*). *Flora* **2009**, *204*, 388–395. [CrossRef]

32. Osman, N.; Dorairaj, D.; Halim, A.; Zelan, N.I.A.; Rashid, M.A.A.; Zakaria, R.M. Dynamics of plant ecology and soil conservation: Implications for cut-slope protection. *Acta Oecol.* **2021**, *111*, 103744. [CrossRef]

33. Haq, S.M.; Khuroo, A.A.; Malik, A.H.; Rashid, I.; Ahmad, R.; Hamid, M.; Dar, G.H. Forest ecosystems of Jammu and Kashmir state. In *Biodiversity of the Himalaya: Jammu and Kashmir State*; Dar, G., Khuroo, A., Eds.; Springer: Singapore, 2020; pp. 191–208. [CrossRef]

34. Wang, J.; Chen, W.; Ma, R.; Baskin, C.C.; Baskin, J.M.; Qi, W.; Chen, X. Role of short-term cold stratification on seed dormancy break and germination of alien species in southeastern China. *Plant Ecol.* **2016**, *217*, 383–392. [CrossRef]

35. Kong, C.; Hu, F.; Xu, X. Allelopathic potential and chemical constituents of volatiles from *Ageratum conyzoides* under stress. *J. Chem. Ecol.* **2002**, *28*, 1173–1182. [CrossRef]

36. Singh, H.P.; Batish, D.R.; Kaur, S.; Kohli, R.K. Phytotoxic interference of *Ageratum conyzoides* with wheat (*Triticum aestivum*). *J. Agron. Crop Sci.* **2003**, *189*, 341–346. [CrossRef]

37. Bhatt, B.P.; Tomar, J.M.S.; Misra, L.K. Allelopathic effects of weeds on germination and growth of legumes and cereal crops of North Eastern Himalayas. *Allelopath. J.* **2001**, *8*, 225–232.

38. Idu, M.; Ovuakporie-Uvo, O. Studies on the allelopathic effect of aqueous extract of *Ageratum conyzoides* (Asteraceae) on seedling growth of *Sesanum indicum* (Pedaliaceae). *Int. J. Environ. Sci. Technol.* **2013**, *2*, 1185–1195.

39. Idu, M. Studies on the allelopathic effect of aqueous extract of *Ageratum conyzoides* Asteraceae L. on seedling growth of *Sorghum bicolor* Linn. (Poaceae). *Acad. J. Agric. Res.* **2014**, *2*, 74–79.

40. Batish, D.R.; Singh, H.P.; Kaur, S.; Arora, V.; Kohli, R.K. Allelopathic interference of residues of *Ageratum conyzoides*. *J. Plant Dis. Prot.* **2004**, *18*, 293–299.

41. Muscolo, A.; Panuccio, M.R.; Sidari, M. The effect of phenols on respiratory enzymes in seed germination. *Plant Growth Regul.* **2001**, *35*, 31–35. [CrossRef]

42. Olofsdotter, M. Rice—A step toward use of allelopathy. *Agron. J.* **2001**, *93*, 3–8. [CrossRef]

43. Kaur, A.; Kaur, S.; Singh, H.P.; Batish, D.R.; Kohli, R.K. Phenotypic variations alter the ecological impact of invasive alien species: Lessons from *Parthenium hysterophorus*. *J. Environ. Manag.* **2019**, *241*, 187–197. [CrossRef]

44. Kaur, A.; Kaur, S.; Singh, H.P.; Batish, D.R. Is intraspecific trait differentiation in *Parthenium hysterophorus* a consequence of hereditary factors and/or phenotypic plasticity? *Plant Divers.* 2022, *in press*. [CrossRef]

45. Kotta, J.C.; Lestari, A.B.S.; Candrasari, D.S.; Hariono, M. Medicinal effect, in silico bioactivity prediction, and pharmaceutical formulation of *Ageratum conyzoides* L.: A review. *Scientifica* **2020**, *2020*, 6420909. [CrossRef] [PubMed]

46. Dores, R.G.R.; Guimarães, S.F.; Braga, T.V.; Fonseca, M.C.M.; Martins, P.M.; Ferreira, T.C. Phenolic compounds, flavonoids and antioxidant activity of leaves, flowers and roots of goat weed. *Hortic. Bras.* **2014**, *32*, 486–490. [CrossRef]
47. Shah, S.; Dhanani, T.; Sharma, S.; Singh, R.; Kumar, S.; Kumar, B.; Srivastava, S.; Ghosh, S.; Kumar, R.; Juliet, S. Development and validation of a reversed phase high performance liquid chromatography-photodiode array detection method for simultaneous identification and quantification of coumarin, precocene-I, β-caryophyllene oxide, α-humulene, and β-caryophyllene in *Ageratum conyzoides* extracts and essential oils from plants. *J. AOAC Int.* **2020**, *103*, 857–864. [CrossRef]
48. Makokha, D.W.; Irakiza, R.; Malombe, I.; Le Bourgeois, T.; Rodenburg, J. Dualistic roles and management of non-cultivated plants in lowland rice systems of East Africa. *S. Afr. J. Bot.* **2017**, *108*, 321–330. [CrossRef]
49. Bora, D.; Kalita, J.; Das, D.; Nath, S.C. Credibility of folklore claims on the treatment of malaria in north-east India with special reference to corroboration of their biological activities. *J. Nat. Remedies* **2016**, *16*, 7–17. [CrossRef]
50. Olubomehin, O.O.; Adeyemi, O.O.; Awokoya, K.N. Preliminary investigation into the alpha-amylase inhibitory activities of *Ageratum conyzoides* (Linn) leaf extracts. *J. Chem. Soc. Niger.* **2016**, *41*, 73–76.
51. Oso, B.J.; Olaoye, I.F. Comparative in vitro studies of antiglycemic potentials and molecular docking of *Ageratum conyzoides* L. and *Phyllanthus amarus* L. methanolic extracts. *SN Appl. Sci.* **2020**, *2*, 629. [CrossRef]
52. Febriansah, R.; Komalasari, T. Co-chemotherapeutic effect of *Ageratum conyzoides* L. chloroform fraction and 5-fluorouracil on hela cell line. *Pharmacogn. J.* **2019**, *11*, 913–918. [CrossRef]
53. Lin, Z.; Lin, Y.; Shen, J.; Jiang, M.; Hou, Y. Flavonoids in *Ageratum conyzoides* L. Exert potent antitumor effects on human cervical adenocarcinoma HeLa cells in vitro and in vivo. *BioMed. Res. Int.* **2020**, *2020*, 2696350. [CrossRef]
54. Adelakun, S.A.; Ukwenya, V.O.; Peter, A.B.; Siyanbade, A.J.; Akinwumiju, C.O. Therapeutic effects of aqueous extract of bioactive active component of *Ageratum conyzoides* on the ovarian-uterine and hypophysis-gonadal axis in rat with polycystic ovary syndrome: Histomorphometric evaluation and biochemical assessment. *Metab. Open* **2022**, *15*, 100201. [CrossRef]
55. Voukeng, I.K.; Beng, V.P.; Kuete, V. Antibacterial activity of six medicinal Cameroonian plants against gram-positive and gram-negative multidrug resistant phenotypes. *BMC Complement. Altern. Med.* **2016**, *16*, 388. [CrossRef]
56. Pintong, A.R.; Ruangsittichai, J.; Ampawong, S.; Thima, K.; Sriwichai, P.; Komalamisra, N.; Popruk, S. Efficacy of *Ageratum conyzoides* extracts against *Giardia duodenalis* trophozoites: An experimental study. *BMC Complement. Med. Ther.* **2020**, *20*, 63. [CrossRef]
57. Kumar, R.; Jinnah, Z.; Khan, A.H.; Arya, D. Impact of alien invasive plant species on crop fields and forest areas of Hawalbag Block of Kumaun Himalaya-people's perceptions. *Imp. J. Interdiscip. Res.* **2016**, *2*, 632–635.
58. Ramasamy, V.; Karthi, S.; Ganesan, R.; Prakash, P.; Senthil-Nathan, S.; Umavathi, S.; Krutmuang, P.; Vasantha-Srinivasan, P. Chemical characterization of billy goat weed extracts *Ageratum conyzoides* (Asteraceae) and their mosquitocidal activity against three blood-sucking pests and their non-toxicity against aquatic predators. *Environ. Sci. Pollut. Res.* **2021**, *28*, 28456–28469. [CrossRef] [PubMed]
59. Moreira, M.D.; Picanço, M.C.; Barbosa, L.C.D.A.; Guedes, R.N.C.; da Silva, E.M. Toxicity of leaf extracts of *Ageratum conyzoides* to Lepidoptera pests of horticultural crops. *Biol. Agric. Hortic.* **2004**, *22*, 251–260. [CrossRef]
60. Ragesh, P.R.; Bhutia, T.N.; Ganta, S.; Singh, A.K. Repellent, antifeedant and toxic effects of *Ageratum conyzoides* (Linnaeus) (Asteraceae) extract against *Helicovepra armigera* (Hübner) (Lepidoptera: Noctuidae). *Arch. Phytopathol. Pflanzenschutz.* **2016**, *49*, 19–30. [CrossRef]
61. Vats, T.K.; Rawal, V.; Mullick, S.; Devi, M.R.; Singh, P.; Singh, A.K. Bioactivity of *Ageratum conyzoides* (L.) (Asteraceae) on feeding and oviposition behaviour of diamondback moth *Plutella xylostella* (L.) (Lepidoptera: Plutellidae). *Int. J. Trop. Insect Sci.* **2019**, *39*, 311–318. [CrossRef]
62. Ndacnou, M.K.; Pantaleon, A.; Tchinda, J.B.S.; Mangapche, E.L.N.; Keumedjio, F.; Boyoguemo, D.B. Phytochemical study and anti-oomycete activity of *Ageratum conyzoides* Linnaeus. *Ind. Crops Prod.* **2020**, *153*, 112589. [CrossRef]
63. Kumar, S.; Sharma, A.K.; Kumar, B.; Shakya, M.; Patel, J.A.; Kumar, B.; Bisht, N.; Chigure, G.M.; Singh, K.; Kumar, R.; et al. Characterization of deltamethrin, cypermethrin, coumaphos and ivermectin resistance in populations of *Rhipicephalus microplus* in India and efficacy of an antitick natural formulation prepared from *Ageratum conyzoides*. *Ticks Tick-Borne Dis.* **2021**, *12*, 101818. [CrossRef]
64. Adjou, E.S.; Dahouenon-Ahoussi, E.; Degnon, R.; Soumanou, M.M.; Sohounhloue, D.C.K. Investigations on bioactivity of essential oil of *Ageratum conyzoides* L. from Benin against the growth of fungi and aflatoxin production. *Int. J. Pharm. Sci. Rev. Res.* **2012**, *13*, 143–148.
65. Kong, C.; Hu, F.; Xu, X.; Zhang, M.; Liang, W. Volatile allelochemicals in the *Ageratum conyzoides* intercropped citrus orchard and their effects on mites *Amblyseius newsami* and *Panonychus citri*. *J. Chem. Ecol.* **2005**, *31*, 2193–2203. [CrossRef] [PubMed]
66. Ojo, O.A.; Oladipo, S.O.; Odelade, K.A. In vitro assessment of fungicidal activity of *Ageratum conyzoides* and *Cymbopogon flexuosus* weed extracts against some phytopathogenic fungi associated with fruit rot of water melon (*Citrullus lunatus* (Thunb)). *Arch. Phytopathol. Pflanzenschutz.* **2014**, *47*, 2421–2428. [CrossRef]
67. Yusnawan, E. The effectiveness of polar and non-polar fractions of *Ageratum conyzoides* L. to control peanut rust disease and phytochemical screenings of secondary metabolites. *J. Trop. Plant Pests Dis.* **2014**, *13*, 159–166. [CrossRef]
68. Shafique, S.; Shafique, S.; Yousuf, A. Bioefficacy of extract of *Ageratum conyzoides* against *Drechslera australiensis* and *Drechslera holmii*. *Pak. J. Phytopathol.* **2015**, *27*, 193–200.

69. Chahal, R.; Nanda, A.; Akkol, E.K.; Sobarzo-Sánchez, E.; Arya, A.; Kaushik, D.; Dutt, R.; Bhardwaj, R.; Rahman, M.H.; Mittal, V. *Ageratum conyzoides* L. and its secondary metabolites in the management of different fungal pathogens. *Molecules* **2021**, *26*, 2933. [CrossRef]
70. Nguyen, C.C.; Nguyen, T.Q.C.; Kanaori, K.; Binh, T.D.; Dao, X.H.T.; Vang, L.V.; Kamei, K. Antifungal activities of *Ageratum conyzoides* L. extract against rice pathogens *Pyricularia oryzae* Cavara and *Rhizoctonia solani* Kühn. *Agriculture* **2021**, *11*, 1169. [CrossRef]
71. Kato-Noguchi, H. Assessment of the allelopathic potential of *Ageratum conyzoides*. *Biol. Plant.* **2001**, *44*, 309–311. [CrossRef]
72. Xuan, T.D.; Shinkichi, T.; Hong, N.H.; Khanh, T.D.; Min, C.I. Assessment of phytotoxic action of *Ageratum conyzoides* L (Billy goat weed) on weeds. *Crop Prot.* **2004**, *23*, 915–922. [CrossRef]
73. Erida, G.; Saidi, N.; Hasanuddin, H.; Syafruddin, S. Herbicidal Effects of Ethyl Acetate Extracts of Billygoat Weed (*Ageratum conyzoides* L.) on Spiny Amaranth (*Amaranthus spinosus* L.) Growth. *Agronomy* **2021**, *11*, 1991. [CrossRef]
74. Kong, C.; Liang, W.; Hu, F.; Xu, X.; Wang, P.; Jiang, Y.; Xing, B. Allelochemicals and their transformations in the *Ageratum conyzoides* intercropped citrus orchard soils. *Plant Soil* **2004**, *264*, 149–157. [CrossRef]
75. Nongmaithem, D.; Pal, D. Effect of weed management practices on soil actinomycetes and fungi population under different crops. *J. Crop Weed* **2016**, *12*, 120–124.
76. Palmer, P.A.; Bryson, J.A.; Clewell, A.E.; Endres, J.R.; Hirka, G.; Vértesi, A.; Béres, E.; Glávits, R.; Szakonyiné, I.P. A comprehensive toxicological safety assessment of an extract of *Ageratum conyzoides*. *Regul. Toxicol. Pharmacol.* **2019**, *103*, 140–149. [CrossRef]
77. Paul, S.; Datta, B.K.; Ratnaparkhe, M.B.; Dholakia, B.B. Turning waste into beneficial resource: Implication of *Ageratum conyzoides* L. In sustainable agriculture, environment and biopharma sectors. *Mol. Biotechnol.* **2022**, *64*, 221–244. [CrossRef]
78. Manandhar, S.; Shrestha, B.B.; Lekhak, H.D. Weeds of paddy field at Kirtipur, Kathmandu. *Sci. World* **2007**, *5*, 100–106. [CrossRef]
79. Devi, C.; Khwairakpam, M. Feasibility of vermicomposting for the management of terrestrial weed *Ageratum conyzoides* using earthworm species *Eisenia fetida*. *Environ. Technol. Innov.* **2020**, *18*, 100696. [CrossRef]
80. Dogra, K.S. Impact of some invasive species on the structure and composition of natural vegetation of Himachal Pradesh. Ph.D. Thesis, Panjab University, Chandigarh, India, 2007.
81. Leke, W.N. Molecular Epidemiology of Begomoviruses that Infect Vegetable Crops in Southwestern Cameroon. Ph.D. Thesis, Acta Universitatis Agriculturae Sueciae, Uppsala, Sweden, 2010.
82. Tahir, M.; Amin, I.; Haider, M.S.; Mansoor, S.; Briddon, R.W. Ageratum enation virus–a Begomovirus of weeds with the potential to infect crops. *Viruses* **2015**, *7*, 647–665. [CrossRef]
83. Serfraz, S.; Amin, I.; Akhtar, K.P.; Mansoor, S. Recombination among Begomoviruses on Malvaceous plants leads to the evolution of okra enation leaf curl virus in *Pakistan*. *J. Phytopathol.* **2015**, *163*, 764–776. [CrossRef]
84. Anwar, I.; Bukhari, H.A.; Nahid, N.; Rashid, K.; Amin, I.; Shaheen, S.; Hussain, K.; Mansoor, S. Association of cotton leaf curl Multan betasatellite and *Ageratum conyzoides* symptomless alphasatellite with tomato leaf curl New Delhi virus in *Luffa cylindrica* in Pakistan. *Australas. Plant Pathol.* **2020**, *49*, 25–29. [CrossRef]
85. Martins, D.D.S.; Ventura, J.A.; Paula, R.D.C.A.L.; Fornazier, M.J.; Rezende, J.A.M.; Culik, M.P.; Ferreira, P.S.F.; Peronti, A.L.B.G.; de Carvalho, R.C.Z.; Sousa-Silva, C.R. Aphid vectors of Papaya ringspot virus and their weed hosts in orchards in the major papaya producing and exporting region of Brazil. *Crop Prot.* **2016**, *90*, 191–196. [CrossRef]
86. Reddy, C.S.; Pattanaik, C. An assessment of floristic diversity of Gandhamardan hill range, Orissa, India. *Bangladesh J. Plant Taxon.* **2009**, *16*, 29–36. [CrossRef]
87. Sengupta, R.; Dash, S.S. A comprehensive inventory of alien plants in the protected forest areas of Tripura and their ecological consequences. *Nelumbo* **2021**, *63*, 163–182. [CrossRef]
88. Semy, K.; Singh, M.R. Changes in plant diversity and community attributes of coal mine affected forest in relation to a community reserve forest of Nagaland, Northeast India. *Trop. Ecol.* **2023**, *in press*. [CrossRef]
89. Yang, Q.; Jin, B.; Zhao, X.; Chen, C.; Cheng, H.; Wang, H.; He, D.; Zhang, Y.; Peng, J.; Li, Z.; et al. Composition, distribution, and factors affecting invasive plants in grasslands of Guizhou Province of Southwest China. *Diversity* **2022**, *14*, 167. [CrossRef]
90. Sharma, M.; Devi, A.; Badola, R.; Sharma, R.K.; Hussain, S.A. Impact of management practices on the tropical riverine grasslands of Brahmaputra floodplains: Implications for conservation. *Ecol. Indic.* **2023**, *151*, 110265. [CrossRef]
91. Maturi, K.C.; Haq, I.; Kalamdhad, A.S. Insights into the bioconversion of *Ageratum conyzoides* into a nutrient-rich compost and its toxicity assessment: Nutritional and quality assessment through instrumental analysis. *Biomass Convers. Biorefin.* **2022**, *in press*. [CrossRef]
92. Hauchhum, R.; Tripathi, S.K. Impact of rhizosphere microbes of three early colonizing annual plants on improving soil fertility during vegetation establishment under different fallow periods following shifting cultivation. *Agric. Res.* **2020**, *9*, 213–221. [CrossRef]
93. Dogra, K.S.; Kohli, R.K.; Sood, S.K.; Dobhal, P.K. Impact of *Ageratum conyzoides* L. on the diversity and composition of vegetation in the Shivalik hills of Himachal Pradesh (Northwestern Himalaya), India. *Int. J. Biodivers. Conserv.* **2009**, *1*, 135–145.
94. Diallo, A.; Eklu-Gadegbeku, K.; Amegbor, K.; Agbonon, A.; Aklikokou, K.; Creppy, E.; Gbeassor, M. In vivo and in vitro toxicological evaluation of the hydroalcoholic leaf extract of *Ageratum conyzoides* L. (Asteraceae). *J. Ethnopharmacol.* **2014**, *155*, 1214–1218. [CrossRef] [PubMed]
95. Gowda, G.; Lakshmi, S.; Parasuramalu, B.G.; Nagaraj, C.; Gowda, B.V.C.; Somashekara, K.G. A study on allergen sensitivity in patients with allergic rhinitis in Bangalore, India. *J. Laryngol. Otol.* **2014**, *128*, 892–896. [CrossRef] [PubMed]

96. Bhatia, H.; Manhas, R.K.; Kumar, K.; Magotra, R. Traditional knowledge on poisonous plants of Udhampur district of Jammu and Kashmir, India. *J. Ethnopharmacol.* **2014**, *152*, 207–216. [CrossRef] [PubMed]

97. Batish, D.R.; Singh, H.P.; Kohli, R.K.; Johar, V.; Yadav, S. Management of invasive exotic weeds requires community participation. *Weed Technol.* **2004**, *18*, 1445–1448. [CrossRef]

98. Kohli, R.K.; Batish, D.R.; Singh, H.P. Allelopathic interactions in agroecosystems. In *Allelopathy: A Physiological Process with Ecological Implications*; Reigosa, M.J., Pedrol, N., González, L., Eds.; Springer: Cham, The Netherlands, 2006; pp. 465–493. [CrossRef]

99. Zelaya, I.A.; Owen, M.D.K.; Pitty, A. Effect of tillage and environment on weed population dynamics in the dry tropics. *Ceiba* **1997**, *38*, 123–135.

100. Paradkar, V.K.; Sharma, T.R.; Sharma, R.C. Water stagnation for control of weeds in *Eucalyptus rostrata*, Schlecht plantation. *World Weeds* **1998**, *5*, 67–68.

101. Basavaraj, P.; Reddy, V.C. Influence of non-chemical weed management practices on weed composition and diversity in transplanted organic finger millet. *Int. J. Agric. Sci.* **2017**, *9*, 3808–3811.

102. Chand, S.; Rana, M.C.; Rana, S.S. Effect of time and method of post-emergence atrazine application in controlling weeds in maize. *Int. J. Adv. Agric. Sci. Technol.* **2016**, *3*, 49–59.

103. Singh, H.P.; Batish, D.R.; Kohli, R.K.; Saxena, D.B.; Arora, V. Effect of parthenin- a sesquiterpene lactone from *Parthenium hysterophorus* on early growth and physiology of *Ageratum conyzoides*. *J. Chem. Ecol.* **2002**, *28*, 2169–2179. [CrossRef]

104. Bali, A.S.; Batish, D.R.; Singh, H.P. Phytotoxicity of *Callistemon viminalis* essential oil against some weeds. *Ann. Plant Sci.* **2016**, *5*, 1442–1445. [CrossRef]

105. Motmainna, M.; Juraimi, A.S.; Uddin, M.K.; Asib, N.B.; Islam, A.K.M.M.; Ahmad-Hamdani, M.S.; Berahim, Z.; Hasan, M. Physiological and biochemical responses of *Ageratum conyzoides*, *Oryza sativa* f. spontanea (weedy rice) and *Cyperus iria* to *Parthenium hysterophorus* methanol extract. *Plants* **2021**, *10*, 1205. [CrossRef] [PubMed]

106. Singh, H.P.; Batish, D.R.; Kohli, R.K. Allelopathic effect of two volatile monoterpenes against bill goat weed (*Ageratum conyzoides* L). *Crop Prot.* **2002**, *21*, 347–350. [CrossRef]

107. Sahid, I.; Yusoff, N. Allelopathic effects of *Chromolaena odorata* (L.) King and Robinson and *Mikania micrantha* H.B.K. on three selected weed species. *Aust. J. Crop Sci.* **2014**, *8*, 1024–1028.

108. Ismail, B.S.; Chong, T.V. Allelopathic effects of *Dicranopteris linearis* debris on common weeds of Malaysia. *Allelopath. J.* **2009**, *23*, 277–286.

109. Singh, H.; Naik, B.; Kumar, V.; Bisht, G.S. Screening of endophytic actinomycetes for their herbicidal activity. *Ann. Agrar. Sci.* **2018**, *16*, 101–107. [CrossRef]

110. Kaur, A.; Batish, D.R.; Kaur, S.; Singh, H.P.; Kohli, R.K. Phenological behaviour of *Parthenium hysterophorus* in response to climatic variations according to the extended BBCH scale. *Ann. Appl. Biol.* **2017**, *171*, 316–326. [CrossRef]

111. Filho, J.S.; Marchão, R.L.; de Carvalho, A.M.; Carmona, R. Weed dynamics in grain *Sorghum* grass intercropped systems. *Rev. Cienc. Agron.* **2014**, *45*, 1032–1039. [CrossRef]

112. Shen, S.C.; Xu, G.F.; Li, D.Y.; Jin, G.M.; Liu, S.F.; Clements, D.R.; Yang, Y.X.; Rao, J.; Chen, A.D., Zhang, F.D.; et al. Potential use of sweet potato (*Ipomoea batatas* (l.) lam.) to suppress three invasive plant species in agroecosystems (*Ageratum conyzoides* l., *Bidens pilosa* l., and *Galinsoga parviflora* cav.). *Agronomy* **2019**, *9*, 318. [CrossRef]

113. Rapp, R.E.; Datta, A.; Irmak, S.; Arkebauer, T.J.; Knezevic, S.Z. Integrated management of common reed (*Phragmites australis*) along the Platte River in Nebraska. *Weed Technol.* **2012**, *26*, 326–333. [CrossRef]

114. Saha, B.; Sathyan, A.; Mazumder, P.; Choudhury, S.P.; Kalamdhad, A.S.; Khwairakpam, M.; Mishra, U. Biochemical methane potential (BMP) test for *Ageratum conyzoides* to optimize ideal food to microorganism (F/M) ratio. *J. Environ. Chem. Eng.* **2018**, *6*, 5135–5140. [CrossRef]

Article

Invasive Trends of *Spartina alterniflora* in the Southeastern Coast of China and Potential Distributional Impacts on Mangrove Forests

Jiaying Zheng [1,2], Haiyan Wei [1,*], Ruidun Chen [1,2], Jiamin Liu [1,2], Lukun Wang [1,2] and Wei Gu [2,3,*]

[1]   School of Geography and Tourism, Shaanxi Normal University, Xi'an 710119, China;
       zhengjiaying@snnu.edu.cn (J.Z.); crd@snnu.edu.cn (R.C.)
[2]   National Engineering Laboratory for Resource Development of Endangered Crude Drugs in Northwest
       China, Shaanxi Normal University, Xi'an 710119, China
[3]   College of Life Sciences, Shaanxi Normal University, Xi'an 710119, China
*   Correspondence: weihy@snnu.edu.cn (H.W.); weigu@snnu.edu.cn (W.G.)

Abstract: Mangrove forests are one of the most productive and seriously threatened ecosystems in the world. The widespread invasion of *Spartina alterniflora* has seriously imperiled the security of mangroves as well as coastal mudflat ecosystems. Based on a model evaluation index, we selected RF, GBM, and GLM as a predictive model for building a high-precision ensemble model. We used the species occurrence records combined with bioclimate, sea–land topography, and marine environmental factors to predict the potentially suitable habitats of mangrove forests and the potentially suitable invasive habitats of *S. alterniflora* in the southeastern coast of China. We then applied the invasion risk index (IRI) to assess the risk that *S. alterniflora* would invade mangrove forests. The results show that the suitable habitats for mangrove forests are mainly distributed along the coastal provinces of Guangdong, Hainan, and the eastern coast of Guangxi. The suitable invasive habitats for *S. alterniflora* are mainly distributed along the coast of Zhejiang, Fujian, and relatively less in the southern provinces. The high-risk areas for *S. alterniflora* invasion of mangrove forests are concentrated in Zhejiang and Fujian. Bioclimate variables are the most important variables affecting the survival and distribution of mangrove forests and *S. alterniflora*. Among them, temperature is the most important environmental variable determining the large-scale distribution of mangrove forests. Meanwhile, *S. alterniflora* is more sensitive to precipitation than temperature. Our results can provide scientific insights and references for mangrove forest conservation and control of *S. alterniflora*.

Keywords: *Spartina alterniflora*; mangrove forests; species distribution models; ensemble model; invasive risk

check for
updates

Citation: Zheng, J.; Wei, H.; Chen, R.; Liu, J.; Wang, L.; Gu, W. Invasive Trends of *Spartina alterniflora* in the Southeastern Coast of China and Potential Distributional Impacts on Mangrove Forests. *Plants* **2023**, *12*, 1923. https://doi.org/10.3390/plants12101923

Academic Editors: Congyan Wang and Hongli Li

Received: 13 March 2023
Revised: 26 April 2023
Accepted: 4 May 2023
Published: 9 May 2023

Correction Statement: This article has been republished with a minor change. The change does not affect the scientific content of the article and further details are available within the backmatter of the website version of this article.

1. Introduction

With the rapid development of China's economy, especially the growth of trade and transportation, biological invasions are occurring very frequently [1,2]. China has become one of the countries in the world with the most serious damage from biological invasions. At the end of 2018, there were nearly 800 invasive alien species in China; 638 species have been confirmed as having invaded agricultural and forestry ecosystems [3]. *Spartina alterniflora* is a perennial herb that originates from the Atlantic coast and Gulf coasts of North America [4] It has successfully invaded coastal wetland areas worldwide through intentional or unintentional introduction by human beings [5–8]. China introduced *S. alterniflora* in 1979 to protect against wind, siltation, reclamation, and to improve beach vegetation cover and productivity [9,10]. Since its introduction in China, *S. alterniflora* has become the most important invasive plant in coastal wetlands due to its high adaptability and reproductive capacity [11]. The outbreak scale of *S. alterniflora* in China is much larger than in other countries and regions of the world [12].

Mangrove forests are mainly found in the intertidal zone of the world's coastal tropics and subtropics, acting as buffer zones between land and sea [13,14]. They are among the most productive ecosystems in the world [15,16], having not only great social, economic, and ecological value, but also stabilizing coastlines to reduce the damaging effects of natural disasters [17]. Mangrove forests also provide food, medicine, fuel, and building materials for people as well as important social and economic services such as forest products for residents [18]. When it comes to maintaining and protecting tropical and subtropical marine biodiversity, mangrove forests play an important role [19,20]. However, mangrove forests are one of the most threatened ecosystems [21], especially in Asia and the Pacific, where 70% of their original habitats have been lost [22]. As an important ecological barrier, the survival and distribution of mangrove forests are influenced by climate and environmental changes [23]. At the same time, human activities, such as urban development, aquaculture, mining, over-exploitation of timber, and invasive alien organisms [24–28], have led to extensive degradation of mangrove forests.

Although *S. alterniflora* has played a role in promoting siltation, land reclamation, and soil improvement, its continuous expansion has also brought about more serious ecological consequences and economic losses. Due to its adaptability and rapid growth, *S. alterniflora* has invaded the native mangrove forests along the southern coast of China and is likely to occupy increasing areas of mangrove forest habitat in the future [27]. Different areas of *S. alterniflora* have rapidly occupied most of the outer edges of mangrove forests and significantly inhibited the regeneration of native mangrove plants [29] while threatening the security of coastal mudflat ecosystems. Current research on mangrove forests has focused on the physiological characteristics of plants, chemical composition extraction, benthic biodiversity distribution, genome identification and evolution, and ecological restoration engineering [30–33]. Studies on *S. alterniflora* have focused on the dynamic responses of *S. alterniflora* to tidal flat systems, native community diversity, food webs, and trophic structure [34,35]; the biogeochemical cycling processes of *S. alterniflora* in salt marshes [36–38]; the effects of climate change on the physiological characteristics, geospatial changes, and expansion rates of *S. alterniflora* [39,40]; and the inter-population interactions of the ecosystems in which *S. alterniflora* is located [41,42]. Previous studies have mainly investigated the mechanisms of mangrove forest invasion by *S. alterniflora* at microscopic scales, including differences in chemical substances in sediments of different plant habitats; physical and chemical responses in ecosystems; intertidal benthic differences at small scales [43–45]; and the suitability distribution of a particular species alone [46,47]. The spatial areas of mangrove forest invasion by *S. alterniflora* at a large scale have been less studied.

Climate is an important environmental determinant of species distribution, and climate change has significant impacts on biodiversity, including species distribution and interspecific interactions [48,49]. The IPCC has finalized the first part of the *Sixth Assessment Report*, which states that the climate system is expected to continue warming by mid-century, with climate change bringing many different combinations of changes to different regions [50]. As a result, an increasing number of studies have focused on how climate change affects species distribution, while also considering a variety of other factors, such as topography, soils, and human activities [51–56]. However, species located in coastal areas are not only subject to climate change, but also the marine environment. Factors such as sea surface temperature, sea surface salinity, photosynthetically available radiation, and water quality can directly or indirectly affect the distribution of coastal wetland plants [15,57].

Species distribution models (SDMs) are widely used to predict the potential distribution of species [58]. SDMs are used to visualize the association of species occurrence records and environmental variables through functional relationships to form various algorithms and predict the potential distribution of target species [59]. Various SDMs have been developed based on the rapid development of computer software and technological innovation. Commonly used SDMs are maximum entropy models (MaxEnt) [60,61], generalized linear models (GLM) [62], generalized add models (GAM) [63], classification tree analysis (CTA) [64], artificial neural networks (ANN) [65], flexible discriminant analyses (FDA) [66], generalized boosting models (GBM) [67], random forests (RF) [68], surface range envelopes (SRE) [69], multiple adaptive regression splines (MARS) [70,71], support vector machines

(SVM) [72], and maximum likelihood (Maxlike) [73]. A single model may lead to variation in suitable habitats for the same species due to multiple factors, resulting in uncertainty in prediction results. The proposed ensemble model (EM) fixes this problem [53]. The advantage of EM is that it reduces or even eliminates the over-fitting caused by some powerful single algorithms in an integrated manner, significantly enhancing modeling accuracy and reducing model-fitting uncertainty [74,75]. Biomod2 is a collection of programs developed in R for building SDMs. It can be used to simulate the potential distribution of species by building a single model or EM to explore the relationship between species spatial distribution and environmental variables and calibrate and evaluate models [76].

In this study, we used 1358 mangrove forest and 1314 *S. alterniflora* occurrence records from the southeastern coast of China to simulate the potential distribution areas using the EM with terrestrial and marine environmental factors. Then, we applied an invasion risk index (IRI) and analyzed the invasive risk of *S. alterniflora* in the mangrove forest distribution area. Our specific objectives are to (1) select multiple single models for simulation and construct the EM by selecting a single model with a high accuracy of evaluation index; (2) identify the most important environmental variables affecting the distribution of mangrove forests and *S. alterniflora*; (3) predict potentially suitable habitats of mangrove forests and potentially suitable invasion habitats of *S. alterniflora*; (4) analyze the risk of invasion by *S. alterniflora* into mangrove forests in the southeastern coast of China. This study is based on the premise hypothesis of SDM construction: Species distribution and the environment are in an equilibrium stage. The locations where species occur represent suitable environmental spaces and those where they do not occur represent unsuitable spaces. The results of this study can provide the scientific reference for mangrove forest conservation and *S. alterniflora* control.

2. Results

2.1. Model Evaluation

We simulated nine single models and selected RF, GBM, and GLM according to their TSS values for modeling. Their TSS values were all superior to the other six single models. RF, GBM, and GLM were combined to build the ensemble model (EM). Then, we evaluated the model accuracy of RF, GBM, GLM, and EM. The results showed that the ranking of the mean value of TSS of each model was EM > RF > GBM > GLM; the ranking of the mean value of AUC of each model was the same as TSS (Figure 1). The simulation accuracy of the EM is better than that of the single model as can be obtained from the results of the two model metrics, i.e., TSS and AUC. From the simulation results, it can be concluded that the integrated EM with a single model of high accuracy has a higher simulation accuracy.

2.2. Analysis of Environmental Variables

We derived the contribution of each variable through EM and analyzed the dominant variables (Table 1). The cumulative contribution rates of bioclimate variables of mangrove forests and *S. alterniflora* were 52.57% and 56.75%; the marine environment variables were 29.44% and 34.18%; and the sea–land topography variables were 17.99% and 9.07%, respectively. Specifically, bioclimate variables played a crucial role in the distribution of mangrove forests and *S. alterniflora*. The role of Ele in sea–land topography variables was relatively large. Among the marine environment variables, only CHL and PAR would play a dominant role in the distribution of mangrove forests and *S. alterniflora*. The contributions of sea surface temperature (SST1, SST2, SST3, SST4, and SST5) and sea surface salinity (SSS) were both low. Among the variables whose cumulative contribution rate reached more than 90%, mangrove forests had 10 key variables, and *S. alterniflora* had 11 key variables (Table 1). Eight variables were common, namely Bio2, Bio12, Bio15, Bio16, Bio19, Ele, CHL, and PAR, but they played different roles in influencing the growth and distribution of mangrove forests and *S. alterniflora* (Figure S2). Among the bioclimate variables, Bio1 played a key role in the prediction of mangrove forests and was the most important temperature variable with a contribution of 25.07% and an optimal threshold of 18.7–25.7 °C. For *S. alterniflora*, the proportion of precipitation reached approximately 3.3 times that of temperature. Bio16 was the highest contributing precipitation variable, with a contribution of 21.27% and an optimal threshold of 438–1226 mm. Among the marine environment variables, CHL was the key variable affecting mangrove forests and *S. alterniflora*: the contribution of CHL was 17.84% and 22.34%, with optimal thresholds of 1.17–13.14 μg/L and 3.04–13.85 μg/L, respectively. The contribution of PAR to mangrove forests and *S. alterniflora* was 3.18% and 5.92%, respectively. The contribution of SST2 to mangrove forests was 5.14%, SST1 to *S. alterniflora* was 2.31%, and the contributions of other marine environment variables were very low. Among the sea–land topography variables, only Ele had a relatively high influence on the distribution of mangrove forests and *S. alterniflora*, with a contribution of 17.86% and 8.29%, respectively. Slop and Aspe had no effect (Table 1).

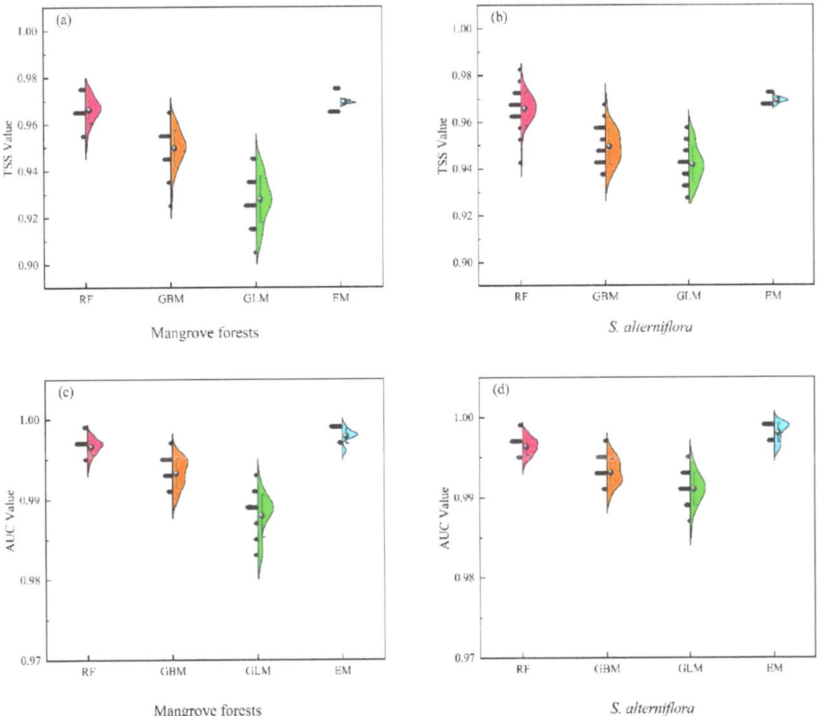

Figure 1. Model evaluation of the single (RF, GBM, GLM) and the ensemble model (EM). The sphere in the graph represents the average value. (**a**) is the TSS value of mangrove forests. (**b**) is the TSS value of *S. alterniflora*. (**c**) is the AUC value of mangrove forests. (**d**) is the AUC value of *S. alterniflora*.

2.3. Simulation Analysis of Potential Distribution for Mangrove Forests and S. alterniflora

In this study, the EM prediction results show that the suitable areas of mangrove forests are mainly distributed in eight provinces and regions along the southeastern coast of China, including Guangxi, Guangdong, Hainan, Fujian, Zhejiang, Taiwan, Hong Kong, and Macau (Figure 2a). Highly suitable habitats for mangrove forests are mainly found in harbors or estuaries that are well covered by waves. Specifically, highly suitable mangrove forest habitats in Guangxi are mainly distributed on the southern coast of Beihai city, the east coast of Fangchenggang city, and the coastal harbors and bays of Qinzhou city. In Guangdong province, they are mainly distributed in the coasts of Zhanjiang, Maoming, Yangjiang, and Jiangmen city; other coastal areas, such as Shenzhen city, Shanwei city, and Shantou city, are sporadically distributed (Figure 2d). Most coastal areas in Hainan province are highly adaptable to mangrove forests, including the coast of Haikou city, where mangrove forest nature reserves, such as Dongzhai Port and Qinglan Port, are located. Mangrove forest highly suitable areas are distributed in all coastal sections of Fujian province, such as the Zhangjiang River estuary, Chiu-lung River estuary, Quanzhou Bay, and Xiamen Bay (Figure 2c). Among these, Ningde city is the northernmost boundary of the natural distribution of mangrove forests in China while the rest of the coast has a sporadic distribution. The highly suitable habitats in Taiwan are mainly distributed in the west coast and concentrated in the coastal areas of Taipei freshwater estuaries and the northern area of Tainan City (Figure 2a). Zhejiang is the northernmost distribution area of mangrove forest introduction and cultivation in China, with relatively few highly suitable habitats areas that are sporadically distributed in and around Ximen Island of Yueqing city and the Rui'an coast of Wenzhou city (Figure 2b). The moderate and low suitable habitats of mangrove forests are concentrated around the highly suitable habitats and spread around from the highly suitable areas, with a wide range that extends to Taizhou, Ningbo city in Zhejiang at the northern end.

Table 1. Contribution rate and cumulative contribution rate of each environmental variable based on the ensemble model.

Mangrove Forests				S. alterniflora			
Variables	Contribution Rate (%)	Cumulative Contribution Rate (%)	Best Suitable Range (Unit)	Variables	Contribution Rate (%)	Cumulative Contribution Rate (%)	Best Suitable Range (Unit)
Bio1	25.07	25.07	18.7–25.7 °C	CHL	22.34	22.34	3.04–13.85 µg/L
Ele	17.86	42.93	−140–16 m	Bio16	21.27	43.61	438–1226 mm
CHL	17.84	60.77	1.17–13.14 µg/L	Ele	8.29	51.90	−149–165 m
Bio16	7.63	68.40	438–1642.76 mm	Bio19	7.05	58.95	69–332 mm
Bio12	7.30	75.70	949–2669 mm	Bio3	6.34	65.29	116–276
SST2	5.14	80.84	25.41–38.4 °C	Bio12	6.08	71.37	932–2491.4 mm
Bio19	3.19	84.03	23.6–566 mm	Bio2	6.06	77.43	25–70
PAR	3.18	87.21	28.13–39.56 E/m^2day	PAR	5.92	83.35	26.25–35.48 E/m^2day
Bio15	2.63	89.84	20–100 mm	Bio18	3.68	87.03	302–1137 mm
Bio2	2.54	92.38	20–62	Bio15	2.61	89.64	32–87 mm
Bio18	1.83	94.21	302–1586.94 mm	SST1	2.31	91.95	8.8–35.4 °C
SSS	1.65	95.86	30.9–33.9%	Bio5	1.91	93.86	28.4–33.3
Bio5	1.31	97.17	29.7–32.5	Bio1	1.75	95.61	16.9–23.8 °C
Bio3	1.08	98.25	150–297	SST2	1.16	96.77	22.1–37.3 °C
SST4	0.89	99.14	13.29–32.03 °C	SSS	1.04	97.81	21.4–33.1%
SST3	0.31	99.45	19.82–36.87 °C	Slop	0.75	98.56	0–90°
SST1	0.29	99.74	16.6–36.5 °C	SST3	0.75	99.31	15.9–36.1 °C
SST5	0.14	99.88	22.39–34.9 °C	SST4	0.56	99.87	8.8–31.1 °C
Slop	0.10	99.98	0–90°	SST5	0.10	99.97	15.88–34.42 °C
Aspe	0.02	100.00	−1–359.82°	Aspe	0.03	100.00	−1–359.82°

The results of the EM prediction show that in this study area, the Zhejiang and Fujian coastal areas are the concentrated distribution areas of highly suitable habitats for *S. alterniflora*. Most of the coasts are distributed, the harbor and estuary areas are concentrated, and the width of the distribution strips is wide (Figure 3a). The highly suitable habitats for *S. alterniflora* are scattered in the Fangchenggang, Qinzhou, and Beihai coastal areas of Guangxi Province, and the Zhanjiang, Maoming, Yangjiang, Jiangmen, and Shantou areas of Guangdong Province (Figure 3b–d). In the Zhejiang and Fujian coastal areas, the moderate and low suitable habitats of *S. alterniflora* are mainly at the periphery of the high suitable habitats and extend towards the sea (Figure 3a). In the coastal areas of Zhuhai and Macau, the highly suitable habitats of *S. alterniflora* have not formed a strip, and the moderate and low suitable habitats have a bigger distribution area (Figure 3c). In

Guangxi, the highly suitable habitats are mainly in the surrounding areas, and the moderate and low suitable habitats are less distributed (Figure 3d).

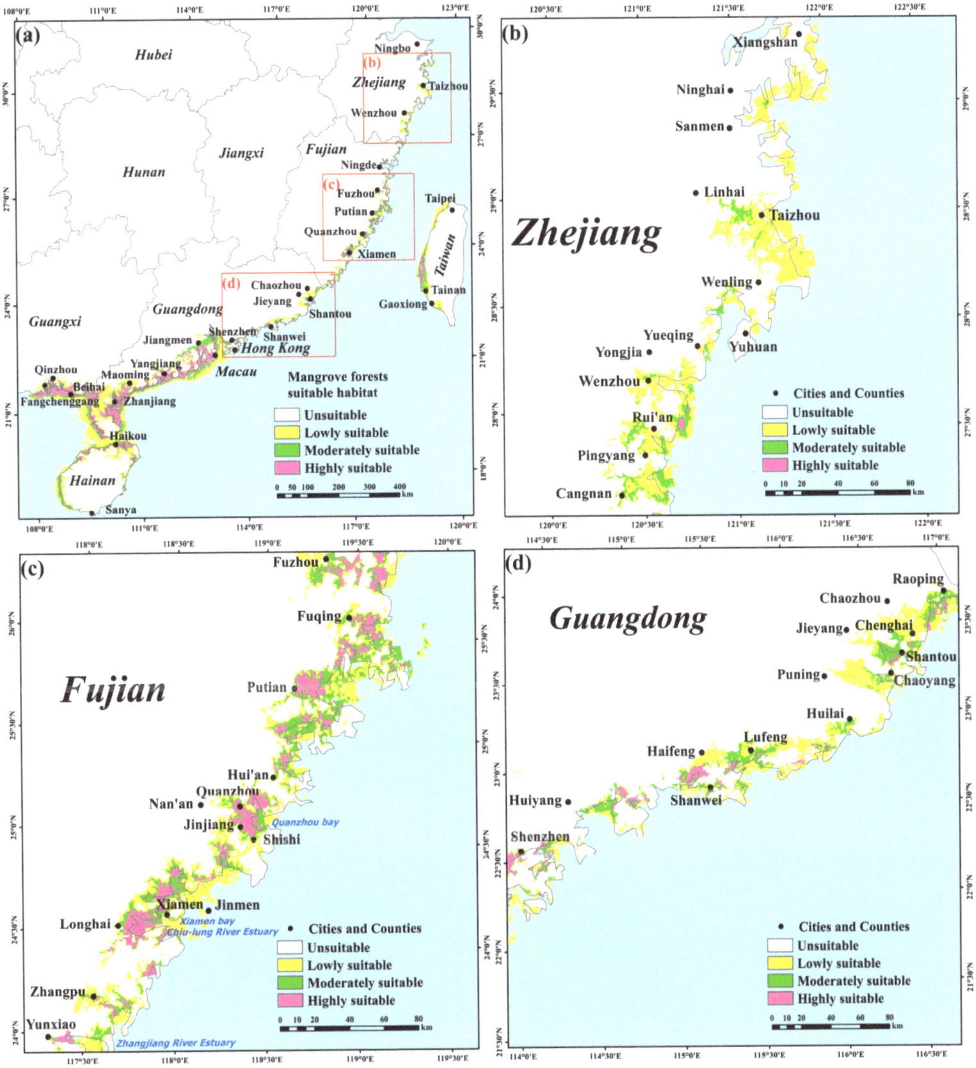

Figure 2. Potential distribution predicted for mangrove forests covering the southeastern coast of China based on the EM. (**b–d**) are local enlargements of (**a**).

2.4. Risk Analysis of S. alterniflora Invades Mangrove Forests along the Southeastern Coast of China

In the coastal areas of Guangdong, Guangxi, and Hainan, the potential suitability of mangrove forests is higher than that of Fujian and Zhejiang, while the high suitability areas of *S. alterniflora* are mostly distributed in the coastal areas of Fujian and Zhejiang. The potential distributions of mangrove forests and *S. alterniflora* have overlapping ecological niches in Zhejiang, Fujian, Guangdong, and Guangxi provinces (Figures 2a and 3a). In Zhejiang and Fujian, the distribution of *S. alterniflora* is very concentrated and the invasive suitability is high. In the range from low to high potential invasion suitability of *S. alterniflora*, the invasive risk to native mangrove forests is relatively high (Figure 4a,b). In Guangdong and Guangxi, the distribution of highly suitable invasive habitats for *S. alterniflora* is small, and the invasive risk to native mangrove forests is relatively low in areas with high potential invasion suitability for *S. alterniflora* (Figure 4c,d). In Hainan and Taiwan, the distribution of highly suitable invasive habitats of *S. alterniflora* and the invasive risk to native mangrove forests are low (Figures 3a and 4e,f). Specifically, the areas with high risk of invasion of *S. alterniflora* are concentrated in the Zhejiang and Fujian coasts (Figure 5a). The risk in northern Zhejiang is higher than that in southern areas (Figure 5b). Compared with Zhejiang, the invasion risk area along Fujian is narrower (Figure 5c). The risk of invasion of *S. alterniflora* is sporadically distributed in Chaozhou, Shantou, and the Jieyang coast of Guangdong (Figure 5d). Meanwhile, the risk of invasion of *S. alterniflora* in the rest of the region is low.

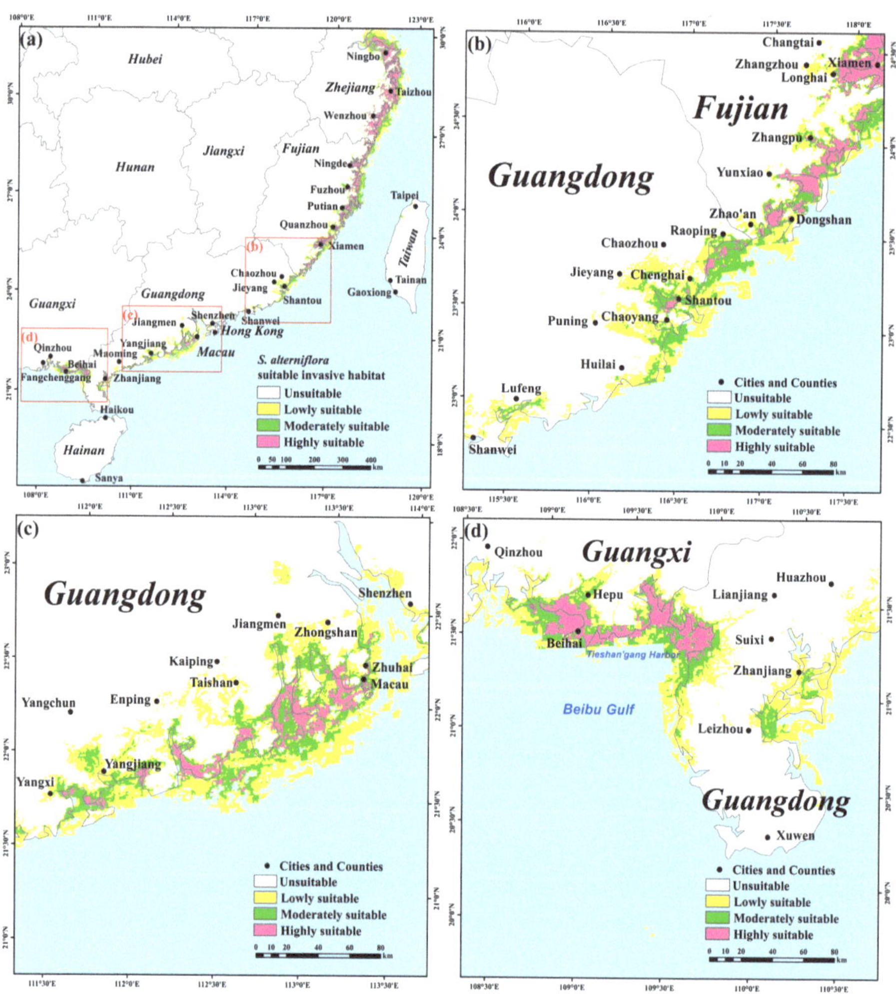

Figure 3. Potential distribution for predicted *S. alterniflora* covering the southeastern coast of China based on the EM. (**b–d**) are local enlargements of (**a**).

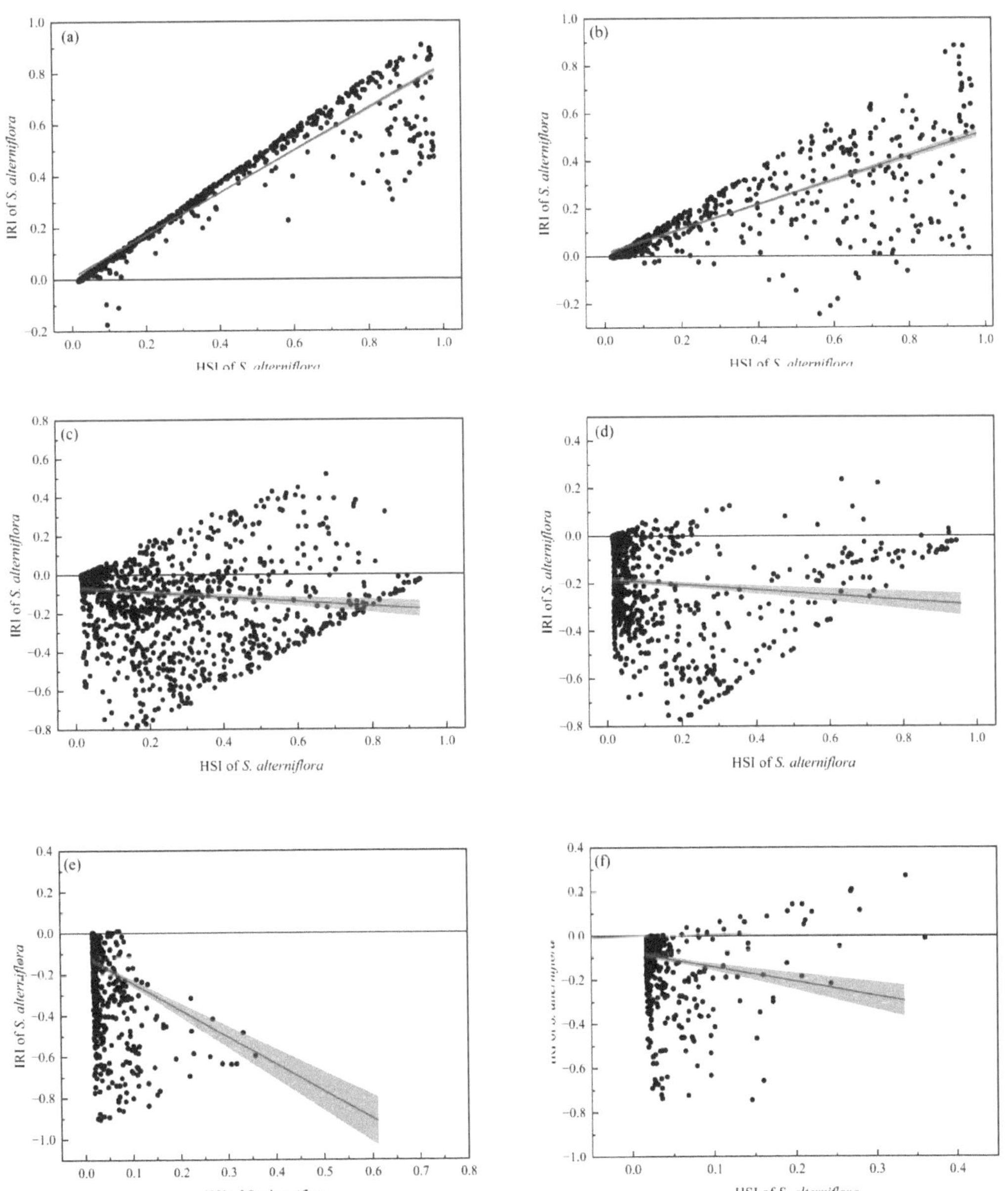

Figure 4. The relationship between invasion suitability and invasion risk of *S. alterniflora* to mangrove forests in the southeastern coast of China. (**a**) Zhejiang, (**b**) Fujian, (**c**) Guangdong, (**d**) Guangxi, (**e**) Hainan, and (**f**) Taiwan. The horizontal axis represents the HSI of *S. alterniflora*, and the vertical axis represents the IRI of *S. alterniflora* to mangrove forests. The red line shows the result of the linear fit; the gray band indicates a 95% confidence band.

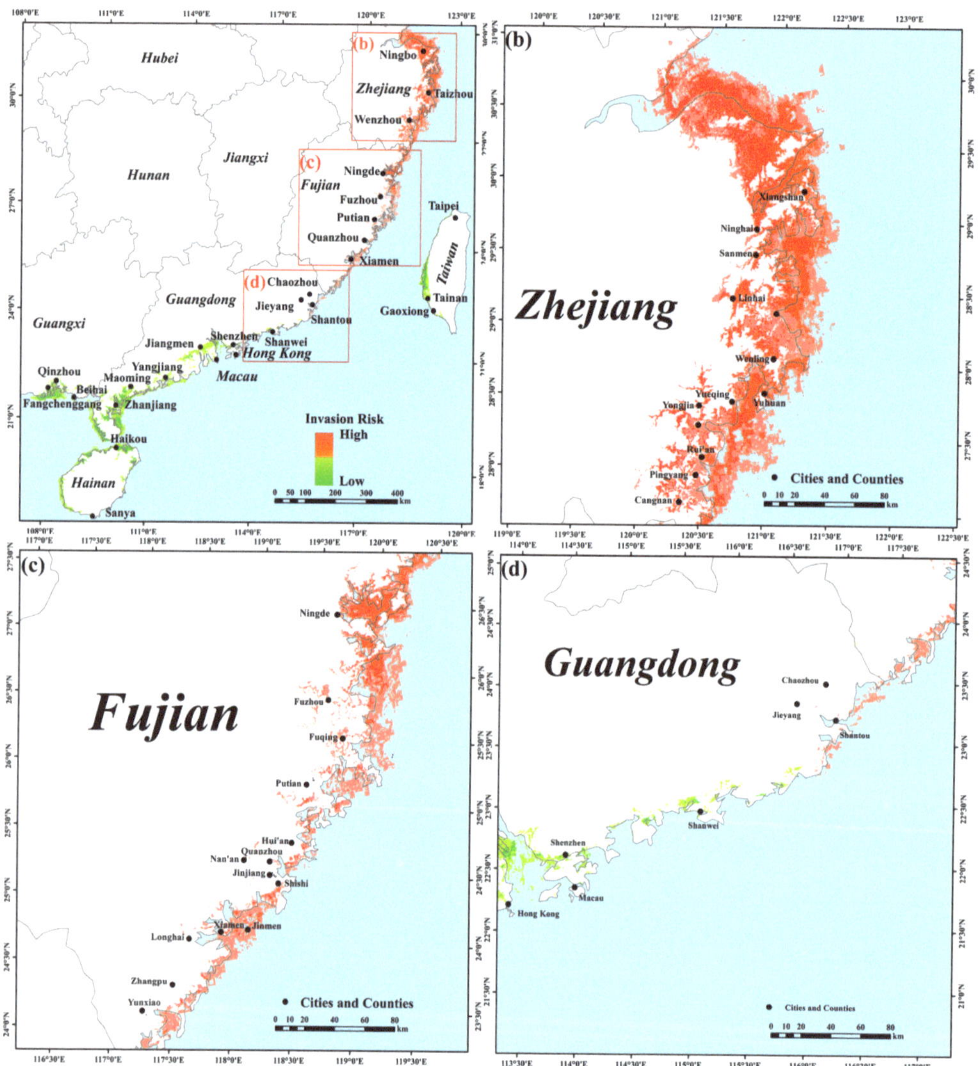

Figure 5. Invasion risk analysis of *S. alterniflora* to mangrove forests in the southeastern coast of China. Red is the area with high invasion risk, and green is the area with low invasion risk. (**b–d**) are local enlargements of (**a**).

3. Discussion

3.1. Applicability of SDMs

SDMs mainly use species distribution data and environmental data to estimate the ecological niches of the species base through specific algorithms. They are then projected onto the landscape to reflect the preference of species for habitats in the form of probabilities [77,78]. In recent years, SDMs have been widely used for the species response to climate change in the context of global change [51]; for potential range prediction of invasive species [52]; for determining the effects of regional climate change on species richness and community stability [79]; for range delineation of protected areas for endangered and rare species; and for determining the impact of human activities on endangered species [80]. In the modeling process for different purposes, species ecological niche characteristics, and modeling database, researchers need to choose different modeling algorithms. In recent years, EM prediction using multiple model information has become a trend in species distribution studies to reduce model uncertainty and increase modeling accuracy [80,81]. Therefore, it is reasonable to use EM in this study.

3.2. Ensemble Model Simulation Accuracy

A single model may result in differences in suitable habitat for the same species due to multiple factors. Different algorithms have different architectures and input data assumptions, such as species ecological characteristics, environmental complexity, and data availability, which can result in uncertainty in prediction results [53,82]. Simulations using a single model inevitably produce under-fitting or over-fitting problems, but the EM can reduce the uncertainty of model fitting [52,83]. The EM can combine the predictive strengths of multiple models, reduce the weaknesses of individual models, improve overall predictive accuracy, and effectively address the uncertainty of model extrapolation [84–86]. In this study, we selected the TSS > 0.9 models to construct the EM by using a weighted average algorithm to predict the suitable habitats of species. The evaluation metrics of the EM were generally higher than those of the three single models (Figure 1). We can conclude that the model integrated by the good single algorithms can be more reliable and overcome the degree of uncertainty in algorithm selection with higher prediction accuracy and better fitting results.

3.3. Selection of Environmental Factors

SDMs is a method for predicting the potential habitats of species by establishing relationships between species distribution points and environmental factors [56]. Therefore, understanding the complexity between species and their environmental factors as well as the selection of appropriate environmental variables are key to building good models [15,86]. In most studies of SDMs, only climatic variables are selected as environmental variables [86], or factors such as climate, topography, and soil are selected as environmental variables [51,82]. Mangrove forests and *S. alterniflora* belong to the intertidal wetland vegetation of coastal mudflats. In this study, two major dual influencing factors, i.e., the terrestrial and marine environments, were considered based on the growth environment of mangrove forests and *S. alterniflora* and previous research results [15,46,57]. Considering the correlation between climate variables, the main climate variables were selected, while the topographic data, as well as sea surface temperature, sea surface salinity, and water quality factors representing the marine ecological environment were selected for predictive simulations. The results of the study show that both terrestrial and marine environmental factors have different degrees of influence on coastal wetland vegetation (Table 2). The response of coastal plants, including mangrove forests and *S. alterniflora*, to factors such as bioclimatic, elevation, and the marine environment, is attributed to the fact that the environmental and material supply required for growth is obtained mainly from both marine and terrestrial sources [15]. Environmental factors, such as temperature, precipitation, salinity, and topography, are the main factors that control the distribution and growth of mangrove forests and *S. alterniflora* in the Chinese coastal zone [87].

3.4. Uncertainty in Species Distribution Model Simulations

In the process of simulating the spatial pattern of species, uncertainties in the simulation results are often caused by a variety of factors, such as species distribution data [88], subjectivity and multi-collinearity of environmental variables [89], selection of species distribution models and setting of parameters [90,91], etc. In this study, we chose to use remote sensing data to obtain species point distributions and selected a single model with high model accuracy to construct EM and run it several times to reduce the uncertainty of model prediction. However, the prediction of species distribution still has a large amount of uncertainty, and the distribution of suitable habitats for species changes due to a variety of factors, such as the physical properties of the environment, resource demand, and human activities. Further exploration on how to reduce the uncertainty of species distribution prediction is needed in future studies.

Table 2. Environmental variables used to simulate predictions of mangrove forests and *S. alterniflora*.

Factors	Variables	Description	Unit	Data Sources
BioClimate	Bio1	Annual mean temperature	°C	CHELSA (https://chelsa-climate.org/ (accessed on 12 March 2023))
	Bio2	Mean diurnal range	°C	
	Bio3	Isothermality	-	
	Bio5	Max temperature of warmest month	-	
	Bio12	Annual precipitation	mm	
	Bio15	Precipitation seasonality	mm	
	Bio16	Precipitation of wettest quarter	mm	
	Bio18	Precipitation of warmest	mm	
	Bio19	Precipitation of coldest quarter	mm	
Sea–land topography	Ele	Elevation	m	National Marine Data Center (http://mds.nmdis.org.cn/ (accessed on 12 March 2023)), Slop and Aspe are extracted from the sea–land topography by ArcGIS.
	Slop	Slope	°	
	Aspe	Aspect	°	
Marine environment	CHL	Chlorophyll concentration	µg/L	NASA MODIS-Aqua Level-3 (http://oceancolor.gsfc.nasa.gov (accessed on 12 March 2023))
	PAR	Photosynthetically available radiation	E/m^{2}day	
	SSS	Annual mean sea surface salinity	%	National Marine Data Center (http://mds.nmdis.org.cn/ (accessed on 12 March 2023))
	SST1	Annual mean sea surface temperature	°C	NASA MODIS-Aqua Level-3 (http://oceancolor.gsfc.nasa.gov (accessed on 12 March 2023))
	SST2	Sea surface temperature of warmest quarter	°C	
	SST3	Sea surface temperature of wettest quarter	°C	
	SST4	Sea surface temperature of coldest quarter	°C	
	SST5	Sea surface temperature of driest quarter	°C	

3.5. Important Variables Affecting Mangrove Forests and S. alterniflora

Previous studies have shown that temperature is a key factor affecting the distribution of mangrove forests, and that temperature affects the growth and reproduction of mangrove forests [46]. The annual average temperature of the mangrove forest distribution area in China is 19–26 °C, the average temperature of the coldest month is 7.4–21 °C, and the average annual precipitation is 1200–2200 mm [92]. This is consistent with the conclusion reached in this study (Table 1). If the temperature rises by 2 °C, the distribution area of mangrove plants in China will likely expand northward by about 2.5°, and the northern boundary of the introgression can reach Hangzhou Bay from the current location in Yueqing County, Zhejiang Province [93,94]. Currently, with the warming climate, the northern boundary of the introduced mangrove forests has reached Wenzhou and Taizhou, Zhejiang Province [93]. In the present study, the northern boundary of the highly suitable distribution area of mangrove forests predicted by the EM can reach Wenzhou and Taizhou, Zhejiang Province (Figure 2). This is consistent with the results of previous studies [93]. In addition to the effect of temperature, Ele, CHL, and precipitation have an influence on the growth of mangrove forests, and studies have shown that mangrove forests are also more suitable to grow in lowland coastal and water body eutrophic level relatively high areas [15,95,96]. Precipitation can regulate nutrient uptake and thus affect mangrove forest productivity [97], and seasonal precipitation can lead to changes in mangrove forest habitat suitability [98].

S. alterniflora is highly suitable and tolerant to climate and environment. Studies have shown that *S. alterniflora* is subject to multiple stresses from precipitation patterns, sea level rise, and nutrient enrichment [99]. Higher nutrient concentrations and increased eutrophication of habitats promote the invasion and expansion of *S. alterniflora* populations [10,100,101]. Precipitation affects the growth of *S. alterniflora* mainly by affecting the salinity of soil pore water, and higher precipitation promotes the growth of *S. alterniflora* [102,103].

3.6. Potential Distribution of Mangrove Forests and S. alterniflora

Mangrove forests are mainly located in the warm and humid subtropical and tropical monsoon regions in some harbors or estuaries with good wave cover. Hainan, Guangdong and Guangxi account for about 96% of China's mangrove forest areas, with small areas in Fujian, Taiwan, Hong Kong, and Macau and no natural distribution of mangrove forests in Zhejiang, which were artificially introduced after the 1950s [24]. From the predicted results, the potentially suitable habitats for mangrove forests are mainly distributed in eight provinces and regions in China, with the northern end reaching Taizhou city, Zhejiang province, and the southern end reaching Sanya city, Hainan province (Figure 2a). This is consistent with previous results [87,92,104–106].

According to the survey, the area of *S. alterniflora* in Jiangsu, Zhejiang, Fujian, and Shanghai accounts for 94% of the total distribution area of the national coastal zone; this is the most concentrated distribution of *S. alterniflora* in China [107]. The Zhejiang and Fujian provinces have winding coastlines and easy to form harbors, which provide favorable conditions for the growth of *S. alterniflora* [11]. In this study, the highly suitable invasive areas for *S. alterniflora* are mainly distributed in the coastal areas of Zhejiang and Fujian. Most of the coasts in these two provinces are distributed, with continuous distribution in the harbor and estuary areas. The width of the distribution strip is wide and extends towards the ocean (Figure 3a).

3.7. S. alterniflora Invades Mangrove Forests along the Southeastern Coast of China

At present, *S. alterniflora* invasion seriously threatens the mangrove forest resources in Fujian and Guangdong, China and replaces the mangrove forest communities in some areas [11,108]. Previous studies have shown that *S. alterniflora* has a wider range of salinity adaptations, greater flood tolerance, and more rapid reproductive dispersal than native mangrove plants [12,27]. Thus, *S. alterniflora* can spread rapidly in mangrove forest communities and form significant competitive exclusion to mangrove plants, affecting the renewal and growth of mangrove forest seedlings. *S. alterniflora* causes not only degradation of mangrove habitats, but also changes the biodiversity and behavioral patterns of mangrove forests [109]. In this study, we used an invasion risk index (IRI) to assess the invasive risk of *S. alterniflora* in mangrove forest distribution areas along the southeastern coast of China. Zhejiang and Fujian are at high risk of invasion, while other southern provinces are at lower risk of invasion (Figure 5a). This is because Zhejiang and Fujian are the areas with the most invasive *S. alterniflora*. Mangrove forests in the southern provinces of Guangdong and Guangxi are relatively intact and can withstand the invasion of *S. alterniflora* [110]. Through the analysis of the IRI results, mangrove forest protection and restoration using *S. alterniflora* control plans and measures can be set more intuitively. This has significance as a practical reference.

3.8. Conservation of Mangrove Forests and Control of S. alterniflora

Mangrove forest wetlands are rich in biodiversity and are one of the most productive ecosystems in the world [15]. Due to its high environmental adaptability and rapid population spread, *S. alterniflora* poses a serious threat to native coastal ecosystems [43] One of the major causes of the extensive destruction of mangrove forests in China since the 1980s is the invasion of *S. alterniflora* [26]. Therefore, mangrove forest conservation and *S. alterniflora* control are urgent. Based on the results of the current study, we propose the following recommendations for future mangrove forest conservation and *S. alterniflora* control along the southeast coast of China.

3.8.1. Rationalize Mangrove Forest Protection Actions

With an increasing number of people recognizing the unique ecological value of mangrove forests, their restoration and protection have received great attention. In recent years, China has made positive progress in mangrove forest restoration and initially reversed the trend of drastic decrease in mangrove forest area [24,111]. In 2020, China's National Forestry and Grassland Administration issued a plan called "Special Action Plan for Mangrove Forest Protection and Restoration (2020–2025)", which required the comprehensive strengthening of mangrove forest protection and restoration work [112]. According to the results of our study, the highly suitable distribution areas of mangrove forests are mainly concentrated in southeastern Guangdong, southern Guangxi, and northern Hainan. Therefore, mangrove forest protection and restoration work should be actively carried out in Guangdong, Guangxi, and Hainan, including the establishment of nature reserves, mangrove forest transplantation, cultivation of new species, and other measures. At present, with the increasing awareness of the social and ecological value of mangrove forests, the introduction of mangrove plants for artificial planting in Zhejiang has also achieved great results, and the northern boundary of the introduction has reached Wenzhou and Taizhou, Zhejiang [92]. At the same time, as the temperature rises, the distribution area of mangrove forest plants will probably expand about 2.5° north, and the northern boundary of the introduced species could reach Hangzhou Bay [92,93]. Mangrove forest priming projects should be planned rationally in the Zhejiang and Fujian areas, using the high suitable distribution area of mangrove forests as reference.

3.8.2. Control *S. alterniflora* According to Local Conditions

The rapid and effective control of *S. alterniflora*, by limiting its proliferation rate and scale and minimizing its ecological damage and impact, have become important issues to be solved in coastal wetland ecosystem and rare species conservation. It was found that low and sparse mangrove forests are vulnerable to the invasion of *S. alterniflora* because it affects their spread. On the other hand, large and lush mangrove forests can shade the growth of *S. alterniflora*, eventually depriving *S. alterniflora* of a suitable environment for survival [113]. At present, various methods have been explored to control *S. alterniflora*, including physical, chemical, and biological methods [114–117]. According to our results, biological control can be carried out in dense mangrove forest areas in Guangdong, Guangxi, and Hainan. This involves the use of a "*Sonneratia apetala* shade" plantation to control the growth and spread of *S. alterniflora*. In areas where *S. alterniflora* invades mangrove forests, fast-growing *S. apetala* suppresses the growth of *S. alterniflora* through shading and chemosensory effects, while promoting the growth of native mangrove plants to restore the mangrove forest community [118]. In Zhejiang and Fujian, where mangrove forests are sparse and *S. alterniflora* is dense, the growth of *S. alterniflora* can be controlled by physical or chemical methods, including artificial removal, mulching for shade, mowing, fire, flooding, and chemical herbicides [119–122].

4. Materials and Methods

4.1. Occurrence Data

In this study, current mangrove forest distribution areas were selected. The study areas in China include Guangdong, Guangxi, Hainan, Fujian, Hong Kong, Macau, Taiwan, and Zhejiang provinces [87]. Mangrove forest occurrence records were obtained from the spatial distribution dataset of Chinese mangrove forests at 30 m resolution in 2015 (National Earth

System Science Data Center (NESSDC), http://www.geodata.cn, accessed on 8 July 2022). Occurrence records of *S. alterniflora* were obtained from the spatial distribution dataset of Chinese *S. alterniflora* at 30 m resolution in 2015 (NESSDC, http://www.geodata.cn, accessed on 22 July 2022). The spatial distribution datasets of mangrove forests and *S. alterniflora* were resampled into the occurrence records with a sampling accuracy of 1 km using the fishnet method in ArcGIS. Then, the collected occurrence records were filtered to eliminate the duplicate records. Finally, 1358 mangrove forest records and 1314 *S. alterniflora* records were obtained for modeling (Figure 6).

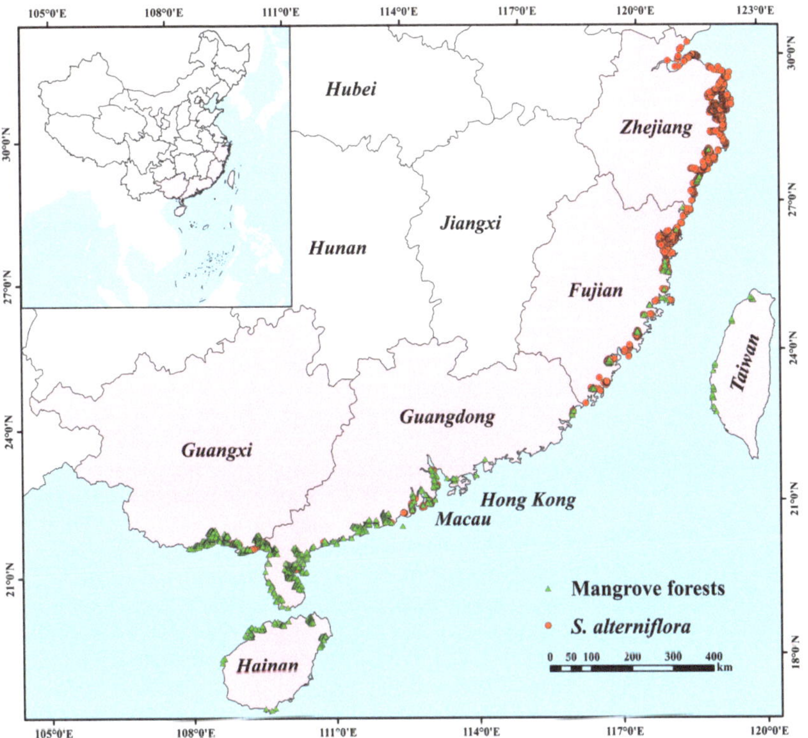

Figure 6. Study area and species occurrence record distribution. The study area mainly includes the provinces of Zhejiang, Fujian, Guangdong, Guangxi, Hainan, Taiwan, Hong Kong, and Macau in China. The green triangles represent the location of mangrove forests. The red dots represent the location of the invasive *S. alterniflora*.

4.2. Environment Data

We selected three categories of environmental factors, including 20 variables in bioclimate, sea–land topography, and marine environment (Table 2). CHELSA (Climatologies at high resolution for the earth's land surface areas, https://chelsa-climate.org/, accessed on 24 July 2022) was selected for the bioclimate data, and the elimination of autocorrelation between predictors helped to avoid prediction errors caused by collinearity in bioclimatic variables. Pearson correlation coefficients of 19 bioclimate variables were calculated (Figure S1); only meaningful variables with $|r| < 0.7$ were retained [43], and 9 bioclimate variables were selected to participate in the modeling. The sea–land topography was obtained from the National Marine Data Center (NMDC, http://mds.nmdis.org.cn/, accessed on 28 July 2022). Elevation (Ele), slope (Slop), and aspect (Aspe) were extracted from sea–land topography by ArcGIS. The photosynthetically available radiation (PAR) and chlorophyll concentration (CHL) represent the marine ecological environment and coastal water quality changes. CHL is one of the key indicators of marine primary productivity; it can visually reflect the degree of seawater eutrophication [123–125]. The PAR, CHL, and sea surface temperature were obtained from NASA MODIS-Aqua Level-3 (NASA Goddard Space Flight Center, Ocean Ecology Laboratory, Ocean Biology Processing Group (OBPG), http://oceancolor.gsfc.nasa.gov, accessed on 26 July 2022). Sea surface temperature included annual mean sea surface temperature (SST1), sea surface temperature of warmest quarter (SST2), sea surface temperature of wettest quarter (SST3), sea surface temperature of coldest quarter (SST4), and sea surface temperature of driest quarter (SST5). The sea surface salinity (SSS) was obtained from the NMDC (http://mds.nmdis.org.cn/, accessed on 28 July 2022) (Table 2). Saga GIS was used to interpolate the environmental variables and make the range in land and marine variables consistent. All the environmental variables were resampled to a resolution of 30″ (about 1 km) for model prediction and analysis.

4.3. Model Construction and Evaluation

This study was completed using the Biomod2 package. We used the ANN, CTA, FDA, GAM, GBM, GLM, MARS, RF, and SRE models in the Biomod2 package on the R platform. To improve the fitting accuracy of the model, three sets of pseudo-absent points were randomly generated. After several experiments, we finally determined that the model had the highest accuracy when generating 5000 pseudo-random points per set. We entered species presence points, pseudo-absence points, and environmental data in the R software environment. Seventy-five percent of the sample data were randomly selected as the training set to build the model, and the remaining twenty-five percent were used for model validation. Each of the nine single models was called and run ten times to reduce uncertainty.

In this study, we evaluated the model using the relative operating characteristic (ROC) curve with the statistics of the AUC value (the area under the ROC curve) and the true skill statistic (TSS). ROC is plotted with 1-specificity (the proportion of species non-occurrence areas correctly predicted) as the horizontal coordinate and sensitivity (the proportion of species occurrence records correctly predicted) as the vertical coordinate [52]. The size of the area under the ROC curve is used to evaluate the ability of the model. A larger value indicates a more accurate prediction, which is one of the indicators for the evaluation of many models [126,127]. TSS (=sensitivity + (specificity-1)) is an improved assay for Kappa that retains the advantages of Kappa [128]. This statistic is one of the default evaluation metrics of the Biomod2 package. In general, when the TSS is greater than 0.8, the AUC is greater than 0.9; this indicates very high accuracy of the model fit [56].

The evaluation results of the single-unit models were obtained, and RF, GBM, and GLM were selected for modeling according to their TSS values which were all higher than other single models. The selected models were run again to obtain 90 single models (3 selected models × 10 repetitions × 3 sets of random datasets). Then, we constructed the EM. All models with TSS values greater than or equal to 0.9 were selected. The weight of the individual model was calculated according to Equation (1), the EM was constructed, and the habitat suitability index [55] of the species was derived using Equation (2).

$$W_i = \frac{a_i}{\sum_{i=1}^{n} a_i}. \tag{1}$$

W_i represents the weight of a single model, a_i represents the TSS value of the ith single model, and n presents the number of models selected.

$$HSI_i = \sum_{j=1}^{n} W_i \times P_{ij}. \tag{2}$$

HSI_i represents the habitat suitability index value of each pixel of model I; W_i represents the weight of the model I; and P_{ij} represents the j pixel value of model i.

The habitat suitability index (0–1) output from the model reflects the probability of species presence [9]. According to the statistical principle of "likelihood" in the presence probability analysis [61], we classified the suitability results into four categories [52,55]: unsuitable ($HSI < 0.3$), low suitability ($0.3 \leq HSI < 0.5$), moderately suitable ($0.5 \leq HSI < 0.7$), and highly suitable ($HSI \geq 0.7$).

4.4. Risk Assessment of S. alterniflora Invading Mangrove Forests

In order to assess the areas where the distribution between mangrove forests and *S. alterniflora* may shift, that is, the invasive risk of *S. alterniflora* within the mangrove forest distribution area, we used the invasive risk index (*IRI*) (Equation (3)) combined with SDMs simulation results of mangrove forests and *S. alterniflora*. We subtracted the habitat suitability value of the native species (i.e., mangrove forests) from the invasive habitat suitability value of the invasive species (i.e., *S. alterniflora*) to obtain the value of *IRI* of *S. alterniflora* in the study area.

$$IRI = HSI_s - HSI_m. \tag{3}$$

IRI represents the invasive risk index with thresholds ranging from -1 to 1; HSI_s represents the result of the SDM simulation of *S. alterniflora*; and HSI_m represents the result of the SDM simulation of mangrove forests. A positive value of *IRI* indicates that the region is more suitable for the growth of *S. alterniflora*. The invasive risk of *S. alterniflora* in this region is high, which may affect the growth and renewal of local mangrove forests. A negative value suggests that the habitat suitability of mangrove forests in this region is higher, and the invasive risk of *S. alterniflora* is lower.

5. Conclusions

S. alterniflora poses a serious threat to native coastal ecosystems due to its high environmental adaptability and rapid population spread. The strong vitality of *S. alterniflora* has led to the disappearance of large areas of mangrove forests because it crowds out mangrove forest growing space. We used EM to simulate the geographic distribution of mangrove forests and *S. alterniflora* in southeastern China and deduced the environmental threshold for their growth. On this basis, we assessed the invasion risk of *S. alterniflora* to mangrove forests along the southeast coast. According to our analysis, mangrove forests are mainly concentrated in southern Guangxi, Guangdong, and *S. alterniflora* is mainly distributed in Zhejiang and Fujian. We should strengthen the control of *S. alterniflora* and the introduction and cultivation of mangrove forests in the coastal areas of Zhejiang, Fujian, Beibu Gulf, and Zhuhai, Guangdong. Mangrove forest restoration projects should be increased in the coastal areas of Guangdong and southern Guangxi. Our study provides a clear understanding of the geographic distribution of mangroves and *S. alterniflora* and the high-risk areas where *S. alterniflora* is more likely to invade mangrove forests. We believe that this study can provide a basis for mangrove forest conservation and *S. alterniflora* control along the southeast coast of China.

Supplementary Materials: The following supporting information can be downloaded at: https://www.mdpi.com/article/10.3390/plants12101923/s1, Figure S1: Correlation analysis of 19 bioclimate variables. The color and size of the circles represent the magnitude of the correlation between the variables. The upper numbers and the lower circles correspond variables with one another; Figure S2: Variables with a cumulative contribution greater than 90% affecting the growth of mangrove forests and *S. alterniflora*. Blue represents the contribution rate of mangrove forests and pink is the contribution rate of *S. alterniflora*.

Author Contributions: J.Z.: Formal analysis, Data curation, Methodology, Software, Writing—original draft. H.W.: Conceptualization, Methodology, Supervision. R.C.: Investigation, Formal analysis, Data curation. J.L.: Investigation, Software, Data curation. L.W.: Investigation, Data curation. W.G.: Conceptualization, Methodology, Writing—review and editing, Supervision. All authors have read and agreed to the published version of the manuscript.

Funding: This work was supported by the National Natural Science Foundation of China (31070293), the Natural Science Basic Research Program of Shaanxi Province (2020JM-277), and the Research and Development Program of Science and Technology of Shaanxi Province (2014K14-01-02).

Data Availability Statement: The data presented in this study are openly available in Mendeley Data at doi: 10.17632/759yn23jb9.1.

Acknowledgments: We are grateful for data provided and help from each of the resource sharing platforms. We appreciate Yunfei Gu, a doctoral student at the University of Southampton for assisting us in revising the paper.

Conflicts of Interest: The authors declare no conflict of interest.

References

1. Ding, J.; Mark, R.N.; Lu, P.; Ren, M.; Huang, H. China's booming economy is sparking and accelerating biological invasions. *Bioscience* **2008**, *58*, 317–324. [CrossRef]
2. Wang, H.; López-Pujol, J.; Meyerson, L.A.; Qiu, J.; Wang, X.; Ouyang, Z. Biological invasions in rapidly urbanizing areas: A case study of Beijing, China. *Biodivers. Conserv.* **2011**, *20*, 2483–2509. [CrossRef]
3. Chen, B.; Sun, Y.; Han, Z.; Huang, H.; Zhang, H.; Li, Y.; Zhang, G.; Liu, W. Challenges in preventing and controlling invasive alien species in China. *J. Biosaf.* **2020**, *29*, 157–163. (In Chinese)
4. Daehler, C.C.; Strong, D.R. Status, prediction and prevention of introduced cordgrass Spartina spp. invasions in Pacific estuaries, USA. *Biol. Conserv.* **1996**, *78*, 51–58. [CrossRef]
5. Partridge, T.R. *Spartina* in New Zealand. *N. Z. J. Bot.* **1987**, *25*, 567–575. [CrossRef]
6. Adams, J.; van Wyk, E.; Riddin, T. First record of *Spartina alterniflora* in southern Africa indicates adaptive potential of this saline grass. *Biol. Invasions* **2016**, *18*, 2153–2158. [CrossRef]
7. Baumel, A.; Ainouche, M.L.; Misset, M.T.; Gourret, J.P.; Bayer, R.J. Genetic evidence for hybridization between the native *Spartina maritima* and the introduced *Spartina alterniflora* (Poaceae) in South-West France: *Spartina* × *neyrautii* re-examined. *Plant Syst. Evol.* **2003**, *237*, 87–97. [CrossRef]
8. Campos, J.A.; Herrera, M.; Biurrun, I.; Loidi, J. The role of alien plants in the natural coastal vegetation in central-northern Spain. *Biodivers. Conserv.* **2004**, *13*, 2275–2293. [CrossRef]
9. Chung, C.-H. Forty years of ecological engineering with *Spartina* plantations in China. *Ecol. Eng.* **2006**, *27*, 49–57. [CrossRef]
10. Deng, Z.; An, S.; Zhi, Y.; Zhou, C.; Chen, L.; Zhao, C.; Fang, S.; Li, H. Preliminary studies on invasive model and outbreak mechanism of exotic species, *Spartiona alterniflora* Loisel. *Acta Ecol. Sin.* **2006**, *26*, 2678–2686. (In Chinese)
11. Zuo, P.; Zhao, S.; Liu, C.a.; Wang, C.; Liang, Y. Distribution of *Spartina* spp. along China's coast. *Ecol. Eng.* **2012**, *40*, 160–166. [CrossRef]
12. Wang, Q.; An, S.; Ma, Z.; Zhao, B.; Chen, J.; Li, B. Invasive *Spartina alterniflora*: Biology, ecology and management. *Acta Phytotaxon. Sin.* **2006**, *44*, 559–588. (In Chinese) [CrossRef]
13. Kesavan, S.; Xavier, K.A.M.; Deshmukhe, G.; Jaiswar, A.K.; Bhusan, S.; Sukla, S.P. Anthropogenic pressure on mangrove ecosystems: Quantification and source identification of surficial and trapped debris. *Sci. Total Environ.* **2021**, *794*, 148677. [CrossRef]
14. Romañach, S.S.; DeAngelis, D.L.; Koh, H.L.; Li, Y.; Teh, S.Y.; Raja Barizan, R.S.; Zhai, L. Conservation and restoration of mangroves: Global status, perspectives, and prognosis. *Ocean Coast. Manag.* **2018**, *154*, 72–82. [CrossRef]
15. Charrua, A.B.; Bandeira, S.O.; Catarino, S.; Cabral, P.; Romeiras, M.M. Assessment of the vulnerability of coastal mangrove ecosystems in Mozambique. *Ocean Coast. Manag.* **2020**, *189*, 105145. [CrossRef]
16. Carugati, L.; Gatto, B.; Rastelli, E.; Lo Martire, M.; Coral, C.; Greco, S.; Danovaro, R. Impact of mangrove forests degradation on biodiversity and ecosystem functioning. *Sci. Rep.* **2018**, *8*, 13298. [CrossRef]
17. Wang, Y.; Gu, J. Ecological responses, adaptation and mechanisms of mangrove wetland ecosystem to global climate change and anthropogenic activities. *Int. Biodeterior. Biodegrad.* **2021**, *162*, 105248. [CrossRef]
18. Inyang, A.I.; Wang, Y. Phytoplankton diversity and community responses to physicochemical variables in mangrove zones of Guangzhou Province, China. *Ecotoxicology* **2020**, *29*, 650–668. [CrossRef]
19. Alongi, D.M. Mangrove forests: Resilience, protection from tsunamis, and responses to global climate change. *Estuar. Coast. Shelf Sci.* **2008**, *76*, 1–13. [CrossRef]
20. Giri, C.; Ochieng, E.; Tieszen, L.L.; Zhu, Z.; Singh, A.; Loveland, T.; Masek, J.; Duke, N. Status and distribution of mangrove forests of the world using earth observation satellite data. *Glob. Ecol. Biogeogr.* **2011**, *20*, 154–159. [CrossRef]
21. Valiela, I.; Bowen, J.L.; York, J.K. Mangrove forests: One of the World's threatened major tropical environments. *Bioscience* **2001**, *51*, 807–815. [CrossRef]
22. Polidoro, B.A.; Carpenter, K.E.; Collins, L.; Duke, N.C.; Ellison, A.M.; Ellison, J.C.; Farnsworth, E.J.; Fernando, E.S.; Kathiresan, K.; Koedam, N.E.; et al. The loss of species: Mangrove extinction risk and geographic areas of global concern. *PLoS ONE* **2010**, *5*, e10095. [CrossRef] [PubMed]
23. Ellison, J.C.; Zouh, I. Vulnerability to climate change of mangroves: Assessment from cameroon, central Africa. *Biology* **2012**, *1*, 617–638. [CrossRef]
24. Wang, W.; Fu, H.; Lee, S.Y.; Fan, H.; Wang, M. Can strict protection stop the decline of mangrove ecosystems in China? From rapid destruction to rampant degradation. *Forests* **2020**, *11*, 55. [CrossRef]

25. Goldberg, L.; Lagomasino, D.; Thomas, N.; Fatoyinbo, T. Global declines in human-driven mangrove loss. *Glob. Chang. Biol.* **2020**, *26*, 5844–5855. [CrossRef] [PubMed]

26. Feng, J.; Huang, Q.; Qi, F.; Guo, J.; Lin, G. Utilization of exotic *Spartina alterniflora* by fish community in the mangrove ecosystem of Zhangjiang Estuary: Evidence from stable isotope analyses. *Biol. Invasions* **2015**, *17*, 2113–2121. [CrossRef]

27. Zhang, Y.; Huang, G.; Wang, W.; Chen, L.; Lin, G. Interactions between mangroves and exotic *Spartina* in an anthropogenically disturbed estuary in southern China. *Ecology* **2012**, *93*, 588–597. [CrossRef]

28. Yu, Z.; Yang, J.; Yu, X.; Liu, L.; Tian, Y. Aboveground vegetation influences belowground microeukaryotic community in a mangrove nature reserve. *Wetlands* **2013**, *34*, 393–401. [CrossRef]

29. Li, Z.; Wang, W.; Zhang, Y. Recruitment and herbivory affect spread of invasive *Spartina alterniflora* in China. *Ecology* **2014**, *95*, 1972–1980. [CrossRef]

30. Rao, K.; Ramanathan, A.L.; Raju, N.J. Assessment of blue carbon stock of Coringa mangroves: Climate change perspective. *J. Clim. Chang.* **2022**, *8*, 41–58. [CrossRef]

31. He, Z.; Feng, X.; Chen, Q.; Li, L.; Li, S.; Han, K.; Guo, Z.; Wang, J.; Liu, M.; Shi, C.; et al. Evolution of coastal forests based on a full set of mangrove genomes. *Nat. Ecol. Evol.* **2022**, *6*, 738–749. [CrossRef] [PubMed]

32. Zamprogno, G.C.; Tognella, M.M.P.; Costa, M.B.d.; Otegui, M.B.P.; Menezes, K.M. Spatio-temporal distribution of benthic fauna in mangrove areas in the Bay of Vitória estuary, Brazil. *Reg. Stud. Mar. Sci.* **2023**, *62*, 102939. [CrossRef]

33. Su, J.; Gasparatos, A. Perceptions about mangrove restoration and ecosystem services to inform ecosystem-based restoration in Large Xiamen Bay, China. *Landsc. Urban Plan.* **2023**, *235*, 104763. [CrossRef]

34. Granse, D.; Suchrow, S.; Jensen, K. Long-term invasion dynamics of *Spartina* increase vegetation diversity and geomorphological resistance of salt marshes against sea level rise. *Biol. Invasions* **2021**, *23*, 871–883. [CrossRef]

35. Zhang, H.; Liu, Y.; Xu, Y.; Han, S.; Wang, J. Impacts of *Spartina alterniflora* expansion on landscape pattern and habitat quality: A case study in Yancheng coastal wetland, China. *Appl. Ecol. Environ. Res.* **2020**, *18*, 4669–4683. [CrossRef]

36. Yang, R.; Yang, F. Impacts of *Spartina alterniflora* invasion on soil inorganic carbon in coastal wetlands in China. *Soil Sci. Soc. Am. J.* **2020**, *84*, 844–855. [CrossRef]

37. Macy, A.; Osland, M.J.; Cherry, J.A.; Cebrian, J. Changes in ecosystem Nitrogen and Carbon allocation with black mangrove (*Avicennia germinans*) encroachment into *Spartina alterniflora* salt marsh. *Ecosystems* **2021**, *24*, 1007–1023. [CrossRef]

38. Zhang, H.; Li, Y.; Pang, M.; Xi, M.; Kong, F. Responses of contents and structure of DOM to *Spartina alterniflora* invasion in Yanghe estuary wetland of Jiaozhou Bay, China. *Wetlands* **2019**, *39*, 729–741. [CrossRef]

39. Liu, W.; Maung-Douglass, K.; Strong, D.R.; Pennings, S.C.; Zhang, Y. Geographical variation in vegetative growth and sexual reproduction of the invasive *Spartina alterniflora* in China. *J. Ecol.* **2016**, *104*, 173–181. [CrossRef]

40. Zhang, D.; Hu, Y.; Liu, M.; Chang, Y.; Sun, L. Geographical variation and influencing factors of *Spartina alterniflora* expansion rate in coastal China. *Chin. Geogr. Sci.* **2020**, *30*, 127–141. [CrossRef]

41. Ge, B.; Jiang, S.; Yang, L.; Zhang, H.; Tang, B. Succession of macrofaunal communities and environmental properties along a gradient of smooth cordgrass *Spartina alterniflora* invasion stages. *Mar. Environ. Res.* **2020**, *156*, 104862. [CrossRef] [PubMed]

42. Noto, A.E.; Hughes, A.R. Genotypic diversity weakens competition within, but not between, plant species. *J. Ecol.* **2020**, *108*, 2212–2220. [CrossRef]

43. Wang, B.; Lin, X. Exotic *Spartina alterniflora* invasion enhances sediment N-loss while reducing N retention in mangrove wetland. *Geoderma* **2023**, *431*, 116362. [CrossRef]

44. Xu, X.; Zhou, C.; He, Q.; Qiu, S.; Zhang, Y.; Yang, J.; Li, B.; Nie, M. Phenotypic plasticity of light use favors a plant invader in nitrogen-enriched ecosystems. *Ecology* **2022**, *103*, e3665. [CrossRef]

45. Nie, S.; Mo, S.; Gao, T.; Yan, B.; Shen, P.; Kashif, M.; Zhang, Z.; Li, J.; Jiang, C. Coupling effects of nitrate reduction and sulfur oxidation in a subtropical marine mangrove ecosystem with *Spartina alterniflora* invasion. *Sci. Total Environ.* **2023**, *862*, 160930. [CrossRef]

46. Hu, W.; Wang, Y.; Zhang, D.; Yu, W.; Chen, G.; Xie, T.; Liu, Z.; Ma, Z.; Du, J.; Chao, B.; et al. Mapping the potential of mangrove forest restoration based on species distribution models: A case study in China. *Sci. Total Environ.* **2020**, *748*, 142321. [CrossRef]

47. Liu, H.; Qi, X.; Gong, H.; Li, L.; Zhang, M.; Li, Y.; Lin, Z. Combined effects of global climate suitability and regional environmental variables on the distribution of an invasive marsh species *Spartina Alterniflora*. *Estuaries Coast.* **2018**, *42*, 99–111. [CrossRef]

48. Thomas, C.D.; Cameron, A.; Green, R.E.; Bakkenes, M.; Beaumont, L.J.; Collingham, Y.C.; Erasmus, B.F.; De Siqueira, M.F.; Grainger, A.; Hannah, L.; et al. Extinction risk from climate change. *Nature* **2004**, *427*, 145–148. [CrossRef]

49. Wu, J.; Lv, J.; Ai, L. The impacts of climate change on the biodiversity:vulnerability and adaptation. *Energy Environ. Sci.* **2009**, *18*, 693–703. (In Chinese) [CrossRef]

50. IPCC. Climate Change 2021: The Physical Science Basis. In *Contribution of Working Group I to the Sixth Assessment Report of the Intergovernmental Panel on Climate Change*; Masson-Delmotte, V., Zhai, P., Pirani, A., Connors, S.L., Péan, C., Berger, S., Caud, N., Chen, Y., Goldfarb, L., Gomis, M.I., et al., Eds.; Cambridge University Press: Cambridge, UK, 2021.

51. Ning, H.; Ling, L.; Sun, X.; Kang, X.; Chen, H. Predicting the future redistribution of Chinese white pine *Pinus armandii* Franch. Under climate change scenarios in China using species distribution models. *Glob. Ecol. Conserv.* **2021**, *25*, e01420. [CrossRef]

52. Fang, Y.; Zhang, X.; Wei, H.; Wang, D.; Chen, R.; Wang, L.; Gu, W. Predicting the invasive trend of exotic plants in China based on the ensemble model under climate change: A case for three invasive plants of Asteraceae. *Sci. Total Environ.* **2021**, *756*, 143841. [CrossRef]

53. Guo, Y.; Li, X.; Zhao, Z.; Nawaz, Z. Predicting the impacts of climate change, soils and vegetation types on the geographic distribution of *Polyporus umbellatus* in China. *Sci. Total Environ.* **2019**, *648*, 1–11. [CrossRef]

54. Zhang, X.; Wei, H.; Zhang, X.; Liu, J.; Zhang, Q.; Gu, W. Non-pessimistic predictions of the distributions and suitability of *Metasequoia glyptostroboides* under climate change using a Random Forest model. *Forests* **2020**, *11*, 62. [CrossRef]
55. Zhang, X.; Wei, H.; Zhao, Z.; Liu, J.; Zhang, Q.; Zhang, X.; Gu, W. The global potential distribution of invasive plants: *Anredera cordifolia* under climate change and human activity based on Random Forest models. *Sustainability* **2020**, *12*, 1491. [CrossRef]
56. Wang, D.; Wei, H.; Zhang, X.; Fang, Y.; Gu, W. Habitat suitability modeling based on remote sensing to realize time synchronization of species and environmental variables. *J. Plant Ecol.* **2021**, *14*, 241–256. [CrossRef]
57. Jayathilake, D.R.M.; Costello, M.J. A modelled global distribution of the seagrass biome. *Biol. Conserv.* **2018**, *226*, 120–126. [CrossRef]
58. Guo, Y.; Li, X.; Zhao, Z.; Wei, H.; Gao, B.; Gu, W. Prediction of the potential geographic distribution of the ectomycorrhizal mushroom *Tricholoma matsutake* under multiple climate change scenarios. *Sci. Rep.* **2017**, *7*, 46221. [CrossRef]
59. Li, G.; Liu, C.; Liu, Y.; Yang, J.; Zhang, X.; Guo, K. Advances in theoretical issues of species distribution models. *Acta Ecol. Sin.* **2013**, *33*, 4827–4835. (In Chinese)
60. Elith, J.; Phillips, S.J.; Hastie, T.; Dudík, M.; Chee, Y.E.; Yates, C.J. A statistical explanation of MaxEnt for ecologists. *Divers. Distrib.* **2011**, *17*, 43–57. [CrossRef]
61. Phillips, S.J.; Anderson, R.P.; Schapire, R.E. Maximum entropy modeling of species geographic distributions. *Ecol. Model.* **2006**, *190*, 231–259. [CrossRef]
62. Guisan, A.; Edwards, T.C.; Hastie, T. Generalized linear and generalized additive models in studies of species distributions: Setting the scene. *Ecol. Model.* **2002**, *157*, 89–100. [CrossRef]
63. Yee, T.W.; Mitchell, N.D. Generalized Additive Models in Plant Ecology. *J. Veg. Sci.* **1991**, *2*, 587–602. [CrossRef]
64. Thuiller, W.; Araújo, M.B.; Lavorel, S. Generalized models vs. classification tree analysis: Predicting spatial distributions of plant species at different scales. *J. Veg. Sci.* **2003**, *14*, 669–680. [CrossRef]
65. Özesmi, S.L.; Özesmi, U. An artificial neural network approach to spatial habitat modelling with interspecific interaction. *Ecol. Model.* **1999**, *116*, 15–31. [CrossRef]
66. Hastie, T.; Tibshirani, R.; Buja, A. Flexible discriminant analysis by optimal scoring. *J. Am. Stat. Assoc.* **1994**, *89*, 1255–1270. [CrossRef]
67. Moisen, G.G.; Freeman, E.A.; Blackard, J.A.; Frescino, T.S.; Zimmermann, N.E.; Edwards, T.C. Predicting tree species presence and basal area in Utah: A comparison of stochastic gradient boosting, generalized additive models, and tree-based methods. *Ecol. Model.* **2006**, *199*, 176–187. [CrossRef]
68. Freeman, E.A.; Moisen, G.G.; Frescino, T.S. Evaluating effectiveness of down-sampling for stratified designs and unbalanced prevalence in Random Forest models of tree species distributions in Nevada. *Ecol. Model.* **2012**, *233*, 1–10. [CrossRef]
69. Booth, T.H.; Nix, H.A.; Busby, J.R.; Hutchinson, M.F.; Franklin, J. bioclim: The first species distribution modelling package, its early applications and relevance to most current MaxEnt studies. *Divers. Distrib.* **2014**, *20*, 1–9. [CrossRef]
70. Verburg, P.H.; Overmars, K.P. Combining top-down and bottom-up dynamics in land use modeling: Exploring the future of abandoned farmlands in Europe with the Dyna-CLUE model. *Landsc. Ecol.* **2009**, *24*, 1167. [CrossRef]
71. Friedman, J.H. Multivariate adaptive regression splines. *Ann. Statist.* **1991**, *19*, 1–67. [CrossRef]
72. Pouteau, R.; Meyer, J.-Y.; Stoll, B. A SVM-based model for predicting distribution of the invasive tree *Miconia calvescens* in tropical rainforests. *Ecol. Model.* **2011**, *222*, 2631–2641. [CrossRef]
73. Hefley, T.J.; Hooten, M.B.; Warton, D. On the existence of maximum likelihood estimates for presence-only data. *Methods Ecol. Evol.* **2015**, *6*, 648–655. [CrossRef]
74. Dietterich, T.G. Ensemble methods in machine learning. In *Multiple Classifier Systems. MCS 2000.*; Lecture Notes in Computer Science; Springer: Berlin/Heidelberg, Germany, 2000; Volume 1857.
75. Zhao, Z.; Wei, H.; Guo, Y.; Gu, W. Potential distribution of *Panax ginseng* and its predicted responses to climate change. *Chin. J. Appl. Ecol.* **2016**, *27*, 3607–3615.
76. Thuiller, W.; Georges, D.; Gueguen, M.; Engler, R.; Breiner, F. Biomod2: Ensemble Platform for Species Distribution Modeling. Available online: https://cran.r-project.org/web/packages/biomod2/index.html (accessed on 8 August 2022).
77. Dyderski, M.K.; Paź, S.; Frelich, L.E.; Jagodziński, A.M. How much does climate change threaten European forest tree species distributions? *Glob. Chang. Biol.* **2018**, *24*, 1150–1163. [CrossRef]
78. Ranc, N.; Santini, L.; Rondinini, C.; Boitani, L.; Poitevin, F.; Angerbjörn, A.; Maiorano, L. Performance tradeoffs in target-group bias correction for species distribution models. *Ecography* **2017**, *40*, 1076–1087. [CrossRef]
79. Ramirez-Villegas, J.; Cuesta, F.; Devenish, C.; Peralvo, M.; Jarvis, A.; Arnillas, C.A. Using species distributions models for designing conservation strategies of Tropical Andean biodiversity under climate change. *J. Nat. Conserv.* **2014**, *22*, 391–404. [CrossRef]
80. Wang, L.; Liu, J.; Liu, J.; Wei, H.; Fang, Y.; Wang, D.; Chen, R.; Gu, W. Revealing the long-term trend of the global-scale Ginkgo biloba distribution and the impact of future climate change based on the ensemble modeling. *Biodivers. Conserv.* **2023**, *32*, 2077–2100. [CrossRef]
81. Grenouillet, G.; Buisson, L.; Casajus, N.; Lek, S. Ensemble modelling of species distribution: The effects of geographical and environmental ranges. *Ecography* **2011**, *34*, 9–17. [CrossRef]

32. Mtengwana, B.; Dube, T.; Mudereri, B.T.; Shoko, C. Modeling the geographic spread and proliferation of invasive alien plants (IAPs) into new ecosystems using multi-source data and multiple predictive models in the Heuningnes catchment, South Africa. *Gisci. Remote Sens.* **2021**, *58*, 483–500. [CrossRef]

33. Marmion, M.; Parviainen, M.; Luoto, M.; Heikkinen, R.K.; Thuiller, W. Evaluation of consensus methods in predictive species distribution modelling. *Divers. Distrib.* **2009**, *15*, 59–69. [CrossRef]

34. Hao, T.; Elith, J.; Guillera-Arroita, G.; Lahoz-Monfort, J.J. A review of evidence about use and performance of species distribution modelling ensembles like BIOMOD. *Divers. Distrib.* **2019**, *25*, 839–852. [CrossRef]

35. Ardestani, E.G.; Ghahfarrokhi, Z.H. Ensembpecies distribution modeling of Salvia hydrangea under future climate change scenarios in Central Zagros Mountains, Iran. *Glob. Ecol. Conserv.* **2021**, *26*, e01488. [CrossRef]

36. Valavi, R.; Shafizadeh-Moghadam, H.; Matkan, A.; Shakiba, A.; Mirbagheri, B.; Kia, S.H. Modelling climate change effects on Zagros forests in Iran using individual and ensemble forecasting approaches. *Theor. Appl. Clim.* **2018**, *137*, 1015–1025. [CrossRef]

37. Liao, B.; Zhang, Q. Area, distribution and species composition of mangroves in China. *Wetl. Sci.* **2014**, *12*, 435–440. [CrossRef]

38. Fei, S.; Yu, F. Quality of presence data determines species distribution model performance: A novel index to evaluate data quality. *Landsc. Ecol.* **2016**, *31*, 31–42. [CrossRef]

39. Yan, Y.; Li, Y.; Wang, W.; He, J.; Yang, R.; Wu, H.; Wang, X.; Jiao, L.; Tang, Z.; Yao, Y. Range shifts in response to climate change of *Ophiocordyceps sinensis*, a fungus endemic to the Tibetan Plateau. *Biol. Conserv.* **2017**, *206*, 143–150. [CrossRef]

40. Tsoar, A.; Allouche, O.; Steinitz, O.; Rotem, D.; Kadmon, R. A comparative evaluation of presence-only methods for modelling species distribution. *Divers. Distrib.* **2007**, *13*, 397–405. [CrossRef]

41. Thibaud, E.; Petitpierre, B.; Broennimann, O.; Davison, A.C.; Guisan, A. Measuring the relative effect of factors affecting species distribution model predictions. *Methods Ecol. Evol.* **2014**, *5*, 947–955. [CrossRef]

42. Lin, P. Ecological notes on mangroves in southeast coast of China including Taiwan Province and Hainan Island. *Acta Ecol. Sin.* **1981**, *3*, 283–290. (In Chinese)

43. Wang, Y. Impacts, challenges and opportunities of global climate change on mangrove ecosystems. *J. Trop. Oceanogr.* **2021**, *40*, 1–14. [CrossRef]

44. Chen, X.; Lin, P. Responses and roles of mangroves in China to global climate changes. *Trans. Oceanol. Limnol.* **1999**, *2*, 11–17. (In Chinese)

45. Wang, H.; Gao, H.; Zhou, H.Y.; Guo, W.D. Preliminary study on the chlorophyll-a and nutrients in seawater of the mangrove area of Qi'ao Island in Zhujiang River estuary. *J. Oceanogr. Taiwan Strait* **2005**, *4*, 502–507. (In Chinese) [CrossRef]

46. He, B.; Wei, M.; Fan, H.; Pan, L.; Cao, Q. Spatio-temporal change of inorganic nitrogen content and the evaluation of eutrophication in the surface seawaters of mangrove area in Guangxi bays. *J. Appl. Oceanogr.* **2014**, *33*, 140–148. [CrossRef]

47. Snedaker, S.C. Mangroves and climate change in the Florida and Caribbean region: Scenarios and hypotheses. *Hydrobiologia* **1995**, *295*, 43–49. [CrossRef]

48. Choudhury, A.K.; Das, M.; Philip, P.; Bhadury, P. An assessment of the implications of seasonal precipitation and anthropogenic influences on a mangrove ecosystem using phytoplankton as proxies. *Estuar. Coast.* **2014**, *38*, 854–872. [CrossRef]

49. Hanson, A.; Johnson, R.; Wigand, C.; Oczkowski, A.; Davey, E.; Markham, E. Responses of *Spartina alterniflora* to multiple stressors: Changing precipitation patterns, accelerated sea level rise, and nutrient enrichment. *Estuaries Coasts* **2016**, *39*, 1376–1385. [CrossRef]

100. Levine, J.M.; Stephen Brewer, J.; Bertness, M.D. Nutrients, competition and plant zonation in a New England salt marsh. *J. Ecol.* **1998**, *86*, 285–292. [CrossRef]

101. Matzke, S.; Elsey-Quirk, T. *Spartina patens* productivity and soil organic matter response to sedimentation and nutrient enrichment. *Wetlands* **2018**, *38*, 1233–1244. [CrossRef]

102. O'Donnell, J.P.R.; Schalles, J.F. Examination of abiotic drivers and their influence on *Spartina alterniflora* biomass over a Twenty-Eight Year period using Landsat 5 TM satellite imagery of the central Georgia coast. *Remote Sens.* **2016**, *8*, 477. [CrossRef]

103. Zhao, X.; Zhao, C.; Liu, X.; Gong, L.; Deng, Z.; Li, J. Growth characteristic and adaptability of *Spartina alterniflora* in different latitudes areas along China coast. *Ecol. Sci.* **2015**, *34*, 119–128. (In Chinese) [CrossRef]

104. Fan, H.; Wang, W. Some thematic issues for mangrove conservation in China. *J. Xiamen Univ. Nat. Sci.* **2017**, *56*, 323–330. (In Chinese)

105. Wang, H.; Ren, G.; Wu, P.; Liu, A.; Pan, L.; Ma, Y.; Ma, Y.; Wang, J. Analysis on the remote sensing monitoring and landscape pattern change of mangrove in China from 1990 to 2019. *J. Ocean Technol.* **2020**, *39*, 1–12. (In Chinese)

106. Fang, B.; Dan, X. On the mangrove resources and its conservation in China. *Contral S. For. Inv. Plan.* **2001**, *3*, 20–30. (In Chinese)

107. Lu, J.; Zhang, Y. Spatial distribution of an invasive plant *Spartina alterniflora* and its potential as biofuels in China. *Ecol. Eng.* **2013**, *52*, 175–181. [CrossRef]

108. Wang, W.; Sardans, J.; Wang, C.; Zeng, C.; Tong, C.; Chen, G.; Huang, J.; Pan, H.; Peguero, G.; Vallicrosa, H.; et al. The response of stocks of C, N, and P to plant invasion in the coastal wetlands of China. *Glob. Chang. Biol.* **2019**, *25*, 733–743. [CrossRef]

109. Chen, Q.; Ma, K. Research overview and trend on biological invasion in mangrove forests. *Chin. J. Plant Ecol.* **2015**, *39*, 283–299. (In Chinese) [CrossRef]

110. Guo, X.; Pan, W.; Chen, Y.; Zhang, W.; He, T.; Liu, Y.; Liu, W.; Zhang, Y. Invasion of *Spartina alterniflora* and protection of mangroves in Guangdong Zhanjiang Mangrove National Nature Reserve and adjacent coastal area. *For. Environ. Sci.* **2018**, *34*, 58–63. (In Chinese) [CrossRef]

111. Li, S.; Xie, T.; Pennings, S.C.; Wang, Y.; Craft, C.; Hu, M. A comparison of coastal habitat restoration projects in China and the United States. *Sci. Rep.* **2019**, *9*, 14388. [CrossRef]

112. National Forestry and Grassland Administration. Special Action Plan for Mangrove Forests Protection and Restoration (2020–2025). 2020. Available online: http://www.gov.cn/zhengce/zhengceku/2020-08/29/content_5538354.htm (accessed on 20 December 2022).
113. Li, H.; Peng, X.; Wu, Z.; Li, J. Investigation on the invasion status of *Spartina alterniflora* to mangrove wetlands. *J. Guangdong Univ. Edu.* **2014**, *34*, 55–60.
114. Hedge, P.; Kriwoken, L.K.; Patten, K. A Review of *Spartina* Management in Washington State, US. *J. Aquat. Plant Manag.* **2003**, *41*, 82–90
115. Grevstad, F.S.; Strong, D.R.; Garcia-Rossi, D.; Switzer, R.W.; Wecker, M.S. Biological control of *Spartina alterniflora* in Willapa Bay, Washington using the planthopper *Prokelisia marginata*: Agent specificity and early results. *Biol. Control* **2003**, *27*, 32–42. [CrossRef]
116. Wu, M.; Hacker, S.; Ayres, D.; Strong, D.R. Potential of *Prokelisia* spp. as biological control agents of English cordgrass, *Spartina anglica*. *Biol. Control* **1999**, *16*, 267–273. [CrossRef]
117. Liu, J.; Zhuang, Z.; Cai, X.; Xu, Y.; Lu, H. Studies on the influence of Micaojing on the plankton in coastal waters. *China Environ. Sci.* **2004**, *6*, 83–85. (In Chinese) [CrossRef]
118. Zhou, T.; Liu, S.; Feng, Z.; Liu, G.; Gan, Q.; Peng, S. Use of exotic plants to control *Spartina alterniflora* invasion and promote mangrove restoration. *Sci. Rep.* **2015**, *5*, 12980. [CrossRef] [PubMed]
119. Zhao, X.; Li, J.; Liu, X.; Gong, L.; Zhao, C. Combined effects of mowing and shading on growth and survival of *Spartina alterniflora*. *Guihaia* **2017**, *37*, 303–307. (In Chinese) [CrossRef]
120. Gu, X.; Liao, B.; Zhu, N.; Guan, W. Effect of sun-shade on the growth of *Spartina alterniflora*. *For. Pest Dis.* **2010**, *29*, 34–36. (In Chinese)
121. Wang, Z.; Zhang, Y.; Pan, X.; Ma, Z.; Chen, J.; Li, B. Effects of winter burning and cutting on aboveground growth and reproduction of *Spartina alterniflora*: A field experiment at Chongming Dongtan, Shanghai. *Biodivers. Sci.* **2006**, *14*, 275–283. (In Chinese) [CrossRef]
122. Chen, Z.; Wang, G.; Liu, J.; Xu, W.; Ren, L. Effects of waterlogging regulation on growth of *Spartina Alterniflora*. *Res. Environ. Sci.* **2011**, *24*, 1003–1007. (In Chinese) [CrossRef]
123. Huang, Q.; Zhao, F. Advances in remote sensing monitoring of lake eutrophication. *Guizhou Sci.* **2017**, *35*, 39–45. (In Chinese)
124. Li, X.; Zhao, C. Application and development of Lidar to detect the vertical distribution of marine materials. *Infrared Laser Eng.* **2020**, *49*, 24–32.
125. Ye, H.; Liao, X.; He, X.; Yue, H. Remote sensing monitoring and variation analysis of marine ecological environment in coastal waters of SriLanka. *J. Geo-Inf. Sci.* **2020**, *22*, 1463–1475. (In Chinese) [CrossRef]
126. Fielding, A.H.; Bell, J.F. A review of methods for the assessment of prediction errors in conservation presence/absence models. *Environ. Conserv.* **1997**, *24*, 38–49. [CrossRef]
127. Manel, S.; Williams, H.C.; Ormerod, S.J. Evaluating presence-absence models in ecology: The need to account for prevalence. *J. Appl. Ecol.* **2001**, *38*, 921–931. [CrossRef]
128. Allouche, O.; Tsoar, A.; Kadmon, R. Assessing the accuracy of species distribution models: Prevalence, kappa and the true skill statistic (TSS). *J. Appl. Ecol.* **2006**, *43*, 1223–1232. [CrossRef]

Article

New Data on Native and Alien Vascular Flora of Sicily (Italy): New Findings and Updates

Salvatore Cambria [1,*], Dario Azzaro [1], Orazio Caldarella [2], Michele Aleo [3], Giuseppe Bazan [4], Riccardo Guarino [4], Giancarlo Torre [5], Antonia Egidia Cristaudo [1], Vincenzo Ilardi [4], Alfonso La Rosa [6], Valentina Lucia Astrid Laface [7], Fabio Luchino [8], Francesco Mascia [9], Pietro Minissale [1,*], Saverio Sciandrello [1], Luca Tosetto [10] and Gianmarco Tavilla [1]

1   Department of Biological, Geological and Environmental Sciences, University of Catania, Via A. Longo 19, 95125 Catania, Italy; azzaro.dario@gmail.com (D.A.); acristau@unict.it (A.E.C.); s.sciandrello@unict.it (S.S.); gianmarco.tavilla@phd.unict.it (G.T.)
2   Independent Researcher, Via Maria SS. Mediatrice 38, 90129 Palermo, Italy; orazio.caldarella@gmail.com
3   Independent Researcher, Via S. Safina, 91100 Trapani, Italy; michele.aleo@libero.it
4   Department of Biological, Chemical and Pharmaceutical Sciences and Technologies (STEBICEF), University of Palermo, 90123 Palermo, Italy; giuseppe.bazan@unipa.it (G.B.); riccardo.guarino@unipa.it (R.G.); vincenzo.ilardi@unipa.it (V.I.)
5   Stiftung Pro Artenvielfalt®, Meisenstraße 65, 33607 Bielefeld, Germany; giancarlotorre@hotmail.it
6   Cooperativa Silene, Via V. D'Ondes Reggio 8/a, 90127 Palermo, Italy; alfonsolarosa@libero.it
7   Department AGRARIA, Mediterranean University of Reggio Calabria, 89122 Reggio Calabria, Italy; vla.laface@unirc.it
8   Independent Researcher, Via Torrente Allume, 6/A, 98027 Roccalumera (ME), Italy; fabio.luchino@tin.it
9   Independent Researcher, Via Vittorio Emanuele III 41, 09020 Villanovaforru (SU), Italy; fr.maxia@gmail.com
10   Independent Researcher, Via Pegorina 548, 35040 Casale di Scodosia (PD), Italy; luca.tosetto89@gmail.com
*   Correspondence: cambria_salvatore@yahoo.it (S.C.); p.minissale@unict.it (P.M.)

Citation: Cambria, S.; Azzaro, D.; Caldarella, O.; Aleo, M.; Bazan, G.; Guarino, R.; Torre, G.; Cristaudo, A.E.; Ilardi, V.; La Rosa, A.; et al. New Data on Native and Alien Vascular Flora of Sicily (Italy): New Findings and Updates. *Plants* **2023**, *12*, 1743. https://doi.org/10.3390/plants12091743

Academic Editors: Congyan Wang and Hongli Li

Received: 16 March 2023
Revised: 19 April 2023
Accepted: 20 April 2023
Published: 23 April 2023

Abstract: In this paper, based on fieldwork and herbaria surveys, new data concerning the presence of 32 native and alien vascular species for Sicily (Italy) are provided. Among the native species, the occurrence of the following *taxa* is reported for the first time or confirmed after many decades of non-observation: *Aira multiculmis*, *Arum maculatum*, *Carex flacca* subsp. *flacca*, *Mentha longifolia*, *Oxybasis chenopodioides*, *Najas minor* and *Xiphion junceum*. Furthermore, we document the presence of three native species (*Cornus mas*, *Juncus foliosus* and *Limonium avei*) that, despite being repeatedly observed in Sicily and reported in the literature, are inexplicably omitted by the most recent authoritative checklists regarding the flora of Italy. Finally, fifteen alien species new to Sicily (including one new to Europe, i.e., *Pyrus betulifolia*) are reported and seven poorly documented allochthonous taxa are confirmed for the island, and for two of them, a status change is proposed. These new or confirmed records allow us to better define the European and national distribution of the targeted *taxa* and offer new insights on the native and alien flora of Sicily.

Keywords: mediterranean flora; biodiversity records; distribution range; exotic species; floristic records; invasive plants; regional flora; species occurrence; biological invasions; alien species management

1. Introduction

During floristic and vegetational research carried out by the authors throughout the whole territory of Sicily, some taxa that were previously never reported or whose presence in Sicily was doubtful were recorded. After the latest checklists of the Sicilian [1,2] and Italian native and alien vascular flora [3–5], several new contributions have been published for Sicilian flora. Some of these refer to newly described endemic species [6–22], while others confirm or record for the first time species native to the island [23–32]. However, the large majority of these contributions consist of reports of exotic species [33–58]. In fact, as highlighted by Guarino et al. [59], Sicily, due to its long-lasting human history and central location in the Mediterranean basin, is still today a territory strongly subject to an intense and frequent introduction of exotic species for various purposes or accidentally. At the same time, although the native flora of Sicily is quite well known and its territory is one of the best-known floristic areas among Italian and Mediterranean regions [60], not all of the island has been investigated with the same level of effort, particularly in the case of the inland areas of central Sicily. Therefore, an extensive exploration of these less studied areas yielded the discovery or recent confirmation of some native *taxa* already recorded in other Italian regions or countries. Moreover, the examination of environments that are still less investigated, such as the numerous artificial basins and tanks found in Sicily, can lead to the discovery of new, mostly alien taxa. A systematic exploration of these environments was recently carried out through the MedIsWet project [61]. At the same time, the study of the taxonomic position of some critical taxa or populations allows us to identify and describe new taxa. Additional floristic data are collected through the activity of amateur botanists and plant enthusiasts on social networks, in some cases acting as veritable aggregators for botanical resources [62–64]. In particular, fundamental support for floristic research is currently provided by the members of the Facebook group "Flora spontanea Siciliana", distributed throughout the regional territory.

The aim of this paper is to inform the scientific community about several unpublished new records, partly retrieved from social networks and validated by experts, related both to the native and alien flora of Sicily. In particular, 17 *taxa* are recorded for the first time in Sicily, while 13 other species, whose status was doubtful at the regional level, are here confirmed.

2. Results

The results bring to light new floristic data concerning 32 taxa for Sicily (Figure 1), of which 10 are native and 22 are non-native vascular species. Overall, from a chorological viewpoint, most species show an Asian and American distribution (31–25%, respectively), while the most represented life forms are therophytes (28%), phanerophytes (25%) and geophytes (22%). If we only consider alien species (22 sp.), the dominant life forms correspond to phanerophytes and therophytes (32–23%, respectively), while the most represented chorotypes are confirmed as Asian and American (45–36%, respectively) (Figure 2). Regarding the native species (10 sp.), the dominant life forms correspond to therophytes and geophytes (40–30%, respectively), while the most represented chorotype is Mediterranean and European (30%). Overall, the highest number of species recorded falls within the Peloritani (22%), Hyblaean (19%) and Agrigento (16%) districts (Figure 3). The phytogeographic districts most affected by alien plant species are the Peloritani (27%) and Hyblaean and Etna (18%) districts, while affected by native species are the Nebrodi (28%) and Agrigento and Hyblaean (18%) districts. The natural habitats most affected by alien plants are uncultivated areas (45%) and wetlands (28%), with the latter affected by the following phytosociological classes: *Nerio-Tamaricetea africanae, Phragmito-Magnocaricetea, Potametea pectinati, Saginetea maritimae, Isoeto-Nanojuncetea* and *Salici-Populetea nigrae*.

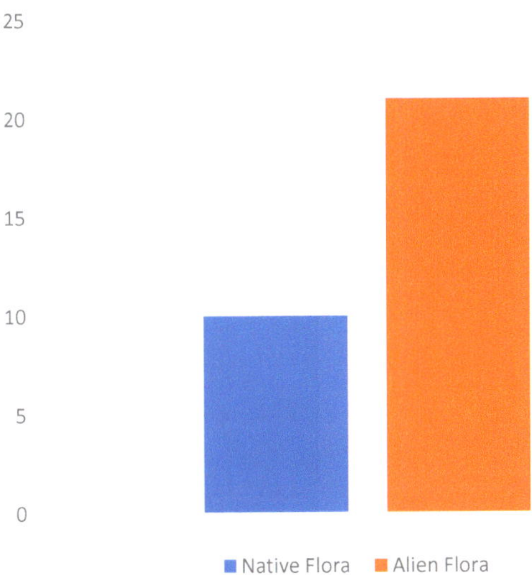

Figure 1. Percentage of native and alien species reported in the paper.

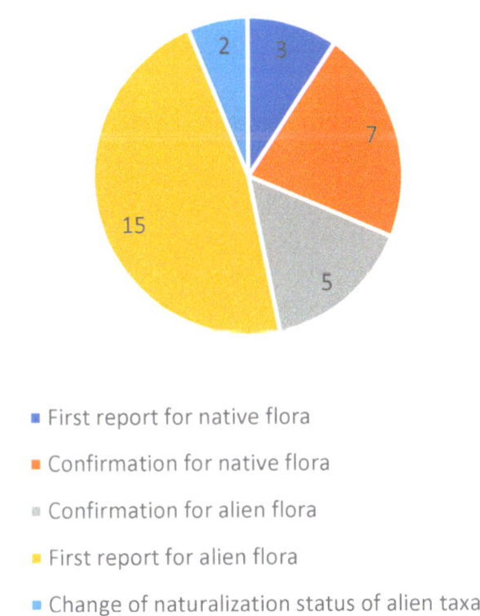

Figure 2. Typology of species recorded in this contribution.

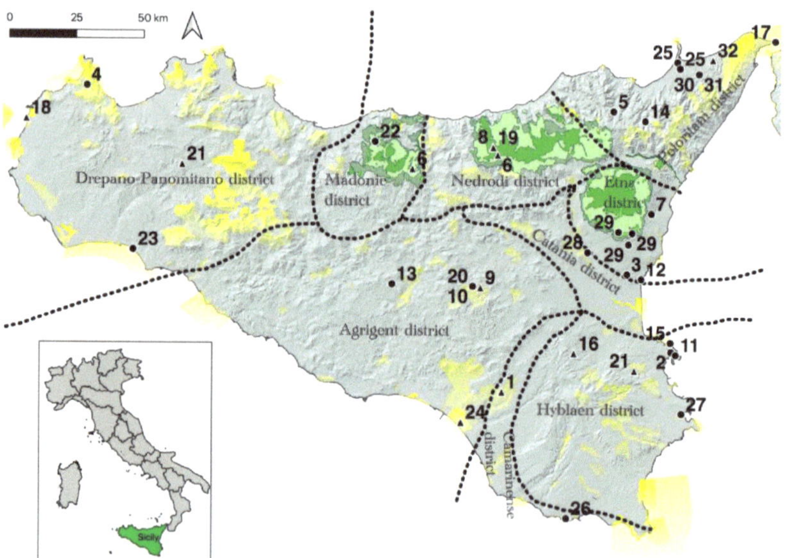

Figure 3. Distribution map of the new records. Black dot, alien species; black triangle, native species; yellow polygons, Natura 2000 areas; green polygons, park boundaries. 1. *Aira multiculmis*. 2. *Allium tuberosum*. 3. *Albuca canadensis*. 4. *Ambrosia artemisifolia*. 5. *Amorpha fruticosa*. 6. *Arum maculatum*. 7. *Bergenia crassifolia*. 8. *Carex flacca* subsp. *flacca*. 9. *Cornus mas*. 10. *Cydonia oblonga*. 11. *Fagopyrum esculentum*. 12. *Grevillea robusta*. 13. *Honorius nutans*. 14. *Hylotelephium spectabile*. 15. *Impatiens balsamina*. 16. *Juncus foliosus*. 17. *Kalanchoe laxiflora*. 18. *Limonium avei*. 19. *Mentha longifolia*. 20. *Morus alba*. 21. *Najas minor*. 22. *Nymphaea x marliacea*. 23. *Oenothera lindheimeri*. 24. *Oxybasis chenopodioides*. 25. *Passiflora morifolia*. 26. *Phacelia tanacetifolia*. 27. *Pontederia crassipes*. 28. *Pyrus betulifolia*. 29. *Secale cereale*. 30. *Solidago gigantea*. 31. *Symphyotrichum x salignum*. 32. *Xiphion junceum*.

Floristic Records

Aira multiculmis Dumort.
Poaceae—T (scap)—Subtrop.—Native.
Confirmation for Sicily
Exsiccatum. Bosco di Santo Pietro (Caltagirone), contrada Ogliastro, sandy soils, 200–250 m a.s.l., 18.04.2022, leg. and det. D. Azzaro, *s.n.* (CAT!).

This is a species related to *Aira caryophyllea* L., which, in Italy, is known for its presence in Tuscany, Campania, Calabria and Sardinia [4]. Its occurrence in Sicily had been mentioned by Lojacono Pojero [65], while more recently, it was excluded by Bartolucci et al. [4] or considered doubtful by Giardina et al. [1]. The species was found in a fairly significant population on the sandy soils occurring in the clearings of the cork oak forest of Santo Pietro near Caltagirone, within the Camarino-Pachinense district, but on the basis of current knowledge, it is very localized. From a phytosociological point of view, the species grows in annual acidophilous meadows referable to the class *Helianthemetea guttate* Rivas Goday & Rivas-Mart. 1963. In the same area, there are two other species of *Aira*, from which *A. multiculmis* is distinguished by certain morphological characteristics. In particular, *A. caryophyllea* L. is characterized by spikelets, generally > 2.6 mm, and by the inferior glume diverging from the peduncle by less than 80° and by a smaller number of culms. *A. cupaniana* Guss. differs instead with its decidedly rough sheaths, smaller spikelets, sharply enlarged peduncles at the apex and its 2–2.5 mm glumes [66]. The population falls within the Nature Reserve "Riserva Naturale Orientata Bosco di Santo Pietro" and the SAC "Bosco di Santo Pietro" (ITA070005).

Albuca canadensis (L.) F.M. Leight (Figure A1G)
Asparagaceae—G (rhiz)—S-Afric.—Alien.
Confirmation for Sicily and Italy
Exsiccatum. Tangenziale di Catania, vicino all'uscita per Misterbianco, roadsides, 100 m
a.s.l., 12.04.2022, leg. and det. S. Cambria, *s.n.* (CAT!).

This South African species was recorded for the first time in Italy by Nicolosi-Roncati [67]
(sub *Albuca altissima* Dryand.), who reported its naturalization from a ruderal stand inside
Catania. This author claims that the species was introduced from the local botanical
garden and from there it spread to other areas of the city. Afterwards, the species was no
longer recollected and was reported as dubious for Sicily and Italy by various authors [1,5].
We found a large population of *A. canadensis* growing on the roadside of the Catania
ring road, consisting of almost 100 individuals. It is often found in dry grasslands with
Hyparrhenia hirta (L.) Stapf., *Isatis tinctoria* L., *Ferula communis* L. and other typical species
of the *Lygeo sparti-Stipetea tenacissimae* Rivas-Martínez 1978 class. Here, the presence of this
taxon is confirmed for Italy and Sicily. We can state that the species does not represent a
threat to the natural habitat 6220* "Pseudo-steppe with grasses and annuals of the *Thero-
Brachypodietea*". The population is circumscribed and does not show an increasing trend to
expand into natural environments.

Allium tuberosum Rottler ex Spreng.
Alliaceae—G (bulb)—Asia—Alien.
New casual alien species for Sicily
Observatum. Roadsides, Augusta (Syracuse), 10 m a.s.l., 09.10.2020, obs. R. Romano, det.
F. Luchino.

A small population of this species was found in the suburb pavements of Augusta (SE
Sicily), probably having escaped from nearby gardens. This species was already known in
several regions in Northern and Central Italy [5].

Ambrosia artemisifolia L.
Asteraceae—T (scap)—N. Amer.—Alien.
New alien species for Sicily
Observatum. Castelluzzo, San Vito Lo Capo (Trapani), 150 m a.s.l., 15.07.2020, obs. S.
Montoleone, det. R. Guarino.

This North American species was observed in very anthropized environments near
San Vito Lo Capo (western Sicily) Although it is still very rare in Sicily, its potential for
further spread should be monitored, as it also represents one of the main causes of seasonal
respiratory allergy in many European countries and in some areas of northern Italy [68].
In Italy, the species is known in most regions, excluding Sardinia, Umbria, Campania and
Basilicata [5].

Amorpha fruticosa L.
Fabaceae—P (caesp)—N. Amer.—Alien.
New alien species for Sicily
Observatum. Valle del Timeto, San Piero Patti-Librizzi, 450 m a.s.l., 23.05.2013, obs. and det.
G. Torre.

This species was already known in all Italian regions as invasive or naturalized,
except for Sicily [5]. Our finding extends its presence to Sicily, where a fairly significant
population was surveyed in the north-eastern part of the island, between Librizzi and
San Pietro Patti (Messina) along the edge of a stream among the riparian vegetation of
Salici purpureae-Populetea nigrae Rivas-Martínez & Cantó ex Rivas-Martínez, Báscones, T.E.
Díaz, Fernández-González & Loidi 2001. The plant community in which *A. fruticosa* grows
is referable to habitat 92A0 "Salix alba and Populus alba galleries".

Arum maculatum L. (Figure A1A)
Araceae—G (rhiz)—Europ.—Native.
New record for Sicily

Exsiccatum. Piano Caterineci (Madonie, Palermo), orophilous vegetation with *Juniperus communis* subsp. *hemisphaerica*, 1590 m a.s.l., 31.05.2016, leg. V. Ilardi and S. Cambria, det. S. Brullo, *s.n.* (CAT!).
Observatum. Bosco della Tassita (Nebrodi, Caronia), Mesophilous woods, 1600 m a.s.l., 23.05.2013, obs. and det. A. La Rosa.

According to Bartolucci et al. [4], *Arum maculatum* occurs in almost all Italian regions, while it is absent in Sicily and doubtful for Sardinia. Giardina et al. [1] report the species as doubtful in Sicily, without providing any precise location. It is possible that in the past the species was confused with the more common *Arum cylindraceum* Gasparr., occurring in nearby localities. However, the two species are distinguished quite easily by the morphology of the underground organs, as *A. maculatum* has an ovoid tuberous rhizome, longer than broad and horizontally developing, while *A. cylindraceum* is characterized by a circular "bulb", from the center of which the scape develops [69]. Pignatti [70] and Pignatti et al. [71–74] reported a peculiar form of *A. maculatum* for Sicily, named *nigro-maculatum* Fiori and indicated it for Caltanissetta. However, according to Sortino [75], it falls within the intraspecific variability of *Arum italicum*. Even the bioclimatic features of the aforementioned locality seem to be favorable to this arrangement, since *A. italicum* is widespread in Sicily at low elevations, while *A. maculatum* was observed by us only within the supra-Mediterranean bioclimatic belt of the Madonie and Nebrodi mountains in the mesophilous beech forests belonging to *Querco roboris-Fagetea sylvaticae* Br.-Bl. & Vlieger in Vlieger 1937 or orophilous shrublands dominated by *Juniperus communis* L. subsp. *hemisphaerica* (J. Presl & C. Presl) Nyman (*Pino sylvestris-Juniperetea sabinae* Rivas-Martínez 1965 class). The two populations fall within protected areas, such as the Madonie regional park and the Nebrodi regional park, as well as in the SAC Pizzo Fau, Monte Pomiere, Pizzo Bidi and Serra della Testa (ITA030014) and Parco delle Madonie (ITA020050).

Bergenia crassifolia (L.) Fritsch. (Figure A2F)
Saxifragaceae—G (rhiz)—Asia—Alien.
New alien species for Sicily
Observatum. Monacella di Santa Venerina (Catania), 450 m a.s.l., 01.02.2022, obs. M. Pappalardo, det. O. Caldarella.

A few individuals of this ornamental species, quite frequent in the gardens of Sicily, were found as casual aliens near the small village of Monacella (Catania), on the eastern flank of Mt. Etna. Previously in Italy, it was reported only in some northern regions [5].

Carex flacca Schreb. subsp. ***flacca*** (Figure A1E)
Cyperaceae—G (rhiz)—Europ.—Native.
New record for Sicily
Exsiccatum. Serra della Testa, hygrophilous vegetation with *Thelypteris palustris*, 1050 m a.s.l., 17.06.2021, leg. and det. S. Sciandrello, P. Minissale, D. Azzaro, S. Cambria, *s.n.* (CAT!).

Carex flacca subsp. *erythrostachys* (Hoppe) Holub is a common species in Sicily; however, as regards the subsp. *flacca*, there are only a few doubtful records mentioned by Lojacono Pojero [65] for various mountain localities and, more recently, by Raimondo [76] for Piano Battaglia (Madonie). However, the species is considered doubtful by Giardina et al. [1] and absent in Sicily by Bartolucci et al. [4]. Our field investigations allowed us to find a new stand in the regional park of Nebrodi mountains, where it grows in a peculiar hygrophilous vegetation dominated by *Thelypteris palustris* Schott. (*Thelypterido palustris-Caricetum paniculatae*, included in the *Caricion gracilis* alliance), recently surveyed by Sciandrello et al. [29].

Cornus mas L.
Cornaceae—P (caesp/scap)—Eurimedit.—Native.
Confirmation for Sicily
Exsiccatum. Valley below Monte Canalotto, Piazza Armerina, riparian woods, 750 m a.s.l., 22.10.2021, leg. D. Azzaro, S. Cambria and det. S. Cambria, G. Tavilla, *s.n.* (CAT!).

The presence of *Cornus mas* in Sicily had already been reported by Giardina et al. [1] in certain localities in central Sicily and in some narrow valleys of the Erei mountains. Similarly, it is reported as "rare and probably only introduced (in reforestations)" in Sicily by Pignatti et al. [71–74]. Despite this, Bartolucci et al. [4] do not report it for Sicily. Our finding testifies to its presence in the previously mentioned area, where it often grows in association with *Cornus sanguinea* L., though much less frequently. From a phytosociological point of view, it grows in the riparian forest of *Salici purpureae-Populetea nigrae* Rivas-Martínez & Cantó ex Rivas-Martínez, Báscones, T.E. Díaz, Fernández-González & Loidi 2001. The surveyed population falls within the SAC "Vallone Rossomanno" (ITA060010) and in the regional Nature Reserve "Riserva Rossomanno-Grottascura-Bellia".

Cydonia oblonga Mill.
Rosaceae—P (scap)—W Asia—Alien.
Confirmation for Sicily as a casual alien
Exsiccatum. Valley below Monte Canalotto, Piazza Armerina, riparian woods, 750 m a.s.l., 22.10.2021, leg. D. Azzaro, S. Cambria and det. S. Cambria, G. Tavilla, *s.n.* (CAT!).

The naturalization of this species in Sicily had already been reported by Terpò [77], Giardina et al. [1] and Pignatti et al. [72]. However, it is not reported in the recent checklists regarding the Italian flora [4,5]. Our findings confirm the presence of this species as a casual alien in Sicily. In fact, a few young trees, most likely seed-generated, have been found within a riparian wood of *Salici purpureae-Populetea nigrae* class (referable to habitat 92A0 "Salix alba and Populus alba galleries"), protected in the regional Nature Reserve "Riserva Rossomanno-Grottascura-Bellia" and in the SAC "Vallone Rossomanno" (ITA060010). However, the few *C. oblonga* individuals do not currently represent a real threat to this habitat.

Fagopyrum esculentum Moench
Polygonaceae—T (scap)—Central Asia—Alien.
Confirmation for Sicily
Observatum. Augusta, ex Saline Regina (Siracusa), Roadsides, 10 m a.s.l., 15.10.2021, obs. and det. R. Romano, A. La Rosa.

Fagopyrum esculentum is not reported for Sicily in any recent work concerning the vascular flora of Sicily [1,4,5,71–74] and historical records regarding its presence are lacking. However, it was recorded in a list of alien species in Sicily by Raimondo et al. [78], without specifying any precise location, as a cultivated and rarely naturalized species. Our findings near Augusta (E Sicily) confirm this species as a casual alien in Sicily. The site falls within the SAC "Saline di Augusta" (ITA090014).

Grevillea robusta A. Cunn. ex R. Br.
Proteaceae—P (scap)—Australia—Alien.
New casual alien species for Sicily
Observatum. Catania, 10 m a.s.l., 24.10.2020, obs. and det. F. Gambilonghi.

This Australian species, commonly used for road tree planting over the last three decades, was found on the sidewalks and retaining walls of Catania. In Italy, it was already recorded as a casual alien in Lazio and Campania [5].

Honorius nutans (L.) Gray (Figure A1C)
Asparagaceae—G (bulb)—W Asia—Alien.
Confirmation for Sicily
Observatum. C.da Serra Pantano, Caltanissetta, olive groves, 470 m a.s.l., 02.04.2018, obs. and det. A. Cristaudo.

The naturalization of this taxon in Sicily was observed only once by Lojacono Pojero [65] near Enna (Central Sicily). The new location was situated on the outskirts of Caltanissetta, where a small population of the species colonized part of an olive grove. The agricultural practices carried out in recent years have probably led to the disappearance of this population, but its presence in other places in the nearby areas cannot be excluded.

Hylotelephium spectabile (Boreau) H. Ohba
Crassulaceae—H (scap)—Eurosiber.—Alien.
New casual alien species for Sicily
Observatum. Oak wood near San Basilio, Novara di Sicilia (Messina) 670 m a.s.l., 08.04.2018, obs. A. Ferrara Currò, det. O. Caldarella.

This species, commonly cultivated as an ornamental plant, was found in a deciduous oak wood (*Querco-Fagetea* class) in the territory of Novara di Sicilia (NW Sicily). In Italy it was already observed as a casual alien in Lombardy, Marche and Umbria [5]. According to the Directive 43/92/EEC, this species has been found in the habitat 91AA* "Eastern white oak woods".

Impatiens balsamina L. (Figure A2H)
Balsaminaceae—T (scap)—Asia—Alien.
New record for Sicily
Observatum. C.da Campolato, Augusta (Syracuse), dry grasslands, 10 m a.s.l., 02.10.2018, obs. C. Arcidiacono, det. O. Caldarella.

This species, often cultivated in gardens as a summer annual, was observed in a coastal grassland dominated by *Hyparrhenia hirta* (L.) Stapf (*Lygeo-Stipetea* class) on calcarenitic substrates near Augusta (SE Sicily). It was reported as a casual alien by Galasso et al. [5] for most of the northern regions as well as Lazio and Puglia. The natural habitat affected by this species is the 6220* "Pseudo-steppe with grasses and annuals of the Thero-Brachypodietea".

Juncus foliosus Desf.
Juncaceae—T (scap)—Stenomedit.—Native.
Confirmation for Sicily
Exsiccatum. Oxena river, near Militello Val di Catania (eastern Sicily), annual hygrophilous vegetation, 310 m a.s.l., 04.07.2021, leg. and det. D. Azzaro, *s.n.* (CAT!).

Bartolucci et al. [4] and Giardina et al. [1] report this species as doubtful for Sicily, while the records by Zodda [79], Lojacono Pojero [65], Barbagallo and Furnari [80] and Raimondo et al. [78] are considered questionable. Recently, Lastrucci et al. [81] considered it to no longer be present in Sicily. However, this species was recorded in several localities in Sicily by Pignatti et al. [71–74] and in the phytosociological relevés by Sciandrello et al. [82] from south-eastern Sicily. Our findings from Oxena river, near Catania, confirm the occurrence of this species in Sicily, where it is linked to the thero-hygrophilous communities belonging to *Isoëto-Nanojuncetea* Br.-Bl. & Tüxen ex Westhoff, Dijk & Passchier 1946 and *Saginetea maritimae* Westhoff, Leeuwen & Adriani 1962.

Kalanchoe laxiflora Baker
Crassulaceae—T (scap)—S.-Afric., Madagascar—Alien.
Confirmation as casual alien for Sicily
Observatum. Via Fortino, Torre Faro, Messina, 10 m a.s.l., 23.10.2022, obs. and det. V.L.A. Laface.

The occurrence of this species as a casual neophyte in Europe and Italy has recently been confirmed by Stinca et al. [83] and Spampinato et al. [84], respectively, for Basilicata and Calabria. Previously, Galasso et al. [5] reported *K. laxiflora* doubtfully for Toscana and Sicily. Our findings confirm the presence of the species in this region, where it is occasionally cultivated. In fact, the observed population, consisting of a few individuals that probably escaped from nearby gardens, colonize wall cracks near the road, within the SPA "Monti Peloritani, Dorsale Curcuraci, Antennamare e area marina dello Stretto di Messina" (ITA030042).

Limonium avei (De Not.) Brullo & Erben (Figure A2C)
Plumbaginaceae—T (scap)—W-Medit.—Native.
Confirmation for Sicily
Exsiccatum. Saline di Nubia presso Paceco (Trapani), 10 m a.s.l., 30.04.2022, leg. and det. M. Aleo (CAT!).

Observatum. Isola Grande dello Stagnone (Marsala), wet clay soils, 5 m a.s.l., 02.03.2022, obs. and det. S. Cambria and G. Tavilla.

The presence of this species has not been recorded in Sicily by Bartolucci et al. [4], despite many authors [1,85–90] repeatedly recording several stands in western Sicily and Lampedusa. Here, we confirm the occurrence of the species in the coastal brackish environments between Trapani and Marsala (western Sicily) and also in the Isola Grande dello Stagnone, inside the regional Nature Reserve "Riserva Saline di Trapani e Paceco" and the SPA "Stagnone di Marsala e Saline di Trapani" (ITA010028), where it is widespread. This annual halophyte prefers soils that has a high salt concentration and is rich in nitrates and grows in markedly xeric environmental conditions, together with several species of the *Limonion avei* alliance (*Saginetea maritimae*).

Mentha longifolia (L.) L. (Figure A2A)
Lamiaceae—H (scap)—Paleotemp.—Native.
New record for Sicily
Exsiccatum. Serra della Testa (Nebrodi), hygrophilous vegetation, 1000 m a.s.l., 17.06.2021, leg. and det. S. Sciandrello, P. Minissale, D. Azzaro, S. Cambria, s.n. (CAT!).

The species is not mentioned by Giardina et al. [1], while it is recorded from Sicily by Conti et al. [3]. Bartolucci et al. [4] reported the reference for Sicily as an incorrect record. Our field surveys have resulted in the discovery of the species in the Nebrodi area, where the species is quite widely distributed and overlooked, perhaps due to the confusion with *M. spicata* L., from which it is mainly distinguished by the presence of only simple straight hairs and lanceolate leaves, with maximum width towards the median part [71–74]. The species is linked to the *Phragmito australis-Magnocaricetea elatae* Klika in Klika & Novák 1941 communities and the surveyed population falls within the Nebrodi Regional Park.

Morus alba L.
Moraceae—P (scap)—E Asia—Alien.
New record for Sicily as casual alien
Exsiccatum. Valley below Monte Canalotto, Piazza Armerina, riparian woods, 750 m a.s.l., 22.10.2021, leg. D. Azzaro, S. Cambria and det. S. Cambria, G. Tavilla, *s.n.* (CAT!).

Despite being commonly cultivated, this species has never been ascribed to the allochthonous flora of Sicily. Based on the observation of some young individuals in the riparian vegetation dominated by *Populus nigra* and *P. alba* (*Salici purpureae-Populetea nigrae* class, habitat 92A0 "Salix alba and Populus alba galleries") colonizing a narrow valley in central Sicily, we report for the first time *Morus alba* as a casual alien for the Sicilian flora. This location is protected in the regional Nature Reserve "Riserva Rossomanno-Grottascura-Bellia" and in the SAC "Vallone Rossomanno" (ITA060010).

Najas minor All. (Figure A1B)
Hydrocharitaceae—I (rad)—Paleotemp.—Native.
New record for Sicily
Exsiccatum. Artificial basin of Cozzo Ogliastri, 381 m a.s.l., 30.07.2021, leg. and det. L. Tosetto, F. Luchino, S. Cambria, s.n. (CAT!).
Observatum. Artificial basin near Roccamena, 470 m a.s.l., 02.04.2018, obs. and det. S. Costa, O. Caldarella and A. La Rosa.

Conspicuous populations of this species were found in two artificial ponds near Roccamena (Palermo) and in a small water reservoir near Melilli (Syracuse). It constitutes a dense, monophytic aquatic vegetation with a summer optimum. Considering that the species has never been reported for Sicily and its presence seems so far restricted to man-made habitats, it is possible to hypothesize its relatively recent arrival on the island. According to Volk [91], the spread of this species is related to water birds and, in particular, to ducks, as the seeds of *Najas minor* partly retain their vitality after digestion and can germinate in polluted waters. In Italy, *Najas minor* is recorded only in northern and central regions, as well as in Sardinia [4].

Nymphaea x marliacea Lat.-Marl. (Figure A1D)
Nymphaeaceae—I (rad)—horticultural origin—Alien.
New record for Sicily as casual alien
Exsiccatum. Artificial pond in Contrada Cella di Fico near Isnello (Palermo), 742 m a.s.l., 14.08.2019, leg. and det. S. Cambria, s.n. (CAT!).

This horticultural hybrid was observed in an artificial pond, built as a water reservoir for the irrigation of nearby agricultural lands in the 1970s and today is in disuse, located within private property occupied by a livestock farm. Over the years, the basin became naturalized and has been colonized by various aquatic and hygrophilous species, such as *Potamogeton pusillus, Scirpoides holoschoenus, Juncus inflexus, Juncus effusus* and *Epilobium hirsutum*, characterizing an aquatic community of *Potametea pectinati* Klika in the Klika & Novák 1941 class. The origin of this *Nymphaea* population is somewhat uncertain, since the site has been abandoned for many years and the distance from the inhabited centers and gardens is quite significant. In any case, the taxon at issue belongs to a complex of horticultural hybrids, whose putative parental species are *N. alba* L., *N. mexicana* Zucc. and *N. odorata* Aiton [92]. Based on multiple years of observation, this population is spontaneously expanding within the basin. In Italy, *N. x marliacea* has been so far reported as a casual alien only in northern regions, such as Lombardia and Trentino-Alto Adige [5,93]. It must be highlighted that the discovered population is located within the Madonie regional park and the SPA "Parco delle Madonie" (Parco delle Madonie).

Oenothera lindheimeri (Engelm. & A. Gray) W.L. Wagner & Hoch (Figure A2G)
Balsaminaceae—H (bienn)—N. Amer.—Alien.
Change of status from casual alien to naturalized
Observatum. Lido Fiori beach near Menfi (Agrigento), 10.08.2022, obs. and det. O. Caldarella.

Oenothera lindheimeri was reported for the first time in Sicily as a casual species by Galasso et al. [39]. Following the observations carried out in the last 5 years on the consistent population occurring near the beach of Lido Fiori near Menfi (Agrigento), the change of status from casual to naturalized is here proposed. Therefore, the species represents a serious threat to the local psammophilous vegetation, referable to the *Euphorbio paraliae-Ammophiletea australis* Géhu & Rivas-Martínez in Rivas-Martínez, Asensi, Díaz-Garretas, Molero, Valle, Cano, Costa & Díaz 2011 (habitat 2120 "Shifting dunes along the shoreline with *Ammophila arenaria* (white dunes)").

Oxybasis chenopodioides (L.) S. Fuentes, Uotila & Borsch.
Chenopodiaceae– T (scap)—Subcosmop.—Native.
Confirmation for Sicily.
Exsiccatum. Biviere di Gela (southern Sicily), lakeshores, 10 m a.s.l., 07.08.2002, leg. and det. S. Sciandrello and S. Brullo, s.n. (CAT!).

This species was reported for the first time in Sicily (sub *Chenopodium botryoides* Sm.) by Brullo and Sciandrello [94] for the lake "Biviere di Gela" (southern Sicily), within the nature reserve Biviere di Gela and the SPA "Torre Manfria, Biviere e Piana di Gela" (ITA050012). However, this record was ignored by Giardina et al. [1], Raimondo et al. [2,95] and Bartolucci et al. [4]. Iamonico [96], while not citing this report for the Biviere di Gela, attributed an herbarium specimen collected in the 19th century near Messina to *Oxybasis chenopodioides* but hypothesized its regional extinction due to the age of the collected specimen and the current dense urbanization of this locality. Therefore, based on our data, it is possible to confirm the current presence of this species in Sicily, where the species is linked to thero-hygrophilous vegetation with the summer–autumn development of *Nanocyperetalia flavescentis* Klika 1935.

Passiflora morifolia Mast. (Figure A2D)
Passifloraceae—P (lian)—S. Amer.– Alien.
New naturalized alien species for Sicily

Exsiccatum. Milazzo, presso l'Istituto Professionale Agrario (IPSAA), 10 m a.s.l., 03.07.2010, leg. and det. A. Cristaudo, *s.n.* (CAT!).
Observatum. Piana di Milazzo, nelle vicinanze dei Vivai Torre, 40 m a.s.l., 22.06.2022, obs. and det. G. Torre.

This species, native to tropical America, is cultivated in gardens for its ornamental qualities. Its ability to disseminate itself is well known among gardening enthusiasts and, in fact, the presence of the species as a casual alien has been already reported in Sardinia by Galasso et al. [36]. According to our observations, it is quite frequent in the irrigated land of the Milazzo plain (Messina), particularly near gardens and nurseries, and its presence has now been well established in the territory for more than 10 years.

***Phacelia tanacetifolia* Benth.**
Hydrophyllaceae—T (scap)—N. Amer.– Alien.
New casual alien species for Sicily
Observatum. Marina di Modica (Ragusa), 5 m a.s.l., 16.04.2020, obs. C. Spadaro, det. R. Guarino.

This North American species, adventitious in almost all regions of central-northern Italy, including Sardinia [5], is sometimes cultivated as an ornamental nectar plant and tends to escape cultivation in disturbed environments. It was found in uncultivated land near Marina di Modica (SE Sicily), where the species is, however, represented only by a small population.

***Pontederia crassipes* Mart. (Figure A2B)**
Pontederiaceae—I (rad)—S. Amer.—Alien.
Change of status from casual alien to invasive
Exsiccatum. Canalizzazioni del Fiume Ciane (Siracusa), 5 m a.s.l., 21.09.2006, leg. and det. R. Guarino, s.n. (PAL!).

This species was recorded for the first time in Sicily by Bartolo et al. [97], who observed it in a channel connecting the Pantano Gariffi and the sea in SE Sicily (Ispica, Ragusa province). Since then, *P. crassipes* has been repeatedly reported as a casual alien in Sicily [1,5,71–74]. Following the observations carried out in the last 16 years on the abundant population extant in the channels of the Ciane River (SE Sicily, Syracuse Province), the change of status from casual to invasive is here proposed. The surveyed site falls within the nature reserve "Fiume Ciane e Saline di Siracusa" and in the SAC "Saline di Siracusa e Fiume Ciane" (ITA090006). This species represents a significant threat to the habitat 3260 "Water courses of plain to montane levels with the *Ranunculion fluitantis* and *Callitricho-Batrachion* vegetation".

***Pyrus betulifolia* Bunge (Figure A1F)**
Rosaceae—P (caesp/scap)—E Asia—Alien.
New record for Sicily and Europe as a casual alien
Exsiccatum. Paternò (Catania), near Pietralunga bridge, scrublands, 89 m a.s.l., 15.10.2019, leg. and det. A. Cristaudo, F. Carruggio, *s.n.* (CAT!).

Pyrus betulifolia is a wild pear native to eastern Asia and, in particular, to north-central and south-east China, Laos, Manchuria and Tibet [98]. According to literature data, the occurrence of this species in the exotic flora of Europe and Italy had never been reported. Outside its original range, the species has been naturalized only in Illinois, United States of America [99]. Moreover, *P. betulifolia* is listed in the Invasive Plant Atlas of the US, but it does not appear to be a species with significant invasive potential [100]. According to Swearingen et al. [101], seeds can be dispersed from planted trees via birds. The surveyed population was small, consisting of only three shrubs that grow within a long-abandoned agricultural area, subject to flooding during the winter and autumn periods. In general, *P. betulifolia* occurs within very scattered thermo-hygrophilous vegetation with *Tamarix africana* Poir. and *T. gallica* L. (*Nerio-Tamaricetea africanae* Br.-Bl. & O. Bolòs 1958 class, habitat 92D0 "Southern riparian galleries and thickets (*Nerio-Tamaricetea* and *Securinegion tinctoriae*)"), which occurs also in the nearby areas next to the Simeto river. The origin of this population can almost certainly be related to the use of this species as rootstock for 'Coscia' pear (*Pyrus communis*), as reported by Stern et al. [102] for Israel, although the species is rarely offered in the catalogs of Italian nurseries. Currently, this type of cultivation has almost disappeared from this area, where citrus groves are predominant, while it is possible that in the past it was more widely practiced. The surveyed plants show different ages and bear fruit regularly. Further investigations are needed to understand the future dynamics of this small population.

Secale cereale L. subsp. ***cereale***
Poaceae—H bienn—Asiat.—Alien.
New record for Sicily as a casual alien
Observatum. Loc. Contrada San Leo, Belpasso, 1000 m a.s.l.; loc. Caldera, Contrada Simita e Tarderia, Pedara, 830–870 m a.s.l.; Contrada Mompilieri, Nicolosi, 640 m a.s.l. (Catania), 27.07.2021, obs. F. Mascia and N. Serafica, det. F. Mascia.

In Sicily, the traditional cultivation of *Secale cereale* has been reported since the end of the XVIII century [103], mainly in the Aeolian islands and the mountain (Etna and Nebrodi) areas of the Catania and Enna provinces. This crop, in particular, a local variety called *Irmana*, had a substantial increase during the Second World War and was later almost completely abandoned [104] and was recently considered probably extinct [1]. In the last decade, the impulse to recover native cereal crops has led to the rediscovery of rye and its renewed cultivation, especially in the province of Catania [105]. In Italy, it is reported as a casual alien for all regions except Molise, Sicily and Val d'Aosta [3,106]. In Sicily, it was found along field edges and roadsides and secondarily as an occasional weed in *Lupinus albus* L. and *Vicia faba* L. fields.

Solidago gigantea Aiton
Asteraceae—H (scap)—N. Amer.—Alien.
New record for Sicily as a casual alien
Observatum. Pace del Mela (Messina), 100 m a.s.l., 2.10.2022, obs. and det. F. Berenato.

This North American species is reported as casual or naturalized in the regions of central-northern Italy, Abruzzo and Calabria [5]. In Sicily, it was observed in seasonal humid environments with vegetation referable to *Scrophulario-Helichrysetea italici* Brullo, Scelsi & Spampinato 1998 near Pace del Mela (Messina), colonizing the edges of ditches and canals. This species was observed within the habitat 3250 "Constantly flowing Mediterranean rivers with *Glaucium flavum*".

Symphyotrichum x salignum (Willd.) G.L. Nesom
Asteraceae—NP—N. Amer.—Alien.
New record for Sicily as a casual alien
Observatum. Pace del Mela (Messina), 100 m a.s.l., 2.10.2022, obs. and det. F. Berenato.

Symphyotrichum x salignum, probably a natural hybrid between *S. novi-belgii* and *S. lanceolatum*, was surveyed in disturbed aspects of *Scrophulario-Helichrysetea* vegetation (habitat 3250 "Constantly flowing Mediterranean rivers with *Glaucium flavum*") near Pace del Mela (Messina). In Italy, it was reported in most of the northern and central regions and Sardinia [5].

Xiphion junceum (Poir.) Parl. (Figure A2E)
Iridaceae—G (bulb)—W-Stenomedit.—Native.
Confirmation for Sicily
Observatum. Olive Groves, Spadafora (Messina), 200 m a.s.l., 14.04.2015, obs. A. Scoglio, det. A. La Rosa.

This species, native to Iberian Peninsula, Northern Africa and Sicily [98], was recorded for the first time in Sicily by Gussone [85] on the southern coast between Palma di Montechiaro and Licata and later confirmed in the same locality by Ponzo [107]. The species was also reported in Riesi [79,108] and Mascali [78]. We confirm the presence of the species on the island and report it for the first time in northern Sicily. In particular, the plant was found near Messina in an olive grove, where the species is represented by a small population.

3. Materials and Methods

3.1. Study Area

With a surface of 25.711 km^2, Sicily represents the largest island in the Mediterranean Sea, located immediately south of the Italian Peninsula [109]. Sicily's mountain ranges are mainly distributed along the northern sector of the island, namely the Madonie (reaching 1979 m a.s.l.), the Nebrodi (1847 m a.s.l.) and the Peloritani (1374 m a.s.l.). In the central and southern sectors, the landscape is mainly characterized by a typical low relief. The highest peak is the Etna volcano (3340 m). This considerable altitudinal heterogeneity encompasses several climate zones, from semi-arid to humid. Annual rainfall varies from 250 to 1400 mm, whereas the average temperature is 18 °C, with values below zero in the inland territory in winter and over 40 °C along the coast in summer. The smaller islands around Sicily are the Aeolian and the Aegadian archipelagos, as well as the Pelagie, Ustica and Pantelleria. According to the classification of Bazan et al. [109], Sicily is divided into the following 11 bioclimatic belts: 1. upper infra-Mediterranean (It 450–515, 0–30 m a.s.l.); 2. lower thermo-Mediterranean (It 400–450, 0–220 m a.s.l.; 3. upper thermo-Mediterranean (It 350–400, 0–450 m a.s.l.); 4. lower meso-Mediterranean (It 285–350, 250–700 m a.s.l.); 5. upper meso-Mediterranean (It 220–285, 620–1030 m a.s.l.); 6. lower supra-Mediterranean (It 150–220, 960–1400 m a.s.l.); 7. upper supra-Mediterranean (It 120–150, 1370–1550 m a.s.l.); 8. lower oro-Mediterranean (Tp 675–900, 1550–2000 m a.s.l.); 9. upper oro-Mediterranean (Tp 450–675, 2000–2400 m a.s.l.); 10. lower cryoro-Mediterranean (Tp 150–450, 2400–3000 m a.s.l.); and 11. upper cryoro-Mediterranean (Tp 1–150, above 3000 m a.s.l.).

According to the phytogeographic classification proposed by Brullo et al. [110] and subsequently modified [111], 14 well-defined districts can be identified within Sicily, distinguished by floristic and vegetational peculiarities, as well by remarkable geological, geomorphological and climatic features [109,112,113]. The research was carried out throughout all the districts of the island (Figure 3).

3.2. Data Sources

The floristic data are based on field investigations carried out from 2015 to 2022, herbaria surveys (CAT, PAL, FI, RO) and bibliographic analysis. The collected or examined plant materials are preserved mainly in CAT herbarium or private herbaria. Nomenclatures, taxonomic concepts and notes on the regional distribution are based mainly on the checklists published by Giardina et al. [1], Bartolucci et al. [4] and Galasso et al. [5], as well as on the online resources "FlorItaly—Portal to the Flora of Italy" (http://dryades.units.it/floritaly/index.php, accessed on 27 August 2022) and "Acta Plantarum" (https://www.actaplantarum.org, accessed on 27 August 2022). Syntaxonomic classification follows Mucina et al. [114].

Specimens were identified according to Pignatti et al. [71–74] or other monographic works cited in the notes for each species. In the results, *taxa* are arranged in alphabetical order. For each taxon, according to Pignatti et al. [71–74], the following data are reported: family, life form, native range, life form, chorology or origin, as well as the herbarium specimens and/or observations, phytosociological reference, phytogeographic district and occurrence in protected areas. The degree of naturalization was evaluated according to Galasso et al. [5]. Details on locality, habitat, altitude, date of collection or observation, names of collector/observer and identifier are also provided. Additionally, other data relating to the floristic record, any previous reports and other information on taxonomy, ecology and distribution are discussed in the notes.

4. Discussion and Conclusions

This paper presents new data concerning 32 taxa of native and non-native vascular species for Sicily (Figure 3). In particular, *Aira multiculmis*, *Arum maculatum* and *Najas minor* are recorded for the first time in the native flora of Sicily, while *Carex flacca* subsp. *flacca*, *Cornus mas*, *Juncus foliosus*, *Limonium avei*, *Mentha longifolia*, *Oxyasis chenopodioides* and *Xiphion junceum* are confirmed for the island, since their presence on the island, as indicated by some authors, was considered doubtful in the recent literature, mainly due to very old or vague reports. In regard to non-native species, the occurrence of *Albuca canadensis*, *Honorius nutans*, *Fagopyrum esculentum*, *Kalanchoe laxiflora* and *Cydonia oblonga* are confirmed after some decades. Finally, *Pontederia crassipes* is recorded as invasive and *Morus alba*, *Nymphaea x merlata*, *Allium tuberosum*, *Grevillea robusta*, *Ambrosia artemisifolia*, *Impatiens balsamina*, *Hylotelephium spectabile*, *Bergenia crassifolia*, *Secale cereale*, *Symphyotrichum x salignum*, *Solidago gigantea*, *Phacelia tanacetifolia*, *Passiflora morifolia*, *Pyrus betulifolia* and *Amorpha fruticosa* are recorded as casual aliens for the first time in the region. The new findings for the native flora confirm the remarkable floristic richness of the island, which, in spite of what can be hypothesized on the basis of the numerous floristic studies recently published on the Sicilian territory, is still worthy of further study. Furthermore, the results obtained in this research confirm the rapid increase in exotic species even in natural environments; thus, in some cases, the careful monitoring of species with greater invasive potential is advised. Particular attention must be paid to protected areas, such as the sites of the Natura 2000 network, mentioned several times here in the new records, where alien species can cause damage or alteration to the habitats of conservation importance [115].

Thus, having a continuous account of new records allows the planning of early preventive actions for their further diffusion, saving on future high management costs [116]. In conclusion, although attention is often focused on invasive alien species (IAS), causal alien species should not be neglected in the management of a territory, as they are quite difficult to control after their spread and change of status [117,118].

Author Contributions: Conceptualization, S.C.; methodology, S.C., P.M., S.S. and G.T. (Gianmarco Tavilla); investigation, S.C., D.A., O.C., M.A., G.B., R.G., G.T. (Giancarlo Torre), A.E.C., V.I., A.L.R., V.L.A.L., F.L., F.M., P.M., S.S., L.T. and G.T. (Gianmarco Tavilla); data curation, S.C., P.M., S.S. and G.T. (Gianmarco Tavilla); writing—original draft preparation, S.C., O.C., R.G., A.L.R., P.M., S.S. and G.T. (Gianmarco Tavilla); writing—review and editing, S.C., P.M., S.S. and G.T. (Gianmarco Tavilla). All authors have read and agreed to the published version of the manuscript.

Funding: This research was financially supported by the project INTERREG V-A ITALIA-MALTA 2014–2020 Axis III—Objective 3.1 FAST—Fight Alien Species Transborder CUP: E99C20000160005, and by collaboration with PIM in the frame of the project MedIsWet, funded by the MAVA Foundation.

Data Availability Statement: Not applicable.

Acknowledgments: We are grateful to many members of the Facebook group "Flora Spontanea Siciliana" for their observations of some species reported in this paper. In particular, we thank Rosario Romano, Francesco Gambilonghi, Antonino Scoglio, Salvatore Montoleone, Caterina Arcidiacono, Amedeo Ferrara Currò, Mauro Pappalardo, Fabio Berenato, Alessandra Cormaci and Carmelo Spadaro. A special thanks goes also to Rosalinda Nicoletti for giving us access to her property and to the farmer Nino Serafica for the fruitful exchange of experiences. Floristic surveys in wetlands were generously supported by the MedIsWet project, funded by MAVA Foundation.

Conflicts of Interest: The authors declare no conflict of interest.

Appendix A

Figure A1. (**A**) *Arum maculatum* (photo by Alfonso La Rosa). (**B**) *Najas minor* (photo by Fabio Luchino). (**C**) *Honorius nutans* (photo by Antonia Cristaudo). (**D**) *Nymphaea x marliacea* (photo by Salvatore Cambria). (**E**) *Carex flacca* subsp. *flacca* (photo by Salvatore Cambria). (**F**) *Pyrus betulifolia* (photo by Antonia Cristaudo). (**G**) *Albuca canadensis* (photo by Salvatore Cambria).

Figure A2. (**A**) *Mentha longifolia* (photo by Salvatore Cambria). (**B**) *Pontederia crassipes* (photo by Riccardo Guarino). (**C**) *Limonium avei* (photo by Michele Aleo). (**D**) *Passiflora morifolia* (photo by Antonia Cristaudo). (**E**) *Xiphion junceum* (photo by Antonino Scoglio). (**F**) *Bergenia crassifolia* (photo by Mauro Pappalardo). (**G**) *Oenothera lindheimeri* (photo by Orazio Caldarella). (**H**) *Impatiens balsamina* (photo by Caterina Arcidiacono).

References

1. Giardina, G.; Raimondo, F.M.; Spadaro, V. A catalogue of plants growing in Sicily. *Bocconea* **2007**, *20*, 5–582.
2. Raimondo, F.M.; Domina, G.; Spadaro, V. Checklist of the vascular flora of Sicily. *Quad. Bot. Amb. Appl.* **2010**, *21*, 189–252.
3. Conti, F.; Abbate, G.; Alessandrini, A.; Blasi, C. *An Annotated Checklist of the Italian Vascular Flora*; Palombi Editori: Roma, Italy, 2005.
4. Bartolucci, F.; Peruzzi, L.; Galasso, G.; Albano, A.; Alessandrini, A.; Ardenghi, N.M.G.; Astuti, G.; Bacchetta, G.; Ballelli, S.; Banfi, E.; et al. An updated checklist of the vascular flora native to Italy. *Plant Biosyst.* **2018**, *152*, 179–303. [CrossRef]
5. Galasso, G.; Conti, F.; Peruzzi, L.; Ardenghi, N.M.G.; Banfi, E.; Celesti-Grapow, L.; Albano, A.; Alessandrini, A.; Bacchetta, G.; Ballelli, S.; et al. An updated checklist of the vascular flora alien to Italy. *Plant Biosyst.* **2018**, *152*, 556–592. [CrossRef]

6. Raimondo, F.M.; Gabrieljan, E.; Greuter, W. The genus *Aria* (*Sorbus* s. l., Rosaceae) in the Sicilian flora: Taxonomic updating, re-evaluation, description of a new species and two new combinations for one Sicilian and one SW Asian species. *Bot. Chron.* **2019**, *22*, 15–37.

7. Raimondo, F.M.; Spadaro, V.; Di Gristina, E. *Centaurea heywoodiana* (Asteraceae), a new species from the Nebrodi Mountains (NE-Sicily). *Flora Mediterr.* **2020**, *30*, 369–376.

8. Raimondo, F.M.; Bajona, E.; Spadaro, V.; Di Gristina, E. Recent and new taxonomic acquisitions in some native genera of Asteraceae from southern Italy and Sicily. *Flora Mediterr.* **2021**, *31*, 109–122.

9. Raimondo, F.M.; Venturella, G.; Domina, G. *Pyrus pedrottiana* (Rosaceae), a new species from the Nebrodi Mountains (N-E Sicily). *Flora Mediterr.* **2022**, *32*, 25–34.

10. Sciandrello, S.; Giusso del Galdo, G.; Salmeri, C.; Minissale, P. *Vicia brulloi* (Fabaceae), a new species from Sicily. *Phytotaxa* **2019**, *418*, 57–78. [CrossRef]

11. Ilardi, V.; Troia, A.; Geraci, A. *Brassica tardarae* (Brassicaceae), a New Species from a Noteworthy Biotope of South-Western Sicily (Italy). *Plants* **2020**, *9*, 947. [CrossRef]

12. Malfa, G.A.; Acquaviva, R.; Bucchini, A.A.E.; Ragusa, S.; Raimondo, F.M.; Spadaro, V. The Sicilian wild cabbages as biological resources: Taxonomic update and a review on chemical constituents and biological activities. *Flora Mediterr.* **2020**, *30*, 245–260.

13. Brullo, S.; Brullo, C.; Cambria, S.; Tavilla, G.; Giusso del Galdo, G.; Bogdanovic, S. Taxonomical and chorological remarks on the Mediterranean *Poa maroccana* (Poaceae) and first record in Italy from Sicilian flora. *Acta Bot. Croat.* **2021**, *80*, 63–73.

14. Brullo, S.; Brullo, C.; Cambria, S.; Tavilla, G.; Pasta, S.; Scuderi, L.; Zimmitti, A. A new subspecies of *Epipactis microphylla* (Orchidaceae; Epidendroideae) from Pantelleria Island (Sicily). *Phytotaxa* **2021**, *512*, 83–96. [CrossRef]

15. Brullo, S.; Cambria, S.; Crisafulli, A.; Tavilla, G.; Sciandrello, S. Taxonomic remarks on the *Centaurea aeolica* (Asteraceae) species complex. *Phytotaxa* **2021**, *483*, 9–24. [CrossRef]

16. Brullo, S.; Brullo, C.; Cambria, S.; Ilardi, V.; Siracusa, G.; Giusso del Galdo, G. *Ephedra aurea* (Ephedraceae), a new species from Sicily. *Phytotaxa* **2022**, *530*, 1–20. [CrossRef]

17. Brullo, S.; Brullo, C.; Cambria, S.; Ilardi, V.; Siracusa, G.; Minissale, P.; Giusso del Galdo, G. *Ephedra strongylensis* (Ephedraceae), a new species from Aeolian islands (Sicily). *Phytotaxa* **2022**, *576*, 250–264. [CrossRef]

18. Brullo, S.; Brullo, C.; Cambria, S.; Minissale, P.; Sciandrello, S.; Siracusa, G.; Tavilla, G.; Tomaselli, V.; Giusso del Galdo, G. Taxonomical remarks on *Solenopsis laurentia* (Campanulaceae) in Italy. *Phytotaxa* **2023**, *584*, 59–88. [CrossRef]

19. Cambria, S.; Brullo, C.; Tavilla, G.; Sciandrello, S.; Minissale, P.; Giusso del Galdo, G.; Brullo, S. *Ferula sommieriana* (Apiaceae), a new species from Pelagie Islands (Sicily). *Phytotaxa* **2021**, *525*, 89–108. [CrossRef]

20. Domina, G.; Barone, G.; Di Gristina, E.; Raimondo, F.M. The *Centaurea parlatoris* complex (Asteraceae): Taxonomic checklist and typifications. *Phytotaxa* **2021**, *527*, 243–247. [CrossRef]

21. Domina, G.; Di Gristina, E.; Barone, G. A new species within the *Centaurea busambarensis* complex (Asteraceae, Cardueae) from Sicily. *Biodivers. Data J.* **2022**, *10*, e91505. [CrossRef]

22. Barone, G.; Domina, G.; Bartolucci, F.; Galasso, G.; Peruzzi, L. A Nomenclatural and Taxonomic Revision of the *Senecio squalidus* Group (Asteraceae). *Plants* **2022**, *11*, 2597. [CrossRef]

23. La Rosa, A.; Cambria, S.; Brullo, S. Considerazioni tassonomiche sulle popolazioni sicule di *Trifolium isthmocarpum* (Fabaceae). *Not. Soc. Bot. Ital.* **2019**, *3*, 19–20.

24. Costa, R.M.S.; Di Gristina, E. *Xeranthemum cylindraceum* (Asteraceae) in Italy: First record for the Sicilian flora. *Flora Mediterr.* **2020**, *30*, 235–243.

25. Azzaro, D.; Calanni-Rindina, M.; Siracusa, G.; Cambria, S. Sulla presenza in Sicilia di *Myosotis congesta* (Boraginaceae) con osservazioni sulle sue caratteristiche morfologiche ed esigenze ecologiche. *Not. Soc. Bot. Ital.* **2021**, *5*, 1–2.

26. Cambria, S.; Brullo, S.; Brullo, C.; Ilardi, V.; Siracusa, G.; Giusso del Galdo, G. Sulla presenza in Sicilia di *Elatine campylosperma* (Elatinaceae), specie critica della flora Mediterranea. *Not. Soc. Bot. Ital.* **2022**, *6*, 5–6.

27. Cambria, S.; Ippolito, G.; Azzaro, D. Nuovi dati sulla presenza di *Scrophularia vernalis* (Scrophulariaceae) in Sicilia. *Not. Soc. Bot. Ital.* **2021**, *5*, 9–10.

28. Caldarella, O.; Lastrucci, L.; Bolpagni, R.; Gianguzzi, L. Contribution to the knowledge of Mediterranean wetland vegetation: *Lemnetea* and *Potamogetonetea* classes in Western Sicily. *Plant Sociol.* **2021**, *58*, 107–131. [CrossRef]

29. Sciandrello, S.; Cambria, S.; Giusso del Galdo, G.; Tavilla, G.; Minissale, P. Unexpected Discovery of *Thelypteris palustris* (Thelypteridaceae) in Sicily (Italy): Morphological, Ecological Analysis and Habitat Characterization. *Plants* **2021**, *10*, 2448. [CrossRef]

30. Ilardi, V.; Troia, A. Reevaluation and typification of *Foeniculum piperitum* (Apiaceae), an underknown medicinal plant and crop wild relative. *Phytotaxa* **2021**, *508*, 197–205. [CrossRef]

31. Di Natale, S.; Lastrucci, L.; Hroudovà, Z.; Viciani, D. A review of *Bolboschoenus* species (Cyperaceae) in Italy based on herbarium data. *Plant Biosyst.* **2022**, *156*, 261–270. [CrossRef]

32. Domina, G.; Uhlich, H.; Barone, G. *Orobanche australis* Moris ex Bertol. the correct name for *O. thapsoides* Lojac. (Orobanchaceae). *Phytotaxa* **2022**, *531*, 91–096. [CrossRef]

33. Badalamenti, E.; La Mantia, T. *Handroanthus heptaphyllus* (Bignoniaceae) in Sicily: A new casual alien to Italy and Europe. *Flora Mediterr.* **2018**, *28*, 331–338.

34. Guarino, R.; Lo Cascio, P.; Mustica, C.; Pasta, S. A new naturalized alien plant in Sicily: *Cotula australis* (Sieber ex Spreng.) Hook. *Nat. Sic.* **2018**, *42*, 125–135.

35. Scafidi, F.; Raimondo, F.M. First record of *Pilea microphylla* (Urticaceae) in Sicily. *Flora Mediterr.* **2018**, *28*, 79–84.

36. Galasso, G.; Domina, G.; Adorni, M.; Angiolini, C.; Apruzzese, M.; Ardenghi, N.M.G.; Assini, S.; Aversa, M.; Bacchetta, G.; Banfi, E.; et al. Notulae to the Italian alien vascular flora: 9. *Ital. Bot.* **2020**, *9*, 47–70. [CrossRef]

37. Galasso, G.; Domina, G.; Andreatta, S.; Angiolini, C.; Ardenghi, N.M.G.; Aristarchi, C.; Arnoul, M.; Azzella, M.M.; Bacchetta, G.; Bartolucci, F.; et al. Notulae to the Italian alien vascular flora: 8. *Ital. Bot.* **2019**, *8*, 63–93. [CrossRef]

38. Galasso, G.; Domina, G.; Andreatta, S.; Argenti, E.; Bacchetta, G.; Bagella, S.; Banfi, E.; Barberis, D.; Bardi, S.; Barone, G.; et al. Notulae to the Italian alien vascular flora: 11. *Ital. Bot.* **2021**, *11*, 93–119. [CrossRef]

39. Galasso, G.; Domina, G.; Ardenghi, N.M.G.; Aristarchi, C.; Bacchetta, G.; Bartolucci, F.; Bonari, G.; Bouvet, D.; Brundu, G.; Buono, S.; et al. Notulae to the Italian alien vascular flora: 7. *Ital. Bot.* **2019**, *7*, 157–182. [CrossRef]

40. Galasso, G.; Domina, G.; Azzaro, D.; Bagella, S.; Barone, G.; Bartolucci, F.; Bianco, M.; Bolzani, P.; Bonari, G.; Boscutti, F.; et al. Notulae to the Italian alien vascular flora: 10. *Ital. Bot.* **2020**, *10*, 57–71. [CrossRef]

41. Galasso, G.; Domina, G.; Angiolini, A.; Azzaro, D.; Bacchetta, G.; Banfi, E.; Barberis, D.; Barone, G.; Bartolucci, F.; Bertolli, A.; et al. Notulae to the Italian alien vascular flora: 13. *Ital. Bot.* **2022**, *10*, 27–44. [CrossRef]

42. Badalamenti, E. Notes about the naturalization in Sicily of *Paulownia tomentosa* (Paulowniaceae) and remarks about its global spread. *Flora Mediterr.* **2019**, *29*, 67–70.

43. Badalamenti, E. First record of *Heptapleurum arboricola* Hayata (Araliaceae) as a casual non-native woody plant in the Mediterranean area. *Bioinvasions Rec.* **2021**, *10*, 805–815. [CrossRef]

44. Campisi, P.; Raimondo, F.M.; Spadaro, V. New floristic data of alien vascular plants from Sicily. *Flora Mediterr.* **2019**, *29*, 263–267.

45. Giordano, M.; Ilardi, V.; Cambria, S. *Iris unguicularis* Poir. subsp. *unguicularis* (Iridaceae): Una nuova specie aliena per la Sicilia. *Nat. Sic.* **2021**, *45*, 29–36.

46. Rosati, L.; Fascetti, S.; Romano, V.A.; Potenza, G.; Lapenna, M.R.; Capano, A.; Nicoletti, P.; Farris, E.; de Lange, P.J.; Del Vico, E.; et al. New Chorological Data for the Italian Vascular Flora. *Diversity* **2020**, *12*, 22. [CrossRef]

47. Collesano, G.; Fiorello, A.; Pasta, S. *Strelitzia nicolaii* Regel & Körn. (Strelitziaceae), a casual alien plant new to Northern Hemisphere. *Webbia* **2021**, *76*, 135–140.

48. von Raab-Straube, E.; Raus, T. *Chenopodium betaceum* Andrz. P. 145. Euro+Med-Checklist Notulae, 13. *Willdenowia* **2021**, *51*, 141–168. [CrossRef]

49. Di Gristina, E.; Domina, G.; Barone, G. The alien vascular flora of Stromboli and Vulcano (Aeolian Islands, Italy). *Ital. Bot.* **2021**, *12*, 63–75. [CrossRef]

50. Di Gristina, E.; Raimondo, F.M. *Muehlenbeckia sagittifolia* (Polygonaceae), a new alien for the Italian flora. *Flora Mediterr.* **2022**, *31*, 477–481.

51. Domina, G.; Mazzola, P. *Wedelia glauca* (Asteraceae) a new naturalized alien to Italy. *Flora Mediterr.* **2022**, *31*, 483–488.

52. Pavone, P.; Raimondo, F.M.; Spadaro, V. New Aloes casual aliens in Sicily. *Flora Mediterr.* **2022**, *31*, 489–493.

53. Venturella, G.; Gargano, M.L. *Araucaria columnaris* (Araucariacee) casual alien in Sicily. *Flora Mediterr.* **2022**, *31*, 515–519.

54. Banfi, E.; Bajona, E.; Raimondo, F.M. *Megathyrsus maximus* var. *maximus* (Poaceae), a new naturalised grass alien to Italy. *Flora Mediterr.* **2022**, *31*, 453–462.

55. Aleo, M.; Cambria, S.; Minissale, P.; Bazan, G. First record of *Sphaeralcea bonariensis* (Cav.) Griseb. (Malvaceae) as a casual alien species in the Mediterranean area. *Bioinvasions Rec.* **2022**, *11*, 338–344. [CrossRef]

56. Cambria, S.; Crisafulli, A.; Giusso del Galdo, G.; Picone, R.M.; Soldano, A.; Sciandrello, S.; Tavilla, G. First record of *Sida rhombifolia* L. (Malvaceae) for Italian flora: Taxonomical and ecological investigation. *Acta Bot. Croat.* **2022**, *81*, 159–167. [CrossRef]

57. Cambria, S.; Giusso del Galdo, G.; Minissale, P.; Sciandrello, S.; Tavilla, G. *Lablab purpureus* (Fabaceae), a new alien species for Sicily. *Flora Mediterr.* **2022**, *32*, 73–78.

58. Brullo, S.; Minissale, P.; Brullo, C.; Tavilla, S.; Sciandrello, S. Considerazioni nomenclaturali e morfologiche su *Cardamine occulta* (Brassicaceae) e sua presenza in Sicilia. *Not. Soc. Bot. Ital.* **2021**, *5*, 7–8.

59. Guarino, R.; Chytrý, M.; Attorre, F.; Landucci, F.; Marcenò, C. Alien plant invasions in Mediterranean habitats: An assessment for Sicily. *Biol. Invasions* **2021**, *23*, 3091–3107. [CrossRef]

60. Raimondo, F.M.; Domina, G.; Bazan, G. Carta dello stato delle conoscenze floristiche della Sicilia. In *Stato delle Conoscenze sulla Flora Vascolare d'Italia*; Scoppola, A., Blasi, C., Eds.; Palombi Editore: Roma, Italy, 2005; pp. 203–207.

61. Fois, M.; Cuena-Lombraña, A.; Araç, N.; Artufel, M.; Atak, E.; Attard, V.; Bacchetta, G.; Cambria, S.; Ben Charfi, K.; Dizdaroğlu, D.E.; et al. The Mediterranean Island Wetlands (MedIsWet) inventory: Strengths and shortfalls of the currently available floristic data. *Flora Mediterr.* **2022**, *32*, 339–349.

62. Martellos, S.; Bartolucci, F.; Conti, F.; Galasso, G.; Moro, A.; Pennesi, R.; Peruzzi, L.; Pittao, E.; Nimis, P.L. FlorItaly—The portal to the Flora of Italy. *PhytoKeys* **2020**, *156*, 55–71. [CrossRef]

63. Marcenò, C.; Cubino, J.P.; Chytrý, M.; Genduso, E.; Gristina, A.S.; La Rosa, A.; Salemi, D.; Landucci, F.; Pasta, S.; Guarino, R. Plant hunting: Exploring the behaviour of amateur botanists in the field. *Biodivers. Conserv.* **2021**, *30*, 3265–3278. [CrossRef]

64. Marcenò, C.; Cubino, J.P.; Chytrý, M.; Genduso, E.; Salemi, D.; La Rosa, A.; Gristina, A.S.; Agrillo, E.; Bonari, G.; del Galdo, G.G.; et al. Facebook groups as citizen science tools for plant species monitoring. *J. Appl. Ecol.* **2021**, *58*, 2018–2028. [CrossRef]

55. Lojacono Pojero, M. *Flora Sicula 3*; Tipo-Litografia Salvatore Bizzarrilli: Palermo, Italy, 1909.
56. Banfi, E. Poaceae (R.Br.) Barnhart. In *Flora d'Italia*, 2nd ed.; Pignatti, S., Guarino, R., La Rosa, M., Eds.; Edagricole, Edizioni Agricole di New Business Media: Bologna, Italy, 2017; Volume 1, pp. 512–781.
57. Nicolosi-Roncati, F. *Albuca altissima* Dryand. in Sicilia. *Boll. Orto. Bot. Napoli.* **1911**, *3*, 1–4.
58. Albertini, R.; Ugolotti, M.; Ghillani, L.; Adorni, M.; Vitali, P.; Signorelli, C.; Pasquarella, C. Aerobiological monitoring and mapping of *Ambrosia* plants in the province of Parma (northern Italy, southern Po valley), a useful tool for targeted preventive measures. *Ann. Ig. Med. Prev. Comunita* **2017**, *29*, 515–528.
59. Fridlender, A. Le genre *Arum* en Corse. *Candollea* **2000**, *55*, 255–267.
60. Pignatti, S. Volume 3: Flora d'Italia & Flora Digitale. In *Flora d'Italia: In 3 Volumi*, 1st ed.; Edagricole: Bologna, Italy, 1982.
61. Pignatti, S.; Guarino, R.; La Rosa, M. Volume 1: Flora d'Italia & Flora Digitale. In *Flora d'Italia: In 4 Volumi*, 2nd ed.; Edagricole-Edizioni Agricole di New Business Media srl: Milano, Italy, 2017.
62. Pignatti, S. Volume 2: Flora d'Italia & Flora Digitale. In *Flora d'Italia: In 4 Volumi*, 2nd ed.; Edagricole-Edizioni Agricole di New Business Media srl: Milano, Italy, 2017.
63. Pignatti, S. Volume 3: Flora d'Italia & Flora Digitale. In *Flora d'Italia: In 4 Volumi*, 2nd ed.; Edagricole-Edizioni Agricole di New Business Media srl: Milano, Italy, 2018.
64. Pignatti, S.; Guarino, R.; La Rosa, M. Volume 4: Flora d'Italia & Flora Digitale. In *Flora d'Italia: In 4 Volumi*, 2nd ed.; Edagricole-Edizioni Agricole di New Business Media srl: Milano, Italy, 2019.
65. Sortino, M. Sull'esatto valore tassonomico di *Arum italicum* Millerf *nigro-maculatum* Fiori, e sua distribuzione. *R. Istit. Bot. Giar. Colon. Palermo* **1968**, *24*, 3–10.
66. Raimondo, F.M. *Carta della Vegetazione di Piano della Battaglia e del Territorio Circostante (Madonie, Sicilia)*; C.N.R.: Roma, Italy, 1980.
67. Terpò, A. Studies on Taxonomy and Grouping of *Pyrus* Species. *Feddes Repert.* **1985**, *96*, 73–87.
68. Raimondo, F.M.; Mazzola, P.; Domina, G. Check-list of the vascular plant collected during Iter Mediterraneum III. *Bocconea* **2004**, *17*, 65–231.
69. Zodda, G. Entità nuove o importanti della flora sicula. *Mem. Reale Acc. Zelanti Acireale* **1908**, *5*, 99–162.
70. Barbagallo, C.; Furnari, F. *Contributo alla Flora del Territorio di Caltanissetta con Osservazioni Sulle Piante Officinali*; Istituto di botanica dell'Università di Catania: Catania, Italy, 1970; pp. 1–61.
71. Lastrucci, L.; Gambirasio, V.; Lazzaro, L.; Viciani, D. Revision of the Italian material of *Juncus* sect. Tenageia in the Herbarium Centrale Italicum: Confirmations and novelties for Italy. *Mediterr. Bot.* **2022**, *43*, e72370. [CrossRef]
72. Sciandrello, S.; Musarella, C.M.; Puglisi, M.; Spampinato, G.; Tomaselli, V.; Minissale, P. Updated and new insights on the coastal halophilous vegetation of southeastern Sicily (Italy). *Plant Sociol.* **2019**, *56*, 81–98.
73. Stinca, A.; Musarella, C.M.; Rosati, L.; Laface, V.L.A.; Licht, W.; Fanfarillo, E.; Wagensommer, R.P.; Galasso, G.; Fascetti, S.; Esposito, A.; et al. Italian vascular flora: New findings, updates and exploration of floristic similarities between regions. *Diversity* **2021**, *13*, 600. [CrossRef]
74. Spampinato, G.; Laface, V.L.A.; Posillipo, G.; Cano Ortiz, A.; Quinto Canas, R.; Musarella, C.M. Alien flora in Calabria (Southern Italy): An updated checklist. *Biol. Invasions* **2022**, *24*, 2323–2334. [CrossRef]
75. Gussone, G. *Supplementum ad Florae Siculae Prodromum*; Regia Typographia: Neapoli, Italy, 1832; pp. 1–166.
76. Brullo, S.; Di Martino, A. Vegetazione dell'Isola Grande dello Stagnone (Marsala). *Boll. R. Orto Bot. Palermo* **1974**, *26*, 15–62.
77. Brullo, S. Miscellaneous notes on the genus *Limonium*. *Willdenowia* **1988**, *17*, 11–18.
78. Raimondo, F.M.; Gianguzzi, L.; Ilardi, V. Inventario delle specie "a rischio" nella flora vascolare nativa della Sicilia. *Quad. Bot. Amb. Appl.* **1994**, *3*, 65–132.
79. Santo, A.; Mattana, E.; Grillo, O.; Sciandrello, S.; Peccenini, S.; Bacchetta, G. Variability on morphological and ecological seed traits of *Limonium avei* (De Not.) Brullo & Erben (Plumbaginaceae). *Plant Species Biol.* **2017**, *32*, 368–379.
80. Brullo, S.; Guarino, R. Limonium Mill. In *Flora d'Italia*; Pignatti, S., Guarino, R., La Rosa, M., Eds.; Edagricole: Milano, Italy, 2018; Volume 2, pp. 18–49.
81. Volk, K.M. Assessing the Invasive Potential of Najas minor in Maine. Bachelor's Thesis, Colby College, Waterville, ME, USA.
82. Dana, D.E.; Verloove, F.; Guillot Ortiz, D.; Rodriguez-Marzal, J.L.; Paredes-Carretero, F.; Juan Bañón, J.L.; Esteban, E.; Garcia de Lomas, J. First record of *Nymphaea* × *marliacea* Lat.-Marl. 'Rosea' in the Iberian Peninsula: Identification based on morphological features and molecular techniques. *Bouteloua* **2017**, *28*, 132–139.
83. Banfi, E.; Galasso, G. *La Flora Esotica Lombarda*; Museo di Storia Naturale di Milano, Regione Lombardia: Milano, Italy, 2010.
84. Brullo, S.; Sciandrello, S. La vegetazione del bacino lacustre "Biviere di Gela" (Sicilia meridionale). *Fitosociologia* **2006**, *43*, 21–40.
85. Raimondo, F.M.; Domina, G.; Spadaro, V.; Aquila, G. Prospetto delle piante avventizie e spontaneizzate in Sicilia. *Quad. Bot. Amb. Appl.* **2010**, *15*, 153–164.
86. Iamonico, D. Taxonomical, morphological, ecological and chorological notes on *Oxybasis chenopodioides* and *O. rubra* (Chenopodiaceae) in Italy. *Hacquetia* **2014**, *13*, 285–296. [CrossRef]
87. Bartolo, G.; Brullo, S.; Marcenò, C. Contributo alla flora sicula. *Boll. Acc. Gioenia Sci. Nat.* **1976**, *12*, 72–78.
88. POWO. Plants of the World Online. Facilitated by the Royal Botanic Gardens, Kew. 2022. Available online: http://www.plantsoftheworldonline.org/ (accessed on 9 February 2022).
89. Mohlenbrock, R.H. *Vascular Flora of Illinois. A Field Guide*, 4th ed.; Southern Illinois University Press: Carbondale, IL, USA, 2014; pp. 1–536.

100. Swearingen, J.; Bargeron, C. Invasive Plant Atlas of the United States. University of Georgia Center for Invasive Species and Ecosystem Health. Available online: http://www.invasiveplantatlas.org/ (accessed on 12 January 2023).
101. Swearingen, J.; Slattery, B.; Reshetiloff, K.; Zwicker, S. *Plant Invaders of Mid-Atlantic Natural Areas*, 4th ed.; National Park Service and U.S. Fish and Wildlife Service: Washington, DC, USA, 2010; 168p.
102. Stern, R.A.; Doron, I.; Redel, G.; Raz, A.; Goldway, M.; Holland, D. Lavi 1—A new *Pyrus betulifolia* rootstock for 'Coscia' pear (*Pyrus communis*) in the hot climate of Israel. *Sci. Hortic.* **2013**, *161*, 293–299. [CrossRef]
103. Sestini, D. *Lettere del Signor Abate Domenico Sestini Scritte dalla Sicilia e dalla Turchia a Diversi suoi amici in Toscana*; Tomo V. Stamperia Carlo Gorgi: Livorno, Italy, 1782.
104. Venora, G.; Blangiforti, S. *I Grani Antichi Siciliani—Manuale Tecnico per il Riconoscimento delle Varietà Locali Dei Frumenti Siciliani*; Le Fate Editore: Ragusa, Italy, 2017; 193p.
105. I.Stat. Coltivazioni (istat.it). Available online: http://dati.istat.it/Index.aspx?DataSetCode=DCSP_COLTIVAZIONI (accessed on 12 January 2023).
106. Celesti-Grapow, L.; Alessandrini, A.; Arrigoni, P.V.; Banfi, E.; Bernardo, L.; Bovio, M.; Brundu, G.; Cagiotti, M.R.; Camarda, I.; Carli, E.; et al. Inventory of non-native flora of Italy. *Plant Biosyst.* **2009**, *143*, 386–430. [CrossRef]
107. Ponzo, A. Escursioni nei dintorni di Licata. *Malpighia* **1902**, *16*, 227–260.
108. Nicotra, L.; Campagna, C. Addenda ad floram siculam nonnulla. *Malpighia* **1908**, *22*, 537–548.
109. Bazan, G.; Marino, P.; Guarino, R.; Domina, G.; Schicchi, R. Bioclimatology and vegetation series in Sicily: A geostatistical approach. *Ann. Bot. Fenn.* **2015**, *52*, 1–18. [CrossRef]
110. Brullo, S.; Minissale, P.; Spampinato, G. Considerazioni fitogeografiche sulla flora della Sicilia. *Ecol. Medit.* **1995**, *21*, 99–117. [CrossRef]
111. Sciandrello, S.; Guarino, R.; Minissale, P.; Spampinato, G. The endemic vascular flora of Peloritani Mountains (NE Sicily): Plant functional traits and phytogeographical relationships in the most isolated and fragmentary micro-plate of the Alpine orogeny. *Plant Biosyst.* **2015**, *149*, 838–854. [CrossRef]
112. Fierotti, G.; Dazzi, C.; Raimondi, S. *Commento Alla Carta Dei Suoli della Sicilia*; Istituto di Agronomia Generale: Palermo, Italy, 1988.
113. Montanari, L. *Geologica Sicula, un Intreccio tra Rocce e Storia*; Collana Studi e Ricerche; ARPA: Rimini, Italy, 2004.
114. Mucina, L.; Bültmann, H.; Dierßen, K.; Theurillat, J.P.; Raus, T.; Čarni, A.; Šumberová, K.; Willner, W.; Dengler, J.; García, R.G.; et al. Vegetation of Europe: Hierarchical floristic classification system of vascular plant, bryophyte, lichen, and algal communities. *Appl. Veg. Sci.* **2016**, *19*, 3–264. [CrossRef]
115. Lazzaro, L.; Bolpagni, R.; Buffa, G.; Gentili, R.; Lonati, M.; Stinca, A.; Acosta, A.T.R.; Adorni, M.; Aleffi, M.; Allegrezza, M.; et al. Impact of invasive alien plants on native plant communities and Natura 2000 habitats: State of the art, gap analysis and perspectives in Italy. *J. Environ. Manag.* **2020**, *274*, 111140. [CrossRef] [PubMed]
116. Haubrock, P.J.; Turbelin, A.J.; Cuthbert, R.N.; Novoa, A.; Taylor, N.G.; Angulo, E.; Ballesteros-Mejia, L.; Bodey, T.W.; Capinha, C.; Diagne, C.; et al. Economic costs of invasive alien species across Europe. *NeoBiota* **2021**, *67*, 153–190. [CrossRef]
117. Zenni, R.D.; McDermott, S.; García-Berthou, E.; Essl, F. The economic costs of biological invasions around the world. *NeoBiota* **2021**, *67*, 1–9. [CrossRef]
118. Brancatelli, G.I.E.; Zalba, S.M. Vector analysis: A tool for preventing the introduction of invasive alien species into protected areas. *Nat. Conserv.* **2018**, *24*, 43–63. [CrossRef]

Article

Chemical Management of *Senecio madagascariensis* (Fireweed)

Kusinara Wijayabandara [1,*], Shane Campbell [1], Joseph Vitelli [2], Sundaravelpandian Kalaipandian [1] and Steve Adkins [1]

[1] School of Agriculture and Food Science, University of Queensland, Gatton Campus, 4343, QLD, Australia
[2] Department of Agriculture and Fisheries, Brisbane 4102, QLD, Australia
* Correspondence: k.wijayabandara@uq.edu.au

Abstract: Fireweed (*Senecio madagascariensis* Poir.) is a herbaceous weed-producing pyrrolizidine alkaloid that is poisonous to livestock. To investigate the efficacy of chemical management on fireweed and its soil seed bank density, a field experiment was conducted in Beechmont, Queensland, in 2018 within a pasture community. A total of four herbicides (bromoxynil, fluroxypyr/aminopyralid, metsulfuron-methyl and triclopyr/picloram/aminopyralid) were applied either singularly or repeated after 3 months to a mix-aged population of fireweed. The initial fireweed plant density at the field site was high (10 to 18 plants m^{-2}). However, after the first herbicide application, the fireweed plant density declined significantly (to ca. 0 to 4 plants m^{-2}), with further reductions following the second treatment. Prior to herbicide application, fireweed seeds in both the upper (0 to 2 cm) and lower (2 to 10 cm) soil seed bank layers averaged 8804 and 3593 seeds m^{-2}, respectively. Post-herbicide application, the seed density was significantly reduced in both the upper (970 seeds m^{-2}) and lower (689 seeds m^{-2}) seed bank layers. Based on the prevailing environmental conditions and nil grazing strategy of the current study, a single application of either fluroxypyr/aminopyralid, metsulfuron-methyl or triclopyr/picloram/aminopyralid would be sufficient to achieve effective control, whilst a second follow-up application is required with bromoxynil.

Keywords: herbicides; plant density; seed bank; fireweed; management

check for updates

Citation: Wijayabandara, K.; Campbell, S.; Vitelli, J.; Kalaipandian, S.; Adkins, S. Chemical Management of *Senecio madagascariensis* (Fireweed). *Plants* **2023**, *12*, 1332. https://doi.org/10.3390/plants12061332

Academic Editors: Congyan Wang and Hongli Li

Received: 16 February 2023
Revised: 8 March 2023
Accepted: 9 March 2023
Published: 15 March 2023

1. Introduction

Madagascar ragwort (*Senecio madagascariensis* Poir.), commonly known as fireweed in Australia, is a short-lived perennial herb that belongs to the Asteraceae family. It is native to southern Africa and has been introduced to several countries, including Argentina, Brazil, Colombia, Uruguay, Japan, Hawaii, and Australia [1]. Fireweed plants produce pyrrolizidine alkaloids (PAs) that are poisonous to livestock, particularly cattle (*Bos taurus* L.) and horses (*Equus ferus caballus* L.). Several general management approaches, including cultural, physical, chemical, biological, or a combination of these, have been used to manage fireweed. It is known to be susceptible to the action of several selective herbicides [2], spanning several 'mode of action' groups [3]. For example, fluroxypyr/aminopyralid (HotShot™, Corteva Agriscience Australia, 67 Albert Avenue, Chatswood, New South Wales, Australia, 2067) and triclopyr/picloram/aminopyralid (Grazon™ Extra, Corteva Agriscience Australia, 67 Albert Avenue, Chatswood, New South Wales, Australia, 2067) are combinations of synthetic auxin herbicides (Group 4) [4]. These disrupt plant cell growth in the newly forming stems and leaves and negatively affect protein synthesis and normal cell division, leading to malformed growth and tissue tumours [4]. Bromoxynil (Bromicide® 200, Nufarm Australia, 103–105 Pipe Road, Laverton North, Victoria, Australia, 3026) acts as a Photosystem II photosynthetic inhibitor (Group 5) while metsulfuron-methyl (Brush-Off®, Bayer CropScience Australia, 391–393 Tooronga Road, Hawthorn East, Victoria, Australia, 3123)—a member of the sulfonylurea group of herbicides (Group 2)—impedes the normal function of acetolactate synthase (ALS), a key enzyme in the pathway of biosynthesis of the branched-chain amino acids isoleucine,

leucine, and valine [4]. A commonplace recommendation for herbicide control of fireweed in Australia is that plants should be sprayed during the early flowering stage of growth with a follow-up treatment often essential 6 months later [2,5]. According to Sindel and Coleman [2], such herbicide applications are best applied to control fireweed populations during April (Autumn) in Australia.

In terms of herbicide efficacy, 2,4-dichlorophenoxyacetic acid (2,4-D) formulations [5,6] dicamba, glyphosate, MCPA, tebuthiuron, triclopyr [6] bromoxynil, fluroxypyr/aminopyralid, triclopyr/picloram/aminopyralid [2,7], and metsulfuron-methyl [7] have all been found to be active in one or more fireweed growth stages (i.e., seedlings, juvenile or mature plants). However, which herbicide and at what rate they can be legally applied may vary between countries and even between jurisdictions within countries (Pest Plants and Animals Act 2005 and Biosecurity Act 2014).

In Australia, 2,4-D amine (3.2 kg ha^{-1}) and 2,4-D sodium salt (2 to 4 kg ha^{-1}) have been reported to give good fireweed control without harming the proximate pasture species, such as blue couch (*Digitaria didactyla* Wild), blady grass (*Imperata cylindrica* (L.) Beauv) and white clover (*Trifolium repens* L.) [8]. Similarly, Motooka et al. [6] suggested that in Hawaiian pastures where forage legumes are mixed with grasses, the amine salt formulation of 2,4-D is preferable because of its mild impact on legumes. In contrast, metsulfuron-methyl at 40 to 80 g ha^{-1} provided effective control of fireweed in an Australian study, but it severely damaged legumes (such as *T. repens*) present within the treated pasture [9].

The seed bank is defined as a collection of viable, non-germinated seeds [9] and is an important component of grazed pastures. To develop a suitable, long-term chemical management strategy for any grazed pasture, it is important to have information on the dynamics of the weed and pasture species' seed banks [10]. The soil seed bank significantly contributes to the regeneration ability and future composition of that pasture community [11]. During a typical Australian Autumn (March–May), with average rainfall (of 94.3 mm in 2019; [12]), most fireweed seeds will germinate in the first 3 months after dispersal from the parent plant, and only a small percentage will remain viable and ungerminated in the seed bank after a year [2]. However, in a relatively drier season, a greater proportion of the seed produced will enter the seed bank and is predicted to maintain its viability there for up to 10 years [2]. In one study, freshly collected fireweed seeds buried 3 cm deep in the soil only lost a small percentage of their viability (from 63 to 54%) in the following 15 months [13]. When Radford [13] assessed the size of the fireweed soil seed bank, they found over 12,000 seeds m^{-2} at a heavily infested site, with most seeds found below 1 cm of depth in the soil profile.

Rapidly reducing or eradicating weed seed banks should be relatively easy if seed production and their placement into the seed bank can be prevented [14]. In addition, determining the seed bank size and structure of grazed pastures will be helpful in determining an effective chemical management approach [14]. To eradicate an invasive weed species like fireweed, it is necessary to not only kill the emerging plants but also deplete the seed bank. Even with high levels of plant mortality, the soil seed bank may still allow populations to reappear in future years [15]. In the estimation of the impact of any chemical management approach in an agroecosystem, knowledge of the germination behaviour of the weed species and its seed bank ecology will be important [16]. By using ecological population indices, such as the Shannon–Weiner index [10], determining the effect of chemical management can be made on the species diversity within treated communities.

Thus, the objectives of this study were to: (1) evaluate the impact of several herbicides, that have all been previously found to be efficacious on fireweed but have different modes of action, on fireweed density, (2) compare a single dose to a repeated dose, and (3) determine their ability to rapidly deplete the fireweed seed bank while maintaining the pasture community species diversity.

2. Results

2.1. Density of Fireweed Plants following Various Chemical Control Approaches

Before applying herbicide treatments, fireweed plant density was relatively high (*ca.* 10 to 18 plants m^{-2}) and not significantly different ($p = 0.56$) between treatment plots (Figure 1). Following the first herbicide application, new seedling recruitment was detected in the bromoxynil-treated sub-plots but not in any of the other herbicide treatments. Two months after the first herbicide application, a significant reduction ($p < 0.05$) in fireweed plant density (*ca.* 0 to 4 plants m^{-2}) was observed in all herbicide-treated sub-plots when compared to the control sub-plots (*ca.* 14 plants m^{-2}; Figure 1).

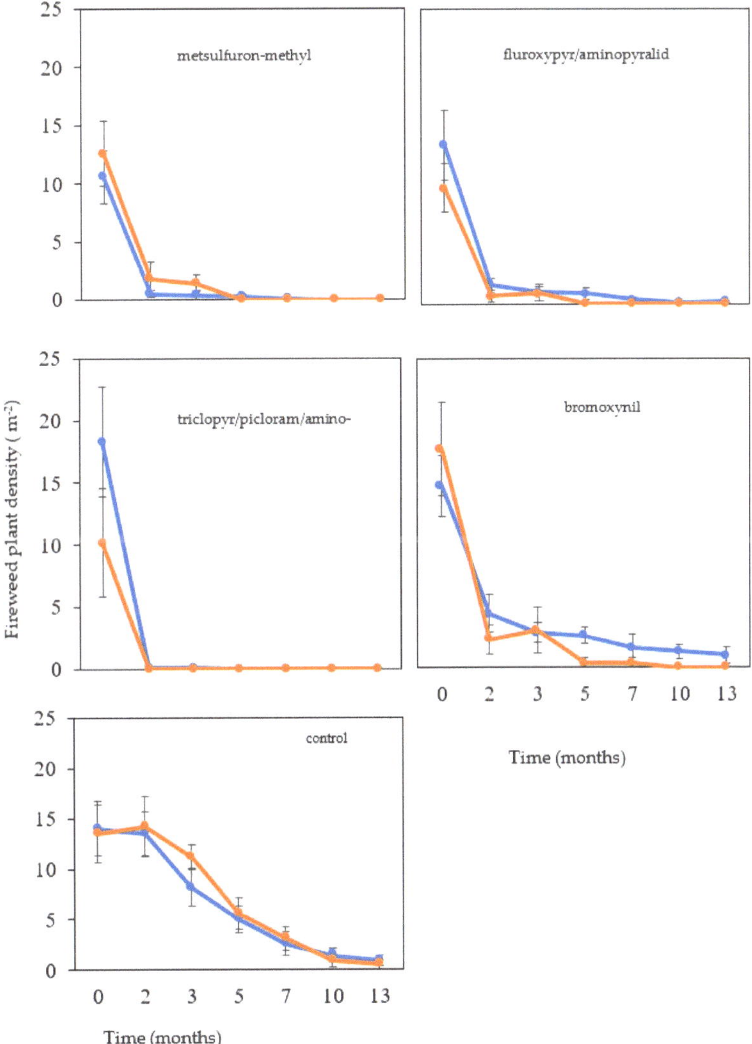

Figure 1. Fireweed plant density (m^{-2}) in two different types of field plots: those plots that had been sprayed once (blue line) and those that had been sprayed twice (orange line) recorded over time following the application of metsulfuron-methyl, fluroxypyr/aminopyralid, triclopyr/picloram/aminopyralid, bromoxynil, and control. The data are the mean ± SEM from five replicate plots.

When comparing the single application plots with the follow-up application plots, the single application had been efficacious for all herbicides except bromoxynil (Figure 1). For bromoxynil, the follow-up application treatment was required to reduce the density of fireweed to zero plants m^{-2}, while the single application treatment averaged three plants m^{-2} (Figure 1).

Ten months after application, there was no significant difference ($p = 0.10$) between the single and the follow-up treatments on fireweed density (Figure 1). In addition, during this time, fireweed density had declined in all plots, including the control. At the end of the experiment (i.e., after 13 months), there was a significant difference ($p < 0.05$) in fireweed density in the herbicide-applied plots as compared to the control; however, there was not a significant difference ($p = 0.15$) between the number of applications (apart from the bromoxynil treated plots) (Figure 1).

For the single application sub-plots for metsulfuron-methyl, fluroxypyr/aminopyralid and triclopyr/picloram/aminopyralid applications significantly reduced the density of fireweed compared to the control sub-plots, while no significant difference was observed between bromoxynil and the control (Figure 1). However, following the second herbicide application, the fireweed plant density dropped to 0 in all herbicide-applied sub-plots (including the bromoxynil sub-plots) at the end of the experiment (13 months; Figure 1).

2.2. Effect of Chemical Control on Seed Bank Structure

Before the initial application of the herbicides, fireweed seeds were found in both the upper (0 to 2 cm) and lower (2 to 10 cm) soil layers. The fireweed seed density in the upper layer (8804 seeds m^{-2}) was significantly ($p < 0.05$) greater than the seed density in the lower layer (3593 seeds m^{-2}) (Table 1). The germinable fireweed seed density across the site varied from 7711 to 10,736 germinable seeds m^{-2} in the upper layer and from 3198 to 3972 m^{-2} in the lower layer (Table 1). Fireweed was the most abundant species in the seed bank and accounted for 90% of the total seed bank in both soil layers (Table 1).

Table 1. Germinable soil seed bank species seed density of a fireweed-dominated kikuyu pasture at Beechmont, Queensland, before the application of metsulfuron-methyl, fluroxypyr/aminopyralid, triclopyr/picloram/aminopyralid or bromoxynil, compared with a non-treated control, at two soil depths: upper (0 to 2 cm deep) and lower (2 to 10 cm deep).

Family and Species	Life Form	Metsulfuron-Methyl		Fluroxypyr Aminopyralid		Triclopyr Picloram Aminopyralid		Bromoxynil		Control	
		Upper	Lower	Upper	Lower	Upper	Lower	Upper	Lower	Upper	Lower
						seeds m^{-2}					
Asteraceae											
* *Senecio madagascariensis*	P/F, W	7710	3769	8740	3494	8128	3535	10,736	3973	8709	3198
* *Conyza bonariensis*	A/F, W	55	10	132	102	143	81	224	112	81	51
Cirsium vulgare	P/F, W	20	20	132	31	81	0	61	10	71	41
Fabaceae											
Trifolium repens	P/F	632	530	407	591	509	580	397	530	519	356
Oxalidaceae											
* *Oxalis dillenii*	P/F, W	61	10	31	31	10	41	20	173	31	10
Poaceae											
* *Pennisetum clandestinum*	P/G, W	1701	2007	2333	1314	2771	4034	3239	2659	3035	2027
Apiaceae											
Cyclospermum leptophyllum	A/F	0	20	0	10	10	10	10	234	10	0
Solanaceae											

Table 1. *Cont.*

Family and Species	Life Form	Metsulfuron-Methyl		Fluroxypyr Aminopyralid		Triclopyr Picloram Aminopyralid		Bromoxynil		Control	
		Upper	Lower	Upper	Lower	Upper	Lower	Upper	Lower	Upper	Lower
* *Solanum nigrum*	PA/S, W	0	0	10	0	0	0	0	0	0	0
Verbeaceae											
Verbena sp.	P/F	0	0	0	0	0	0	0	0	10	51
Gentianaceae											
* *Centaurium erythraea*	A/F, W	784	1345	550	540	1151	1579	1385	1436	784	937
Araliaceae											
Hydrocotyle sp.	P, W	20	81	0	10	0	31	0	0	10	0
Phytolaccaceae											
* *Phytolacca americana* L.	P/F	0	20	0	0	41	61	0	10	0	31
Cyperaceae											
* *Cyperus rotundus* L.	P/F	0	183	20	51	122	346	387	51	71	244
Juncaceae											
Juncus sp.	P/G	0	0	0	0	0	0	10	10	0	0
Unknown sp.		10	0	31	31	0	31	31	0	0	20
Total		11,193	7995	12,386	6205	12,966	10,329	16,500	9198	13,331	6966
Grand total		19,188		18,591		23,295		26,698		23,297	

Life form details for longevity (A = annual or biennial, P = perennial) and for life form (F = forb, G = graminoid, S = shrub) Weed status (W = weed) and * Introduced species.

There were about 15 species recorded from the pre-treatment seed bank (Table 1). Fireweed, followed by kikuyu (*Pennisetum clandestinum* Hochst. ex Chiov.), had the highest seed density (ranged from 1314 to 4034 seed m^{-2}), which is consistent with the study site being a kikuyu-sown pasture. The total seed density of all species, including fireweed in both layers, varied from 18,591 to 26,698 seeds m^{-2}. The Shannon–Weiner index for pre-treatment at the top layer (from -0.47 to 1.37) was less than the bottom layer (-0.29 to 1.40) (Table 2).

Following two rounds of herbicide application, the germinable seed bank of fireweed was significantly ($p < 0.05$) reduced (Table 3) due to the absence of any remaining reproductively active plants. Before application, fireweed seeds were found in both upper and lower soil layers, with a significantly ($p < 0.05$) greater portion (7710 to 10,736) in the upper layer (Table 1). When assessed 5 months after the second herbicide application, the germinable fireweed seeds in the upper layer varied from only 601 (fluroxypyr/aminopyralid) to 1263 m^{-2} (Control), while in the lower layer, germinable fireweed seeds varied from 519 (fluroxypyr/aminopyralid) to 866 m^{-2} (bromoxynil) (Table 3). Not only in herbicide-treated sub-plots but also in the control sub-plots, fireweed seed density had dropped by 28%, indicating that loss in seed viability was occurring. Kikuyu was the dominant species in the seed bank, with only 11 other species observed in the seed bank post-herbicide application (Table 3). In addition, nutgrass (*Cyperus rotundus* L.) had increased following herbicide application except in the triclopyr/picloram/aminopyralid treated sub-plots (Table 3). New plant species also appeared (little hogweed, *Portulaca oleracea* L.; burclover, *Medicago polymorpha* L.; Carolina bristle mallow, *Modiola caroliniana* L. G. Don), possibly due to seed dispersal into the site. Whilst some species—which had been present in the first seed bank assessment—were lost from the second seed bank assessment (e.g., white clover, *Trifolium repens* L.; marsh parsley, *Cyclospermum leptophyllum* (Pers.) Sprague ex Britton and P.Wilson; black nightshade, *Solanum nigrum* L.; vervain, *Verbena* spp.; pennywort,

Hydrocotyle sp. and American pokeweed, *Phytolacca americana* L.), possibly due to short term seed longevity in the seed bank.

Table 2. Shannon—Wiener index for different treatments metsulfuron-methyl (A), fluroxypyr/ aminopyralid (B), triclopyr/picloram/ aminopyralid (C), bromoxynil (D), and Control (E).

Seed Bank	Layer	Treatment	Shannon Wiener Index
Before spray	Top	A	1.035048
		B	0.933172
		C	1.119198
		D	−0.47028
		E	1.025117
	Bottom	A	1.375892
		B	1.295681
		C	1.428605
		D	−0.299021
		E	1.408365
After spray	Top	A	1.415585
		B	1.647513
		C	1.176353
		D	0.852067
		E	1.281637
	Bottom	A	1.165063
		B	0.967562
		C	1.17289
		D	1.027754
		E	−0.59443

Table 3. Species density and change in seed density of germinable soil seed bank species found in a kikuyu pasture at Beechmont, Queensland, after spraying twice with herbicides metsulfuron-methyl, fluroxypyr/aminopyralid, triclopyr/picloram/aminopyralid, bromoxynil, and non-treated control, at two soil depths: upper (0 to 2 cm deep) and lower (2 to 10 cm deep).

Family and Species	Life Form	Metsulfuron-Methyl		Fluroxypyr Aminopyralid		Triclopyr Picloram Aminopyralid		Bromoxynil		Control	
		Upper	Lower	Upper	Lower	Upper	Lower	Upper	Lower	Upper	Lower
		seeds m^{-2}									
Asteraceae											
* *Senecio madagascariensis*	P/F, W	1059 (−39%)	591 (−25%)	601 (−43%)	519 (−35%)	856 (−34%)	805 (−11%)	1070 (−46%)	866 (−26%)	1263 (−28%)	662 (−28%)
* *Cirsium vulgare*	P/F, W	20 (0)	20 (+1%)	20 (0)	0 (0)	0 (−1%)	10 (0)	0 (0)	10 (0)	20 (0)	(−1%)
* *Conyza bonariensis*	A/F, W	112 (+1%)	31 (+1%)	20 (0)	0 (−2%)	0 (−1%)	10 (0)	0 (−1%)	10 (−1%)	20 (0)	0 (−1%)
Poaceae											
* *Pennisetum clandestinum*	P/G, W	1599 (+29%)	1711 (+40%)	1049 (+30%)	1538 (+43%)	1569 (+32%)	2149 (+23%)	4105 (+54%)	3494 (+42%)	1742 (+28%)	2720 (+46%)
Oxalidaceae											
* *Oxalis dillenii*	P/F, W	31 (0)	10 (0)	41 (2%)	10 (0)	10 (0)	51 (+1%)	71 (+1%)	112 (0)	41 (+1%)	61 (+2%)
Portulacaceae											
* *Portulaca oleracea*	A/F, W	20	20	10	20	10	20	0	20	10	31
Fabaceae											
* *Medicago polymorpha*	A/F, W	744	224	316	265	387	377	183	367	326	112
Malvaceae											

Table 3. *Cont.*

Family and Species	Life Form	Metsulfuron-Methyl		Fluroxypyr Aminopyralid		Triclopyr Picloram Aminopyralid		Bromoxynil		Control	
		Upper	Lower	Upper	Lower	Upper	Lower	Upper	Lower	Upper	Lower
* *Modiola caroliniana*	P/F	10	20	31	0	112	20	0	20	0	51
Juncaceae											
Juncus sp.	P/G	0 (0)	10 (0)	10 (0)	0 (0)	0 (0)	20 (0)	0 (+1%)	20 (0)	0 (0)	0 (0)
Cyperaceae											
Cyperus rotundus.	P/F	173 (+5%)	112 (+2%)	411 (+19%)	10 (0)	10 (−1%)	112 (0)	194 (+1%)	132 (+2%)	306 (+8%)	81 (−1%)
Gentanaceae											
* *Centaurium erythraea*	A/F, W	0 (−7%)	(−17%)	(−4%)	(−9%)	51 (−9%)	(−15%)	(−7%)	102 (−16%)	(−6%)	(−11%)
Total		3595	2637	2149	2413	2964	3462	5562	4929	3422	3637
Grand total		6232		4562		6426		10,491		7059	

Life form entails longevity (A= annual or biennial, P = perennial) and lifeform (F = forb, G = graminoid, S = shrub), Weed status (W = weed) and * introduced species. The percentage reduction (−) and increase (+) of species compared to before spray is shown in parentheses.

The total seed density of the seed bank varied from 4562 to 10,491 seeds m^{-2} (Table 3).

2.3. Effect of Chemical Control on Seed Bank Vertical Distribution

There was a significant difference ($p < 0.05$) between the density of the germinable seed banks before and after herbicide application and between the two depths of the soil samples (0 to 2 and 2 to 10 cm), and interaction of these two factors (herbicide application with soil depth) on fireweed seed density in the seed bank. However, there was no significant difference ($p = 0.12$) between the herbicide treatments and the interaction of seed banks with soil depth or with treatment ($p = 0.67$) (Table 4). The Shannon–Weiner Index for post-treatment in the top layer (from 0.85 to 1.64) was higher than in the bottom layer (−0.59 to 1.17) (Table 2).

There was no significant difference in the fireweed seed density within each of the upper layers and lower layers prior to herbicide application in all sub-plots, but the seed density was higher in the upper layer (0 to 2 cm) when compared to the lower layer (2 to 10 cm). However, there was a significant reduction of fireweed density after herbicide application in every treatment as well as in the control sub-plots (Table 4). As compared to the control, the highest reduction of fireweed seed numbers (−18%) was observed in the upper layer of the bromoxynil-treated plots (Table 4).

Table 4. Germinable seed density of fireweed from a kikuyu pasture at Beechmont, Queensland, before and after herbicide treatments compared with total seed density, the percent of total seed density, and the percent reduction of fireweed seed number after herbicide treatment, and reduction in seed density as compared with non-treated control plot. The measurements were taken before and after herbicide treatment and then determined for two soil layers (0 to 2 cm depth and 2 to 10 cm).

Treatment	Timing	Soil Layer	All Species	Fireweed		
			Seed m^{-2}	Total (%)	Reduction (%)	Control (%)
metsulfuron-methyl	Before	Upper	11,193	7710 [a]	69	−40	−12
	After		3595	1059 [cde]	29		
	Before	Lower	7995	3769 [b]	47	−25	+3
	After		2637	591 [efg]	22		

Table 4. *Cont.*

Treatment	Timing	Soil Layer	All Species		Fireweed		
			——Seed m^{-2}——		Total (%)	Reduction (%)	Control (%)
fluroxypyr/aminopyralid	Before	Upper	12,386	8740 [a]	71	−43	−15
	After		2149	601 [fg]	28		
	Before	Lower	6205	3494 [b]	56	−34	−6
	After		2413	519 [g]	22		
triclopyr/picloram/aminopyralid	Before	Upper	12,966	8128 [a]	63	−34	−6
	After		2964	856 [cdef]	29		
	Before	Lower	10,329	3535 [b]	34	−11	+17
	After		3462	805 [cdefg]	23		
bromoxynil	Before	Upper	16,500	10,736 [a]	65	−46	−18
	After		5562	1070 [d]	19		
	Before	Lower	9198	3973 [b]	43	−25	+3
	After		4929	866 [cdef]	18		
Control	Before	Upper	13,331	8709 [a]	65	−28	
	After		3422	1263 [c]	37		
	Before	Lower	6966	3198 [b]	46	−28	
	After		3637	662 [defg]	18		

Seed densities in each row, followed by different letters, are significantly different ($p < 0.05$).

3. Discussion

In the present study, single applications of all the tested herbicides with different modes of action were successful in controlling fireweed growth, except bromoxynil which required a second follow-up application. The bromoxynil treatment was not as effective in a pasture setting, as it is a contact PSII inhibitor herbicide, whereas the other herbicides used were translocatable and thus more effective in a dense grass community. Bromoxynil is most active on smaller fireweed plants, whereas the efficacy of the other translocatable herbicide treatments is less limited by plant size and age. Additionally, bromoxynil seems to be active on a wider range of species, resulting in reduced pasture diversity.

Initially, the fireweed plant density at the Beechmont field site was relatively high (10 to 18 plants m^{-2}). However, following the implementation of treatments, all four herbicides rapidly reduced the density of fireweed. The density remained low thereafter, except for the bromoxynil treatment, which showed some evidence of plant regrowth and seedling recruitment. Consequently, implementing a follow-up application of bromoxynil was advantageous, reducing the fireweed density to zero plants m^{-2}, while the single application treatments contained three plants m^{-2}, 5 months after the first application of bromoxynil.

Watson et al. [5] suggested that a follow-up herbicide treatment was often necessary for effective fireweed management. Similarly, Sindel and Coleman [2] recommended the follow-up of an initial herbicide treatment with spot spraying with one of the registered herbicides, such as triclopyr/picloram/aminopyralid or fluroxypyr/aminopyralid in the Spring. The results from the current study suggest that a second application is not necessary, but if applied, the timing between the two treatments could vary greatly depending on which herbicide is used. For example, follow-up control using bromoxynil may need to be undertaken much sooner than fluroxypyr/aminopyralid, metsulfuron-methyl, or triclopyr/picloram/ aminopyralid (Figure 1). These or similar herbicides have been shown to give some residual control of seedling recruitment for other herbaceous Asteraceae weeds such as florestina (*Florestina tripteris* auth.) [17]. In future, we recommend performing herbicide residual studies on the pasture feed to determine the active substances present in the feed.

Since all the tested herbicides are selective in their mode of control, they are considered not to damage the pasture grass species in the field. However, other components of the pasture community, including the broadleaved pasture legumes, may be damaged (Table 3). This aspect of fireweed management with selective herbicides needs to be further studied. According to Anderson and Panetta [8], clover (*Trifolium* sp.) was severely damaged by metsulfuron-methyl, clopyralid, triclopyr and triclopyr and picloram. Although the 2,4-D formulations damaged neither blue couch (*Digitaria didactyla* Willd.) nor clover, atrazine plus 2,4-D caused severe damage to both species.

In addition, the four tested herbicides have different withholding periods, which may also influence which herbicide is selected for use. Bromoxynil has an 8-week withholding period [2]; therefore, grazing or cutting for stock feed should be avoided during that period (APVMA, 2022). Triclopyr/ picloram/aminopyralid [18] and fluroxypyr/aminopyralid [19] have a 7-day withholding period, and even after death, many plants endure toxicity and more palatable thus, the stock should not be grazing for 7 days [20,21]. However, metsulfuron-methyl has no withholding period [22].

According to the first seed bank analysis, fireweed seeds can be present in the soil in moderately large numbers (3000 to 10,000; Table 2). Ragweed parthenium (*Parthenium hysterophorus* L.), another invasive Asteraceae weed, can form seed banks as large as ca. 45,000 seeds m^{-2} in a similar habitat to that studied in Southeast Queensland (SEQ) [11]. The results for the fireweed seed bank in this study were like that of Sindel et al. [23], who had undertaken studies in two different locations in New South Wales. A recent study undertaken by Karem [24] has indicated that fireweed seeds collected from the same Beechmont site in SEQ had an indicative short life of <1 year in the seed bank. Since a single fireweed plant can produce up to ca. 30,000 seeds [25] that can be effectively dispersed by wind, the invasive strategy of this weed must be seen as a balance between high seed production with rapid dispersal and producing a medium-sized seed bank of short-lived seeds. Land managers should therefore focus more on preventing seed set and dispersal and, to a lesser extent, on preventing the formation of seed banks. The present study has shown that most of the fireweed seed is to be found in the upper soil layers (0 to 2 cm), while a significant decrease occurs in both layers over time after herbicide application. In the present trial, undertaken in the absence of grazing livestock, this could be due to the use of management practices that do not disturb the soil surface. The field trial site used in this study was not grazed for ca. 10 months following the herbicide treatments, with no cattle to disturb the soil surface, the movement of fireweed seeds deeper into the seed bank of a healthy kikuyu pasture was negligible (Table 3).

According to the Shannon-Weiner index, there was no reduction in the seed community biodiversity in the topsoil layers after herbicide application; however, there was an Index reduction in the bottom layers of bromoxynil-treated sub-plots and in the control sub-plots. This indicates that except for bromoxynil, all other herbicides did not reduce the biodiversity of the pasture seed community. Bromoxynil, being a contact herbicide and effective on seedlings [8], may have reduced biodiversity more than other herbicides. Being a contact herbicide, only those parts of the plant that come directly in contact with the herbicide are killed, and the plant will often regrow from the unaffected parts. Significant fireweed seedling recruitment after spraying is often observed [7]. Anderson and Panetta [8] reported that bromoxynil (3 L ha^{-1}) was unsuccessful in controlling mature fireweed plants, with substantial regrowth occurring five months after spraying.

Through natural seed decline and increased competition from the kikuyu pasture, a rapid decline in the soil seed bank is expected to follow. The key to this rapid decline will be the elimination of reproductive plants, a critical component of an effective management strategy for fireweed. Compared to the control, the reduction in seed densities of fireweed in the herbicide-treated sub-plots was high (Table 3). Among tested herbicides, the highest seed density reduction percentage was observed in bromoxynil-treated sub-plots (-18%) (Table 3). Kikuyu was the dominant seed bank species (Table 1) before spraying the herbicides, and there was a dramatic increase of kikuyu seed in the soil seed bank after the herbicides had been applied, indicating the herbicides did not affect kikuyu seed populations or any other grasses or sedges (Table 3). Interestingly, even after the application of herbicide, some new species (*P. oleraceae, M. polymorpha* and *M. carolinia*) appeared from the seed bank while some species declined (*T. repens, C. leptophyllum, S. nigrum, Verbena* sp., *Hydrocotyle* sp., *P. americana*) (Table 3). This may be due to the seasonal variation in the seed bank.

From the present study, when herbicides were used to control mature fireweed plants, the established kikuyu pasture was able to significantly suppress the recruitment of new fireweed plants from the seed bank. However, even without herbicide application (as seen in the control plots; Table 3), a dominant kikuyu pasture can reduce fireweed recruitment and seed input. Perennial pasture species [*viz.* setaria (*Setaria sphacelata* Schum.), kikuyu (*Pennisetum clandestinum* Hochst. ex Chiov.), paspalum (*Paspalum dilatatum* Poir.), and Rhodes grass (*Chloris gayana* Kunth.)], that are competitive through late Summer and Autumn, will help to prevent the establishment of fireweed seedlings in the Autumn and Winter months [22]. Therefore, in a field situation, a well-established pasture community, given time, may well prevent further fireweed seedlings from emerging if stocking rates are sensibly managed [26].

4. Materials and Methods

4.1. Field Experiments

4.1.1. Study Site

A field site at Beechmont ($28°5'32.61''$ S; $153°13'11.14''$ E) in Southeast Queensland (SEQ), containing a dense infestation of fireweed (more than 90% of plants were adult flowering plants), was selected for this study (density of ca. 10 to 18 plants m^{-2}). The soil was a well-drained red ferrosol with a clay loam-to-clay texture and was dominated by kikuyu grass (*Pennisetum clandestinum* Hochst. ex Chiov). Several other key species were observed (Table 2). The climate was warm temperate, with rainfall averaging 656.5 mm annually [12]. The mean annual maximum temperature was 22.5 °C while the mean annual minimum temperature was 14.0 °C in the year of study. At the beginning of the experiment, monthly rainfall was 16.4 mm, and minimum and maximum monthly temperatures averaged 3.8 and 21.6 °C, respectively [12].

4.1.2. Experimental Design

The experiment was established in June 2018 using a split-plot design, with herbicide treatments allocated to main plots, several applications allocated to sub-plots and each treatment replicated five times. Herbicide treatments comprised the application of either bromoxynil, fluroxypyr/aminopyralid, metsulfuron-methyl or triclopyr/picloram/aminopyralid, plus an untreated control (without herbicide). All herbicides were applied at their recommended rates for the management of fireweed (Table 5). Sub-plot treatments were either a one-off herbicide application (July 2018) or two applications whereby the herbicide treatment was repeated 3 months after the first application (i.e., in October 2018). To set up the trial, five treatment blocks were established parallel to each other, with main plots 10×10 m in size and divided into two 5×5 m sub-plots. The site was fenced, and livestock was excluded for the full duration of the study (ca. 10 months).

Table 5. Herbicide treatments, trade names, and rates applied to control fireweed. All treatments included surfactant; 2% (*v/v*) Pulse Penetrant® (1 kg L^{-1}) organo-modified poly dimethyl siloxane, Nufarm Australia Limited, Laverton, North Victoria, Australia.

Herbicide	Trade Name	Rate	Product Rate
		(g Active Ingredient ha^{-1})	(25 m^{-2})
bromoxynil	Bromicide® 200, Nufarm Australia Limited, Victoria, Australia	560	7.00 mL
metsulfuron-methyl	Brush-Off®, Bayer, Victoria, Australia	24	0.10 g
fluroxypyr/aminopyralid	HotShot™, CortevaTM Agriscience, New South Wales, Aus-tralia	210/15	3.75 mL
triclopyr/picloram/aminopyralid	Grazon™ Extra, Cor-tevaTM Agriscience, New South Wales, Australia	900/300/24	7.50 mL

4.1.3. Herbicide Application

The herbicides were applied using a Makita EVH2000 24.5 cm^3, four-stroke petrol backpack sprayer equipped with a 2.5 m swath hand-held boom containing four nozzles (spaced 50 cm apart) and delivering a carrier spray volume of 800 L ha^{-1} (2.0 L 25 m^{-2} plot). A single pass of the boom was undertaken when spraying plots, with the height maintained at 0.5 m above the soil level by attaching a weighted 50 cm vertical guide to the boom. A pressure gauge mounted on the handle of the boom provided confirmation of an operating pressure of 120 kPa, and a portable metronome ensured a constant walking speed of 4.0 km hour^{-1} was maintained. The Turbo Twinjet flat spray nozzles (TTJ60-11002) used in this experiment were supplied by Spraying Systems (Wheaton, IL, USA). All solutions contained a 2% (*v/v*) Pulse Penetrant® (1.0 kg L^{-1} organo-modified polydimethylsiloxane) obtained from Nufarm Australia Limited (Laverton, North Victoria, Australia).

4.1.4. Plant Density

To determine the effect of the herbicides on fireweed density over time, two quadrats (1.0 × 1.0 m) were placed randomly within each sub-plot, and the density of fireweed plants was determined at 0, 2, 3, 5, 7, 10, and 13 months after herbicide application. To identify dead plants, the outer epidermal layer of the stem of the fireweed plants was scraped away to reveal the inner tissues of the stem. Live stems had a green cambium layer immediately beneath the epidermal layer and green or white tissue inside, whereas dead tissues appeared a distinct brown colour.

4.1.5. Seed Bank Density

Before applying the herbicides to the trial site, soil seed bank samples were taken from each subplot designated to receive the follow-up treatment, as well as the untreated control plots, in July 2018. To take soil samples, two 1 × 1 m quadrats were randomly placed within each of the designated sub-plots, and five cylindrical soil cores (5 cm in diameter and 10 cm deep) were extracted from each quadrat (one from each of the four corners and one from the centre of the quadrat) using a soil corer. Soil cores were then separated into two different depths, 0 to 2 and 2 to 10 cm deep.

The soil seed bank samples, taken from the same depths and from both quadrats, were pooled into one sample for each sub-plot. The two soil samples from the two quadrats were placed into separate plastic bags, sealed, and stored at ambient temperature for 2 to 3 days. They were then spread thinly over a 2 cm layer of a Gatton media compost (Osmocote 8–9 M, Osmocote 3–4 M, Nutricote 7 M, coated iron, moisture aid, dolomite, and Osmoform) contained within shallow germination trays (20 × 25 × 6 cm; w/l/d). Then, all trays were distributed randomly to the top of a greenhouse bench at the University of Queensland, Gatton, in July 2018. The temperature in the greenhouse was maintained close to the outside ambient temperature (mean annual maximum of 26.9 °C, with a mean annual minimum of 13.0 °C). Two control trays were placed among the experimental trays to check for compost or greenhouse seed contaminants. All trays were watered daily to maintain soil moisture at or close to the field capacity. The trays were observed regularly for newly emerging seedlings, and when observed, seedlings were marked with a cocktail stick and initially recorded as either being 'fireweed' or 'other species. Once seedlings were large enough to be identified, then, they were counted and removed. If they could not be identified, representative individuals were planted into pots and grown to maturity for further taxonomic identification using the appropriate literature [27,28]. When seedling emergence ceased, the soil in the trays was dried for 2 weeks before being stirred, rewetted, and inspected for any further seedling emergence over a further 9 months [11].

Soil sampling at the field site was undertaken again in March 2019, 5 months after the follow-up herbicide treatments had been applied in October 2018. The same procedure as described above was used to collect soil samples (from follow-up subplots) and to monitor for seedling emergence in the greenhouse. At this time, the temperature in the greenhouse was maintained close to the outside ambient temperature (the mean maximum during March 2019 was 30.0 °C, with a mean minimum during March 2019 of 17.5 °C).

The species diversity of the soil seed bank was calculated using the Shannon–Weiner Index:

$$H' = -\sum_{i-1}^{s} P_i \log_e P_i$$

where S is the number of species and Pi is the proportion of the total of all species' individuals per quadrat represented by the *i*th species [10].

4.2. Statistical Analysis

Plant density data were subjected to analysis of variance (ANOVA) to compare the plant density of different treatments. Means comparisons were through the use of the least significant difference (LSD) test ($p = 0.05$).

An ANOVA was also performed to compare the fireweed seed densities between the two seed banks (before and after spraying) once the data had been transformed to a logarithmic scale. Comparison of treatment means was conducted by the LSD ($p = 0.05$) test. All the data were analyzed using the R statistical software (Version 3.6.3).

5. Conclusions

Herbicides fluroxypyr/aminopyralid, bromoxynil, metsulfuron-methyl and triclopyr/picloram/aminopyralid are effective in controlling fireweed plants. Fireweed seeds dominated both the upper (0 to 2 cm) and lower (2 to 10 cm) soil layers, with the highest density of fireweed seeds observed in the upper layer. Even with a high seed load being produced at the Beechmont site, several of the herbicides tested were effective in controlling plants and reducing the density of fireweed seeds entering the soil seed bank. According to the Shannon—Weiner index, except for bromoxynil, all other herbicides did not reduce the biodiversity of the pasture seed community. However, the implementation of a follow-up application of bromoxynil was more effective than the single application. For other tested herbicides, even one application was sufficient to control fireweed plants at all stages of development at the Beechmont site, but this was when a healthy pasture stand was present, and the stocking rate was zero. Thus, for the effective management of fireweed, the application of a single dose of one of three different herbicides (fluroxypyr/aminopyralid, metsulfuron-methyl and triclopyr/picloram/ aminopyralid) when simultaneously ungrazed for a minimum of 10 months can be efficacious. However, given the presence of fireweed in the soil seed bank, subsequent follow-up herbicide control may still be necessary at some stage, depending on prevailing environmental conditions.

Further research is now needed to evaluate the effect of pasture competition on fireweed establishment and to determine pasture density thresholds that can prevent recruitment. Further research is also needed to evaluate the impact of increased grazing pressure, the effect of different kinds of grazing stock (cattle, sheep, or goats) on established populations of fireweed, and how bare and disturbed ground can affect the establishment of fireweed. In addition, besides selecting the most effective herbicide, cost considerations will also need to be included in the decision-making process.

Author Contributions: Conceptualization: K.W., S.C., J.V. and S.A., Writing—original draft: K.W., Data analysis, interpretation, review, and editing: K.W., S.C., J.V., S.K. and S.A. All authors have read and agreed to the published version of the manuscript.

Funding: We thank the University of Queensland and the Department of Agriculture and Fisheries, Brisbane (RM 2018000131) for providing the funding for part of this project.

Acknowledgments: We also thank the Land holders Irene and Bruce Mills, for giving access to their property at Beechmont for the field research.

Conflicts of Interest: The authors declare no conflict of interest.

References

1. Sindel, B.M.; Michael, P.W.; McFadyen, R.E.; Carthew, J. The continuing spread of fireweed (*Senecio madagascariensis*)—The hottest of topics. In Proceedings of the Sixteenth Australian Weeds Conference, Brisbane, QLD, Australia, 18–22 May 2008; pp. 47–49.
2. Sindel, B.; Coleman, M. *Fireweed—A Best Practice Management Guide for Australian Landholders*; University of New England: Armidale, NSW, Australia, 2012. Available online: https://www.une.edu.au/__data/assets/pdf_file/0004/52366/2012.-Fireweed-A-Best-Practice-Management-Guide-for-Australian-Landholders.pdf (accessed on 4 October 2020).
3. Duke, S.O. Overview of herbicide mechanisms of action. *Environ. Health Perspect.* **1990**, *87*, 263–271. [CrossRef] [PubMed]
4. Sherwani, S.I.; Arif, I.; Khan, H.A. Modes of action of different classes of herbicides. In *Herbicides, Physiology of Action, and Safety*; Price, A., Kelton, J., Sarunaite, L., Eds.; IntechOpen Ltd.: London, UK, 2015; pp. 165–177.
5. Watson, R.; Allan, H.; Munnich, D.; Launders, T.; Macadam, J. *Fireweed, Agfact P7.6.26*, 2nd ed.; New South Wales Agriculture: Orange, NSW, Australia, 1994.
6. Motooka, P.; Nagai, G.; Onuma, K.; DuPone, M.; Kawabata, A.; Fukumoto, G.; Powley, J. *Control of Madagascar Ragwort (Aka Madagascar Fireweed, Senecio Madagascariensis*; College of Tropical Agriculture, University of Hawaii: Honolulu, Hawaii, 2004. Available online: https://scholarspace.manoa.hawaii.edu/bitstream/10125/12611/1/WC-2.pdf (accessed on 17 November 2017).
7. Department of Agriculture and Fisheries. Fireweed (*Senecio madagascariensis*). Available online: https://www.daf.qld.gov.au/__data/assets/pdf_file/0009/67167/IPA-Fireweed-PP31.pdf (accessed on 27 July 2017).
8. Anderson, T.M.D.; Panetta, F.D. Fireweed response to boom spray application of different herbicides and adjuvants. *Plant Prot. Q.* **1995**, *10*, 152–153.
9. Baskin, C.C.; Baskin, J.M. *Seeds: Ecology, Biogeography, and Evolution of Dormancy and Germination*; Academic Press: San Diego, CA, USA, 1998.
10. Nguyen, T.L.T.; Bajwa, A.A.; Navie, S.C.; O'Donnell, C.; Adkins, S.W. The soil seedbank of pasture communities in Central Queensland invaded by *Parthenium hysterophorus* L. Rangel. *Ecol. Manag.* **2017**, *70*, 244–254.
11. Navie, S.C.; Panetta, F.D.; McFadyen, R.E.; Adkins, S.W. Germinable soil seed banks of central Queensland rangelands invaded by the exotic weed *Parthenium hysterophorus* L. *Weed Biol. Manag.* **2004**, *4*, 154–167. [CrossRef]
12. Bureau of Meteorology. Australia in Autumn 2022. Available online: http://www.bom.gov.au/climate/current/season/aus/summary.shtml (accessed on 10 July 2022).
13. Radford, I.J. Impact Assessment for the Biological Control of *Senecio madagascariensis* Poir (Fireweed). Ph.D. Thesis, University of Sydney, Sydney, NSW, Australia, 1997.
14. Buhler, D.D.; Hartzler, R.G.; Forcella, F. Implications of weed seedbank dynamics to weed management. *Weed Sci.* **1997**, *45*, 329–336. [CrossRef]
15. Bear, J.L.; Giljohann, K.M.; Cousens, R.D.; Williams, N.S.G. The seed ecology of two invasive *Hieracium* (Asteraceae) species. *Aust. J. Bot.* **2012**, *60*, 615–624. [CrossRef]
16. Bajwa, A.A.; Chauhan, B.S.; Adkins, S.W. Germination ecology of two Australian biotypes of Ragweed Parthenium (*Parthenium hysterophorus*) relates to their invasiveness. *Weed Sci.* **2018**, *66*, 62–70. [CrossRef]
17. McKenzie, J.; Brazier, D.A.; Campbell, S.; Vitelli, J.; Anderson, A.E.; Mayer, R.J. Foliar herbicide control of sticky Florestina (*Florestina tripteris* DC.). *Rangel. J.* **2014**, *36*, 259–265. [CrossRef]
18. Brooks, S.J.; Setter, S.D. Increased options for controlling mikania vine (*Mikania micrantha*) with foliar herbicide. In Proceedings of the 19th Australasian Weeds Conference, Hobart, TAS, Australia, 1–4 September 2014.
19. Australian Pesticides and Veterinary Medicines Authority. Evaluation of the New Active Aminopyralid in the Product Hotshot Herbicide. Available online: https://apvma.gov.au/sites/default/files/publication/13581-prs-aminopyralid.pdf (accessed on 23 May 2021).
20. Corteva Agriscience. Grazon®Extra. Available online: https://www.corteva.com.au/content/dam/dpagco/corteva/au/au/en/products/files/Grazon-Extra-Herbicide-Label.pdf (accessed on 7 March 2023).
21. Dow AgroSciences. HotShot. Available online: http://www.herbiguide.com.au/Labels/AMINFLUR_59173-0906.PDF (accessed on 7 March 2023).

22. Department of Primary Industries. Fireweed Department of Primary Industries NSW. Available online: http://www.dpi.nsw.gov.au/__data/assets/pdf_file/0007/49840/Fireweed.pdf (accessed on 19 March 2019).
23. Sindel, B.; Coleman, M.; Barnes, P. Fireweed Control Research (DAFF 179/10). Available online: https://www.une.edu.au/__data/assets/pdf_file/0005/38633/fireweed-report-biocontrol.pdf (accessed on 12 March 2018).
24. Karem, S.P.R. Can Fireweed (*Senecio madagascariensis* Poir.) Seed Persistence in the Field be Determined by a Laboratory Controlled Ageing Test? Master's Thesis, University of Queensland, Gatton, QLD, Australia, 2020.
25. Egli, D.; Olckers, T. Abundance across seasons of insect herbivore taxa associated with the invasive *Senecio madagascariensis* (Asteraceae), in its native range in KwaZulu-Natal, South Africa. *Afr. Entomol.* **2015**, *23*, 147–156.
26. Sindel, B.M.; Michael, P.W. Growth, and competitiveness of *Senecio madagascariensis* Poir. (fireweed) in relation to fertilizer use and increases in soil fertility. *Weed Res.* **1992**, *57*, 99–406. [CrossRef]
27. Stanley, T.D.; Ross, E.M. *Flora of South-East Queensland*; Queensland Government: Brisbane, QLD, Australia, 1989; Volume 3.
28. Plantnet. National Herbarium of New South Wales PlantNET—FloraOnline. Available online: https://www.nsw.gov.au/ (accessed on 21 November 2017).

plants

Article

Phenology and Diversity of Weeds in the Agriculture and Horticulture Cropping Systems of Indian Western Himalayas: Understanding Implications for Agro-Ecosystems

Shiekh Marifatul Haq [1], Fayaz A. Lone [2], Manoj Kumar [3], Eduardo Soares Calixto [4], Muhammad Waheed [5], Ryan Casini [6], Eman A. Mahmoud [7] and Hosam O. Elansary [8,*]

1   Department of Ethnobotany, Institute of Botany, Ilia State University, Tbilisi 0162, Georgia
2   Department of Botany, Government Degree College (Women), Kupwara 193222, India
3   GIS Centre, Forest Research Institute (FRI), PO New Forest, Dehradun 248006, India
4   Institute of Food and Agricultural Sciences, University of Florida, Gainesville, FL 32962, USA
5   Department of Botany, University of Okara, Okara 56300, Pakistan
6   School of Public Health, University of California, Berkeley, 2121 Berkeley Way, Berkeley, CA 94704, USA
7   Department of Food Industries, Faculty of Agriculture, Damietta University, Damietta 34511, Egypt
8   Department of Plant Production, College of Food & Agriculture Sciences, King Saud University, P.O. Box 2460, Riyadh 11451, Saudi Arabia
*   Correspondence: helansary@ksu.edu.sa

Citation: Haq, S.M.; Lone, F.A.; Kumar, M.; Calixto, E.S.; Waheed, M.; Casini, R.; Mahmoud, E.A.; Elansary, H.O. Phenology and Diversity of Weeds in the Agriculture and Horticulture Cropping Systems of Indian Western Himalayas: Understanding Implications for Agro-Ecosystems. *Plants* **2023**, *12*, 1222. https://doi.org/10.3390/plants12061222

Academic Editors: Congyan Wang and Hongli Li

Received: 14 February 2023
Revised: 3 March 2023
Accepted: 3 March 2023
Published: 8 March 2023

Abstract: Weeds are a major threat to agriculture and horticulture cropping systems that reduce yield. Weeds have a better ability to compete for resources compared to the main crops of various agro-ecosystems and act as a major impediment in reducing overall yield. They often act as energy drains in the managed agroecosystems. We studied weed infestation for five different agro-ecosystems in the part of Indian Western Himalayas represented by paddy, maize, mustard, apple and vegetable orchards. Systematic random sampling was done to record flowering phenology and diversity of weeds during the assessment period 2015–2020. We recorded 59 weed species, taxonomically distributed among 50 genera in 24 families. The Asteraceae family has the most species (15% species), followed by Poaceae (14% species), and Brassicaceae (12% species). The Therophytes were the dominant life form followed by Hemicryptophytes. The majority of the weeds were shown to be at their most blooming in the summer (predominantly from June to July). The Shannon index based diversity of weeds ranged from 2.307–3.325 for the different agro-ecosystems. The highest number of weeds was in the horticulture systems (apple > vegetable) followed by agriculture fields (maize > paddy > mustard). Agriculture and horticulture cropping systems were distinguished using indicator species analysis, which was supported by high and significant indicator values for a number of species. *Persicaria hydropiper, Cynodon dactylon, Poa annua, Stellaria media,* and *Rorippa palustris* had the highest indicator value in agriculture cropping systems, while *Trifolium repens, Phleum pratense,* and *Trifolium pratense* had the highest indicator value in horticulture cropping systems. We found that eleven weed species were unique to apple gardens followed by nine in maize, four in vegetables, two in mustard and one in paddy fields. Spatial turnover (βsim) and nestedness-resultant components (βsne) of species dissimilarity revealed dissimilarity lower than 50% among the five cropping systems. The study is expected to assist in formulating an appropriate management strategy for the control of weed infestation in the study region.

Keywords: agriculture; weed survey; weed management; beta diversity; indicator species analysis

1. Introduction

The vegetated land can broadly be categorized into forested and agricultural landscapes. In vegetated lands, weed invasion has been referred to as one of the prominent threats after climate change [1–4]. Weeds are a group of specialized plants that have evolved along temporal and spatial scales because of their large seed production, aggressive reproduction, high regenerative capacity, and large phenotypic plasticity [5]. The weed plants are considered unwanted and undesirable at a particular site [6,7]. Among the various biological stresses, weeds are known as one of the most detrimental to crop production [8,9]. In addition, these weeds provide shelter to numerous pests of crop plants and, thus, indirectly become a cause of many crop diseases [10]. Weeds are notorious for decreasing crop yield and are found to be economically more detrimental than bacteria, fungi, insects or other crop pests in many situations. For instance, it has been estimated that weeds result in about USD 11 billion of economic losses every year in India [11].

Weeds often invade human-controlled settings, such as agricultural fields, orchards, parks, and lawns. The presence of weeds in agricultural settings is well-known from the beginning of civilization; however, their role in agriculture and horticulture is recognized only in recent decades [12]. The broad ecological amplitude of weeds facilitates them to invade a broad range of habitats and ecological niches. The presence of weeds in the agricultural and horticulture fields compete with the native plants to affect yield [10]. Although horticulture is considered one of the subdivisions of agriculture, they differ in management practices and the kind of plants grown. Thus, it might be possible that similar land under a different regime of management for growing different crops may influence selective weed infestation. This provides an opportunity to investigate the weed invasion in these two different land use classes. At the same time, the information on the phenological events of weed is important for formulating effective control and management [13,14].

The documentation of weeds in different crop fields is important for the management and control of weeds [15,16]. The identification and documentation of weed invasion are important for formulating various strategies for managing the weeds [17]. Management decisions for agriculture and horticulture fields are dependent upon the phenological information of crops for improving yield [13] while the phenological calendar of weeds is important for formulating their effective eradication strategy. In the study landscape, most of the weeds are invasive and not native to the place. There is a need to understand the type, pattern, and impacts of weeds and invasive species for improving the targeted yields of the selected cropping system. However, there is a wide knowledge gap in invasive species research in developing countries and it acts as a decisive impediment to managing invasion. Developing countries lag far behind developed countries in invasive species research [18]. Specifically, Asian countries are represented poorly in the scientific literature on invasive species. Such a lag and data deficit must not be considered an indicator of low invasion risk or low intensity of invasion in developing countries. Rather, this indicates less effort being made to explore the invasive species, less documentation, and insufficient action on data availability.

For supplementing the existing knowledge base on invasive weeds, we present here a comprehensive assessment to achieve the objectives of (i) documenting various weeds in agriculture (paddy, maize, and mustard) and horticulture (apple and vegetable) systems of the study region, (ii) identification as native vs. exotic, (iii) diversity assessment of weed under different cropping systems, and (iv) understanding flowering phenology of weeds. We also attempt to address research queries on crop specificity of weeds to a given cropping system and to assess whether a specific cropping system has more diverse weeds than another. The findings will supplement existing knowledge on invasive weeds amid a dearth of knowledge for the study region. At the same time, a similar approach can be adopted for collecting vital information on weeds for other study regions.

2. Results

The study recorded 59 weed species, taxonomically distributed among 50 genera in 24 families (Table 1). The perennial weeds were 28 in number constituting 48% of the total encountered weeds in all of the cropping systems. The other life span categories were annual (39%), annual–biennial (7%), and annual–biennial–perennials (3%) having 23, 4, and 2 weed species, respectively (Table 1). The distribution of species among 24 families is lopsided, with 4 families accounting for half of the species and 20 families for the other half. 15 families were represented by the presence of just single species. The Asteraceae was the dominant family with 9 species (15%) followed by Poaceae with 8 species (14%), Brassicaceae with 7 species (12%), and Plantaginaceae with 5 species (8%). The rest of the species was represented by Fabaceae, Lamiaceae, Polygonaceae, Ranunculaceae, Amaranthaceae, and other families (Table 1). The monotypic families are Apiaceae, Boraginaceae, Asparagaceae, Caryophyllaceae, Chenopodiaceae, Cyperaceae, Euphorbiaceae, Fumariaceae, Geraniaceae, Rosaceae. Further enumeration is represented in Table 1.

Table 1. Description of weeds in the part of Indian Western Himalayas (IWH), Kupwara district, Jammu & Kashmir, represented by their scientific name, family, dominance in different cropping systems shown by the Important Value Index (IVI), origin (native/exotic), flowering phenology and Raunkiaer life form. (Life span: A = annual, B = biennial, P = perennial, A-B = annual-biennial, A-B-P = annual-biennial-perennials. Dominant weeds in either of the cropping systems have been highlighted in bold along with their top three IVI values).

| Scientific Name | Family | Life Span | Agriculture | | | Horticulture | | Origin | Flowering | Raunkiaer |
			Paddy	Maize	Mustard	Apple	Vegetable		Phenology	Lifeform
Achillea millefolium Linn.	Asteraceae	P	0	0	0	2.74	0	Native	July–Sept.	Hemicryptophyte
Amaranthus caudatus L.	Amaranthaceae	P	0	0	0	2.74	11.20	Exotic	June–Aug.	Therophyte
Amaranthus viridis L.	Amaranthaceae	A	7.45	0	7.85	0	0	Exotic	June–Aug.	Therophyte
Anthemis cotula L.	Asteraceae	A	8.35	0	0	8.32	12.13	Exotic	May–June	Therophyte
Arabidopsis thaliana (L.) Heynh.	Brassicaceae	A	0	0	0	7.23	0	Exotic	April–June	Therophyte
Avena sativa L.	Poaceae	A	0	0	0	11.13	14.24	Exotic	June–Aug.	Therophyte
Brassica rapa L.	Brassicaceae	A	0	0	13.26	0	0	Exotic	April–June	Hemicryptophyte
Capsella bursa-pastoris (L.) Medik.	Brassicaceae	A	0	0	0	6.78	9.34	Native	March–July	Therophyte
Cardamine hirsuta L.	Brassicaceae	B	11.01	0	0	13.02	6.05	Native	March–May	Therophyte
Carex fedia Nees	Cyperaceae	P	0	20.63	0	0	0	Native	March–June	Hemicryptophyte
Cerastium cerastoides (L.)	Caryophyllaceae	A	3.76	0	7.19	0	0	Native	May–Aug.	Therophyte
Chenopodium album L.	Chenopodiaceae	A	0	0	0	5.05	12.61	Exotic	July–Sept.	Therophyte
Cirsium arvense (L.) Scop.	Asteraceae	P	0	0	0	2.08	0	Exotic	May–Aug.	Hemicryptophyte
Convolvulus arvensis L.	Convolvulaceae	P	0	0	0	3.37	0	Exotic	April–Aug.	Therophyte
Cynodon dactylon (L.) Pers.	Poaceae	P	0	21.21	0	13.32	11.47	Native	May–Aug.	Hemicryptophyte
Daucus carota L.	Apiaceae	B	0	2.97	0	4.18	8.41	Exotic	June–Sept.	Hemicryptophyte
Duchesnea indica (Andrews) Teschem.	Rosaceae	P	0	14.51	0	0	7.41	Native	March–Oct.	Hemicryptophyte
Erigeron canadensis L.	Asteraceae	A-B	5.17	16.73	5.62	7.28	12.61	Exotic	April–Sept.	Therophyte
Euphorbia helioscopia L.	Euphorbiaceae	A	0	0	0	0	3.29	Exotic	April–July	Therophyte
Fumaria indica (Hausskn.) Pugsley	Papaveraceae	A	0	5.64	0	0	0	Native	April–June	Therophyte
Gagea lutea (L.) Ker Gawl.	Liliaceae	P	0	0	0	3.37	0	Native	April–June	Geophyte
Galinsoga parviflora Cav.	Asteraceae	A	0	0	0	3.75	0	Exotic	June–Aug.	Therophyte
Geranium rotundifolium L.	Geraniaceae	P	3.29	0	3.56	10.51	0	Native	June–July	Therophyte
Iris kashmiriana Baker	Iridaceae	P	0	6.39	0	0	0	Native	April–Jun	Geophyte
Lactuca serriola L.	Asteraceae	A	0	0	0	4.62	5.63	Native	July–Sept.	Therophyte
Mazus pumilus (Burm.f.) Steenis	Mazaceae	A	6.67	0	0	0	0	Native	April–May	Therophyte
Medicago polymorpha L.	Fabaceae	A	0	0	7.92	9.81	0	Exotic	April–June	Therophyte

Table 1. *Cont.*

Scientific Name	Family	Life Span	Agriculture			Horticulture		Origin	Flowering	Raunkiaer
			Paddy	Maize	Mustard	Apple	Vegetable		Phenology	Lifeform
Mentha arvensis L.	Lamiaceae	P	0	0	0	5.35	0	Native	July–Oct.	Hemicryptophyte
Mentha longifolia (L.) Huds.	Lamiaceae	P	7.47	0	0	5.92	0	Exotic	July–Sept.	Hemicryptophyte
Myosotis arvensis (L.) Hill	Boraginaceae	P	0	18.93	0	0	7.38	Exotic	May–June	Therophyte
Nepeta cataria L.	Lamiaceae	P	0	0	0	3.37	0	Native	June–Aug.	Hemicryptophyte
Oenothera rosea L'Hér. ex Aiton	Onagraceae	P	0	0	0	6.35	0	Native	April–Aug.	Therophyte
Persicaria hydropiper Delarbre	Polygonaceae	A	56.71	14.92	3.91	9.36	6.05	Native	July–Sept.	Therophyte
Phleum pratense L.	Poaceae	P	0	0	0	0	25.15	Native	July–Aug.	Hemicryptophyte
Plantago lanceolata L.	Plantaginaceae	A-B-P	7.47	0	7.92	5.05	9.38	Native	May–Aug.	Therophyte
Plantago major L.	Plantaginaceae	P	6.31	5.07	6.92	7.61	6.51	Native	May–Sept.	Hemicryptophyte
Poa annua L.	Poaceae	A	95.64	42.03	99.22	34.75	21.26	Native	March–Aug.	Therophyte
Poa bulbosa L.	Poaceae	P	0	0	0	10.54	0	Native	March–July	Hemicryptophyte
Polygonum aviculare L.	Polygonaceae	A	4.96	18.55	5.28	9.39	10.27	Native	June–Aug.	Therophyte
Polypogon fugax Nees ex Steud.	Poaceae	P	0	19.19	0	0	0	Native	May–June	Therophyte
Ranunculus arvensis L.	Ranunculaceae	P	0	16.21	0	2.77	0	Native	March–April	Hemicryptophyte
Ranunculus muricatus L.	Ranunculaceae	A-B-P	0	0	0	4.62	10.14	Exotic	April–June	Therophyte
Ranunculus sceleratus L.	Ranunculaceae	P	15.53	5.91	16.13	0	0	Exotic	May–July	Therophyte
Rorippa palustris (L.) Besser	Brassicaceae	A-B	8.64	21.17	0	0	0	Native	June–Aug.	Geophyte
Rorippa indica (L.) Hiern.	Brassicaceae	A-B	0	5.76	0	0	0	Native	April–June	Geophyte
Rumex dentatus L.	Polygonaceae	P	13.87	4.71	15.12	2.75	10.28	Native	June–July	Hemicryptophyte
Senecio vulgaris L.	Asteraceae	A-B	0	0	0	2.08	0	Exotic	April–Sept.	Therophyte
Sisymbrium loeselii L.	Brassicaceae	A	11.01	0	11.84	0	0	Exotic	June–Aug.	Therophyte
Sonchus oleraceus L.	Asteraceae	A	0	0	0	0	5.62	Exotic	May–July	Therophyte
Sorghum halepense (L.) Pers.	Poaceae	P	0	8.61	0	0	0	Exotic	June–Sept.	Geophyte
Stellaria media (L.) Vill.	Poaceae	A	0	0	58.93	0	0	Native	March–Oct.	Therophyte
Taraxacum officinale Weber	Asteraceae	P	0	3.31	0	5.73	7.41	Native	March–Aug.	Hemicryptophyte
Trifolium pratense L.	Fabaceae	P	9.49	8.41	10.07	19.63	21.28	Native	May–Aug.	Hemicryptophyte
Trifolium repens L.	Fabaceae	P	11.21	5.86	12.57	33.19	10.14	Native	May–Sept.	Hemicryptophyte
Urtica dioica L.	Urticaceae	P	0	0	0	4.96	0	Exotic	June–Aug.	Therophyte
Veronica anagallis-aquatica L.	Plantaginaceae	P	0	8.12	0	0	0	Native	June–Sept.	Hemicryptophyte
Veronica peregrina L.	Plantaginaceae	A	0	0	0	0	13.47	Exotic	May–Oct.	Therophyte
Veronica persica Poir.	Plantaginaceae	A	0	0	0	6.64	14.24	Native	March–July	Therophyte
Vicia sativa L.	Fabaceae	A	5.21	4.71	6.31	0	0	Exotic	May–June	Therophyte

2.1. Functional Traits Including Flowering Phenology

The analysis of floristic distribution using Raunkiaer's life form revealed that the therophytes with 35 species forming 59% of the plant community were the dominant life form in the study region, followed by hemicryptophytes with 19 (32%), and geophytes with 5 (9%) (Table 1). The phytogeographical analysis revealed that the maximum weed species (34 in number, 58%) recorded is native, while many (25 in number, 42%) are alien (Table 1). The phenological spectrum of weed flora was presented mainly by the flowering period of each species. The present study's weed flora displayed a wide range of blooming phenology (Table 1, Figure 1). Different species flower through various seasons. We observed that most of the weeds (77%) (e.g., *Achillea millefolium, Amaranthus viridis, Avena sativa, Daucus carota,*

Galinsoga parviflora, Geranium rotundifolium, Nepeta cataria, Polygonum aviculare, Achillea millefolium, Lactuca serriola, Persicaria hydropiper and *Phleum pratense*) flowered from May to August (Table 1, Figure 1). A few weeds showed flowers in other months of the year, such as *Brassica rapa, Capsella bursa-pastoris, Cardamine hirsuta, Ranunculus arvensis, Mazus pumilus, Rumex hastatus, Viola odorata, Tussilago farfara* and *Sigesbeckia orientalis* which bloomed between September and April. This observed variance in the phenological response of blooming across different weeds was ascribed to seasonal temperature variations. The majority of weeds exhibited their peak blooming in the summer, primarily in June and July. The blossoming season typically began at the beginning of spring and lasted until the end of summer (Table 1, Figure 1). The clustering of weeds based on flowering phenology is presented in Figure 1, where weeds grouped in one limb are more similar in flowering timings and show proximity to each other.

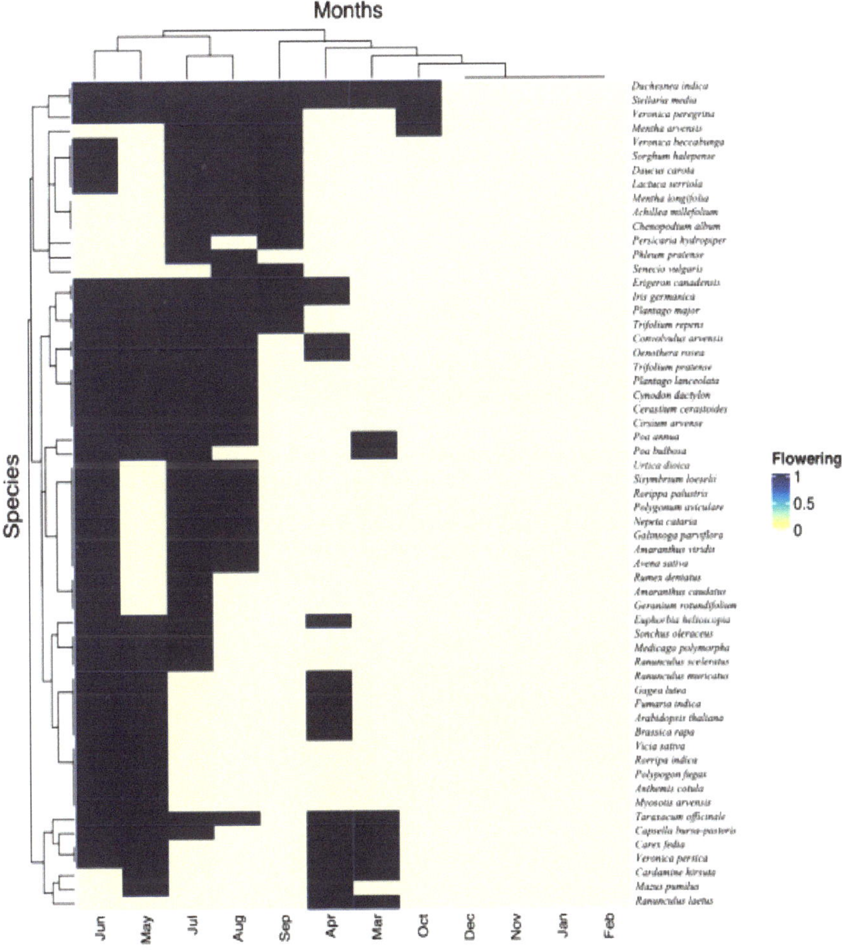

Figure 1. Two-way cluster analysis based on Sorenson's similarity index and flowering phenology of weeds.

2.2. Weed Distribution and Their Diversity in Agriculture and Horticulture Systems

The horticulture management interventions in the apple and vegetable orchards were found to have more weeds compared to the agriculture systems. Among agriculture systems, maize has the highest number of weeds followed by mustard and paddy. The apple garden had significantly higher weed diversity (3.325) than the mustard (2.307). The value of weed dominance based on the Simpson index ranges from 0.835 to 0.955 (Table 2). The other indices of weed diversity in five cropping systems are presented in Table 2.

Table 2. Presence of weeds in the agriculture and horticulture systems of Kupwara, Jammu and Kashmir, India presented as different indices.

Indices	Agriculture			Horticulture	
	Paddy Field	Maize Field	Mustard Field	Apple Garden	Vegetable Garden
Species number	20	24	18	37	27
Dominance	0.153	0.062	0.164	0.048	0.045
Shannon	2.412	2.955	2.307	3.325	3.197
Simpson	0.847	0.937	0.835	0.951	0.955
Evenness	0.557	0.801	0.558	0.751	0.905

2.3. Indicator Species Analysis

The indicator species analysis showed the separation between the paddy, maize, and mustard from apple garden and vegetable fields, as evidenced by high and substantial indicator values for several species. In the paddy field, *Persicaria hydropiper* and *Poa annua* had the highest indication value, while *Cynodon dactylon*, *Poa annua*, and *Rorippa palustris* had the highest indicator value for maize fields. In mustard fields, the indicator species with significant p-value were *Stellaria media* and *Poa annua*. *Trifolium repens* and *Poa annua* had the highest indicator value in the apple garden, while in vegetable fields, the *Phleum pratense* and *Trifolium pratense* had the highest indicator value. *Poa annua* was the indicator species in all four types of agriculture fields excluding vegetable fields (Figure 2).

It was observed that eleven weed species were unique to the apple garden followed by nine species in the maize field, four species in the vegetable garden, two species in the mustard field, and one species in the paddy field (Figure 3). However, eight species were common in all habitat types that included *Erigeron canadensis* L., *Persicaria hydropiper* Delarbre, *Plantago major* L., *Poa annua* L., *Polygonum aviculare* L., *Rumex dentatus* L., *Trifolium pratense* L., *Trifolium repens* L. Similarly, nine species were common between apple and vegetable orchards. Three weed species were common between apple, vegetable, and maize cropping systems. One weed species was common between a maize field and a vegetable orchard. The Venn diagram depicted in Figure 4 shows the number of weed species unique to a specific cropping system and common among different cropping systems. PCA analysis showed four distinct groups based on the composition and IVI of the weed species. Apple, vegetables, and maize were distinctly separated from each other, while mustard and paddy had a similar composition of weeds (Figure 4).

The species *Achillea millefolium*, *Arabidopsis thaliana*, *Cirsium arvense*, *Convolvulus arvensis*, *Gagea elegans*, *Oenothera rosea*, *Poa bulbosa*, *Nepeta cataria*, *Senecio vulgaris*, *Medicago polymorpha*, *Urticadioica* are unique to apple growing fields and species *Carexfedia*, *Fumaria indica*, *Sorghum halepense*, *Rorippaindica*, *Iris germanica*, *Polypogon fugax*, *Veronica anagallis-aquatica*, *Myosotis arvensis* were unique to vegetable fields. Weed plants such as *Carex fedia*, *Fumaria indica*, *Sorghum halepense*, *Rorippaindica*, *Iris germanica*, *Polypogon fugax*, *Veronica anagallis-aquatica*, *Myosotis arvensis* are specific to the maize field. The species linked to paddy and mustard fields are *Lactuca serriola*, *Mentha arvensis*, *Amaranthus caudatus*, *Avena sativa*, *Capsella bursa-pastoris*, *Ranunculus muricatus*, *Chenopodium album*, *Veronica persica*, *Galinsoga parviflora* (Figure 4).

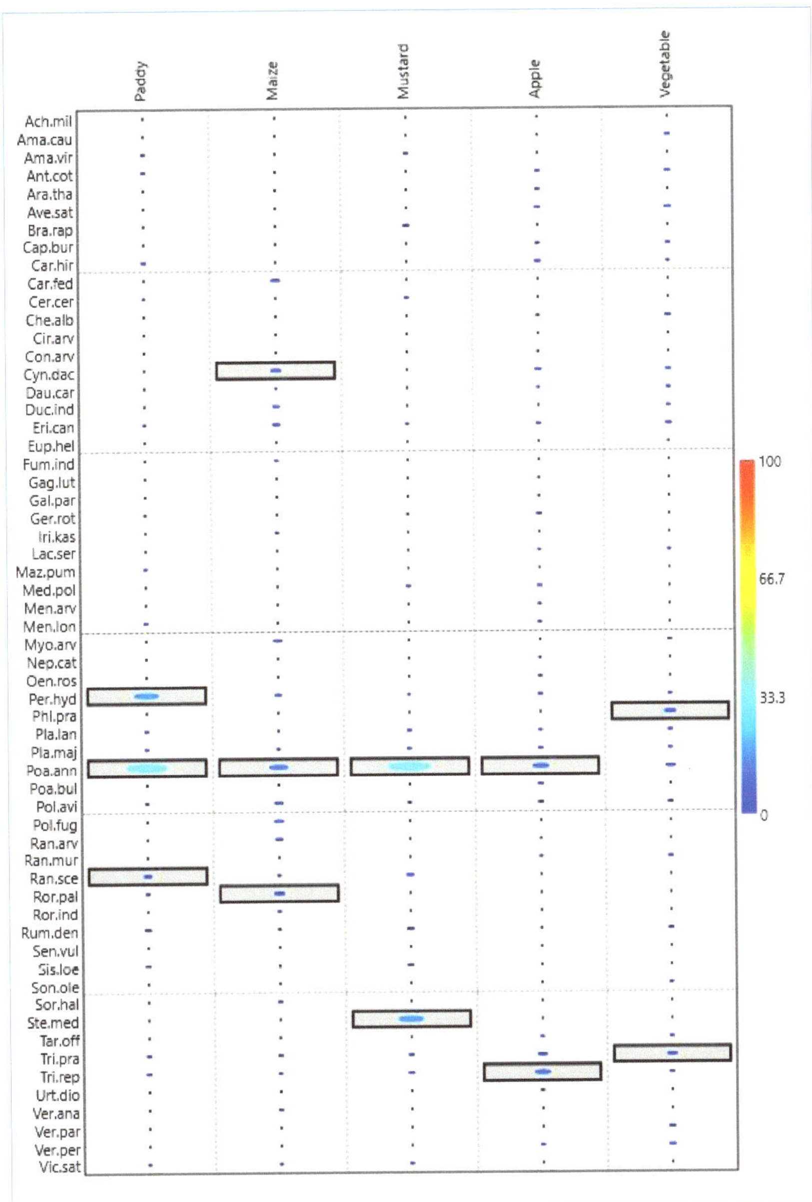

Figure 2. Indicator species analysis diagram showing the most significant indicator species in the agriculture and horticulture systems (paddy, maize, mustard, apple, vegetable). Horizonal lengths show contribution of individual species as indicator values based on combining values for relative abundance, relative density and relative frequency, and *p*-values are shown with bar colour (0 to 100) representing Monte Carlo test of significance of the observed maximum indicator value for each species. The highest indicator values are shown within rectangles. The importance value index for indicator species were as *Poa annua* (292), followed by *Persicaria hydropiper* (91), *Trifolium repens* (73), *Trifolium pratense* (69), *Stellaria media* (59), *Cynodon dactylon* (46), *Ranunculus sceleratus* (38), *Rorippa palustris* (30), and *Phleum pratense* (25). The bar in the figure shows the indicator species with a significant *p*-value. The bigger the dot, the more significant the *p*-value of the species.

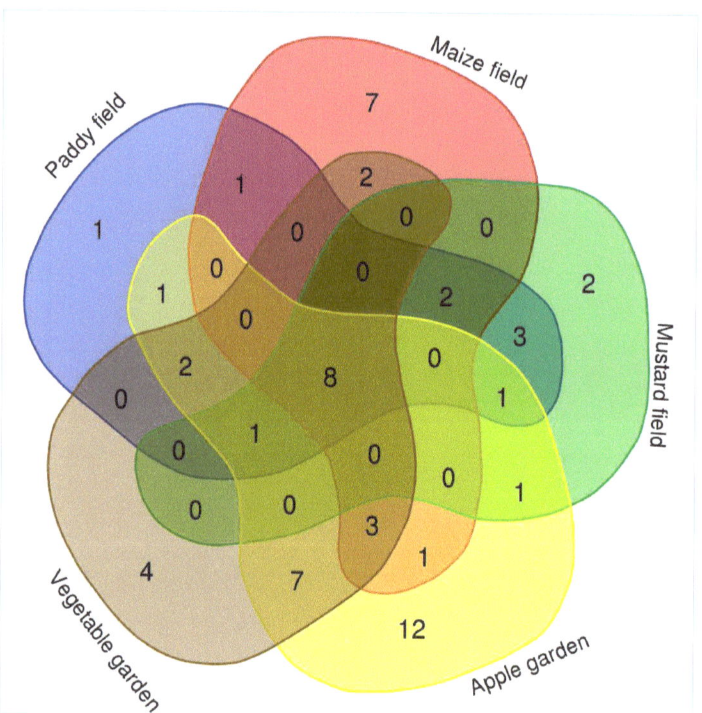

Figure 3. Venn diagram showing the distribution of the number of weed species occurring in different cropping systems of Kupwara district, Jammu and Kashmir, India.

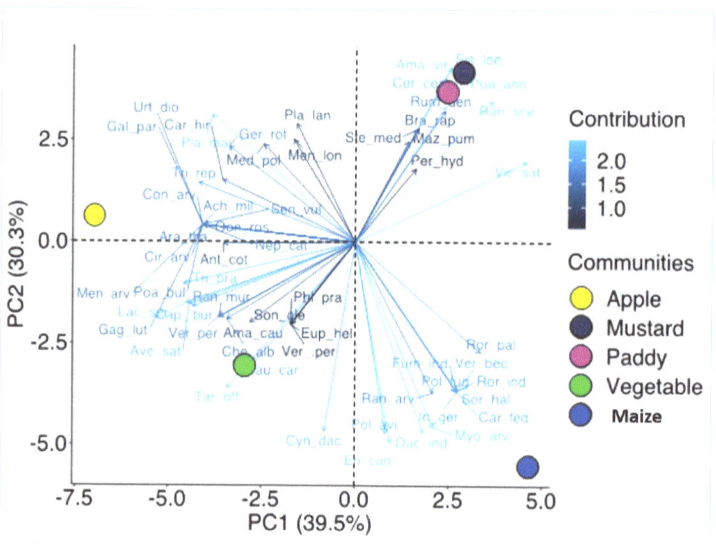

Figure 4. Principle Component Analysis (PCA) illustrating the relationship of weed species and the five cropping systems evaluated for the Kupwara district, Jammu and Kashmir, India. The contribution scale shows how much each weed species contributes to the PCA. (Full name of the species can be referred to in Table 1).

Spatial turnover (βsim) and nestedness-resultant components (βsne) of species dissimilarity revealed dissimilarity lower than 50% among the five crops (Figure 5). In the βsim cluster, we observed three distinct clusters; one made up of apple and vegetable, another of paddy and mustard, and the last one only of maize. Maize showed a difference of 44–50% in weed composition when compared to the other four crops. Apple crops showed a difference in weed composition of 35% and 38% when compared to paddy and mustard, respectively. In addition, vegetable crops had 45% and 50% different weed species when compared to paddy and mustard, respectively. On the other hand, the highest value in the βsne cluster was 21%, showing that the number of species between crops is similar, differing by less than 21%. This highest value was found between apple and mustard. Moreover, the number of weed species in apple crops was 19% different from paddy. The other relationships between crops regarding βsne had less than a 12% difference in weed species number.

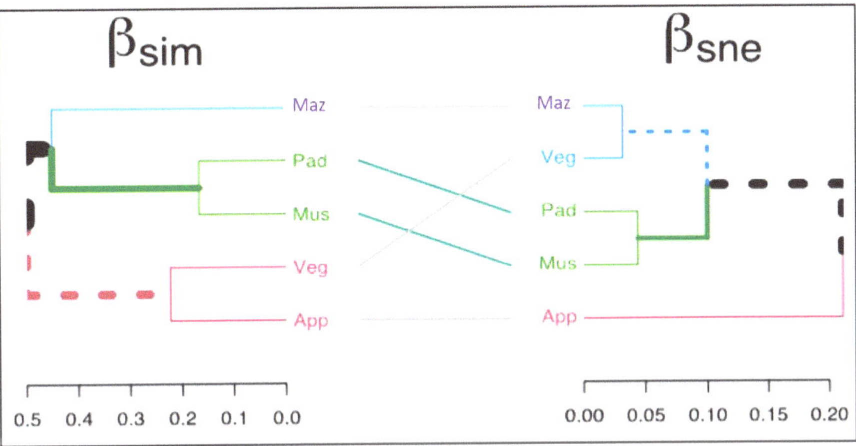

Figure 5. Dissimilarity cluster based on spatial turnover (βsim) and nestedness-resultant components (βsne) of beta diversity components of weeds in five different cropping systems of Kupwara district, Jammu and Kashmir, India. (App—Apple, Veg—Vegetable, Mus—Mustard, Pad—Paddy, Maz—Maize).

3. Discussion

The first step in dealing with the problem of weeds is their identification and documentation not just for a specific location or region but also for the specific cropping systems. The current study would help in formulating management strategies in dealing with weeds. We could find only a limited such studies for the part of IWH while there is a lack of information on weed for most of the regions of India. The findings will help in an improved understanding of weed infestation under the different agroecosystems. As the foremost and important step in weed management is the proper identification of different weeds, their phenological attributes, and habitat preference; the present study provides essential insight for effective weed management.

The presence of weeds in agricultural settings drains nutrients and moisture from the soil and prevents sunlight to reach the plants [19]. As a result, it decreases the yield of crops significantly [20]. Thus, it is imperative to understand the distribution pattern and diversity of weeds in different cropping systems for their effective management to sustain crop yield [10]. The number of species found in the present study is more than what other researchers in other Himalayan locations have discovered. A total of 35 weeds from 33 genera and 18 families were recorded by Khan et al. [21] from the Ochawala valley in the Pakistani district of Charsadda. According to Haq et al. [22], the Pakistani district of Nowshera's onion crop had a total of 21 species. In the Pakistani Himalayas' Mohmand

Agency, Ali et al. [23] discovered 63 weed species. We could not find such studies for our study region, i.e., the Kupwara district of Jammu and Kashmir, India. A higher number of weed species was reported from the families belonging to Asteraceae, Poaceae, and Brassicaceae. The Asteraceae family shows habitat diversity because of its wide ecological amplitude. Khan et al. [21] reported Poaceae as the dominant weed family from Pakistan Himalayas. Several researchers reported Asteraceae as the dominant family of weeds from other regions [24–26]. Interestingly, other Himalayan regions were also dominantly occupied with the above-mentioned families as reported in previous studies [21,27]. The current analysis reveals an uneven distribution of species among families, and 9 of those families were monotypic. These results are fairly equivalent to prior reported values from other western Himalayan locations [21,24,28].

The comparative percentage of dissimilar life form to an existing flora in a specific region or spot is called its biological spectrum [29]. The biological spectrum reveals how plants have changed to cope with their micro- and macro-climates, and it contains significant physiognomic traits that are frequently used in vegetation research [30,31]. The most common group of biological forms were therophytes, followed by hemicryptophytes and geophytes. Therophytes, which are the dominating life form in the studied region, is a sign of significant biotic disturbances on the habitat caused by activities such as grazing, farming, road building, etc. This life form is typically linked to adverse dry environmental conditions [32–35]. Our findings agree with the previous studies [36–38]. Therophytes dominating in such disturbed habitat zones increase the number of species through the introduction of alien annual weedy forbs such as *Anthemis cotula, Amaranthus caudatus, Galinsoga parviflora*, etc. [33,39]. The plausible reason for the predominance of *hemicryptophytes* is due to the available open space and high rate of reproduction [40]. Due to their deep perennial portions, the geophytes only emerge during a brief spring and stay dormant during adverse seasons [41].

The paddy, maize, and mustard were separated from apple orchards and vegetable fields according to the indicator species analysis, which was supported by high and significant indicator values for a number of species. Interesting, each Venn diagram created for each type of habitat across the cropping systems showed a similar pattern (Figure 3). The results from the Venn diagram depicted that eleven species were unique to the apple garden followed by nine species in the maize field, four species in the vegetable garden, two species in the mustard field and one species in the paddy field. The consequence of differential habitat selection results from evolutionary adjustment of species to environmental variables [42]. Species establish under the prolonged prevailing circumstances to function better in a cropping habitat relative to the other habitats [43].

4. Materials and Methods

The study involved field-based surveys for five consecutive years (2015–2020) to collect information on weed listing, diversity assessment under different cropping systems, and flowering phenology. The standard taxonomic procedure was followed for the collection of plant specimens. The specimens were identified using relevant taxonomic literature and were further authenticated by matching the plant specimens kept in the herbarium of the Centre for Biodiversity and Taxonomy, University of Kashmir. The herbarium of the center has been recognized by the International Bureau for Plant Taxonomy and Nomenclature, New York with acronym KASH (http://taxonomy.uok.edu.in/Main/AboutUs.aspx/ accessed on 8 May 2022). Details on sampling and analysis are discussed in this section ahead.

4.1. Study Area

The study was conducted for the Kupwara district of Jammu and Kashmir state, located in the Indian Western Himalayas (IWH). The IWH represents one of the most sensitive ecosystems (http://www.knowledgeportal-nmshe.in/ accessed on 8 May 2022) with very rich biodiversity where many species are rare and endemic [3]. At the same time, the region faces the threat of weed invasion making it vulnerable to ill effects of invasion while the other factors making the region vulnerable include proximity to streams, slopes, and elevations [44]. District Kupwara is the northernmost geographical part of the Kashmir Himalaya in the IWH. Its geographical area is 2379 km^2, situated at a height of around 1616 m amsl. The region extends between the latitudes of 34.17–34.21 E and the longitudes of 73.10–73.16N (Figure 6). It is the backward frontier of the Kashmir Himalayas, with a line of control abutting it to the northwest. The river Kishenganga, originates in the Himalayas and flows from east to west, through the outer areas of Kupwara. The region harbours a large number of weeds because of a broad range of physical land features having large climatic variations across the region. In addition, this part of Himalaya has diverse cropping pattern with a large variety of cultivated crops where few weeds are recognised as crop specific weed [45]. The dominant vegetation of the area is represented by Himalayan dry temperate forests, Himalayan moist temperate forests and Sub-Alpine forests [46]. Scrub forest vegetation intersperse the different forest types at higher altitudes. Patches of grassland meadows are also quite common here (Figure 6).

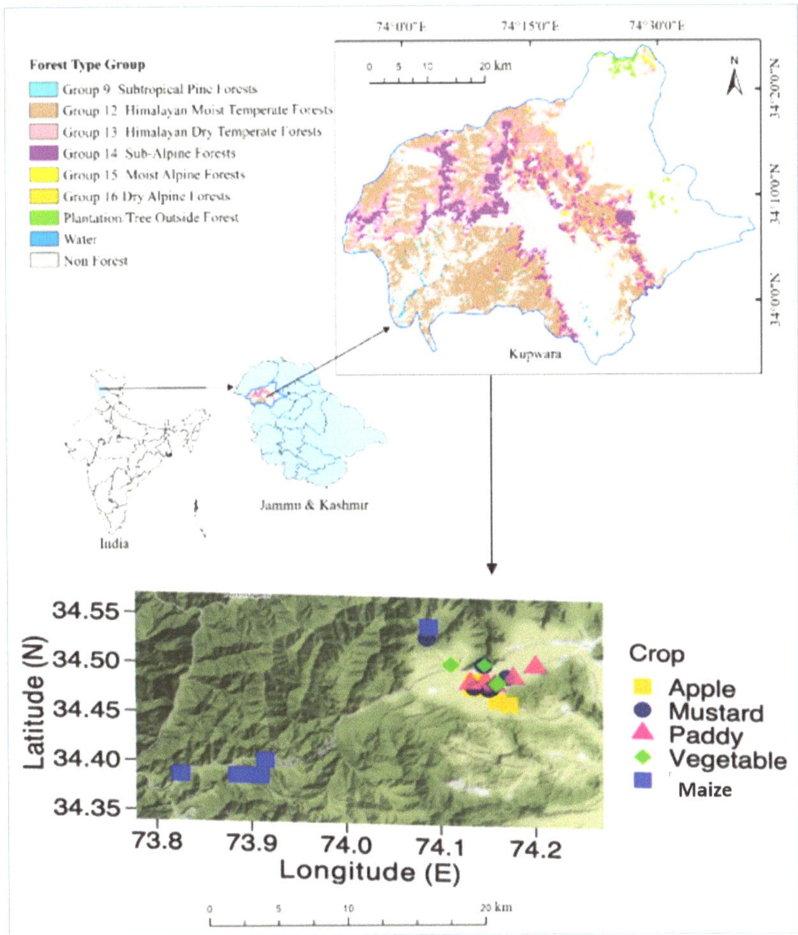

Figure 6. Map of the study area showing the location with dominant forest types in upper panel and lower panel showing sampling sites in the Kupwara district, Jammu and Kashmir state, India. The locations of apple, mustard, paddy, vegetable, and maize sites are indicative of sampling sites that included winter, summer and rainy season sampling and they do not represent the number of samples.

4.2. Sampling Design and Analysis

Preliminary field surveys were carried out to obtain an understanding about the nature of the terrain, vegetation, distribution and accessibility in the study area. After that, five cropping types in two distinct systems of agriculture and horticulture were identified for the assessment. The agriculture system constituted of paddy, maize and mustard cultivation while the horticulture system consisted of apple and vegetable gardens. Systematic random vegetation sampling was carried out to record floral diversity in the different habitats during the year 2015–2020. In total, 250 quadrates of size 5 m × 5 m were laid down for sampling. This constituted 50 plots in each of the five selected cropping systems. The sampling was done to ensure field enumeration for all of the three distinct seasons (i.e., winter, summer, rainy) as few of the weed species were not perennial and their occurrence was reported only during a specific season **(Figure 7).** The selection of plot size was carried out by drawing species vs. area curve, where we found that the number of weed species presence counted was usually saturated at the selected size of 5 m × 5 m size. Species listing of weeds was simply taken as the count of the total number of different weeds occurring in all of the sampled study quadrates for a selected cropping system. The dominating weed species present in a particular agricultural system were identified using the Importance Value Index (IVI) of weeds (Supplementary Materials). The IVI was calculated by adding the relative frequency, relative abundance, and relative density of each weed species in a given cropping system [47]. Field notebooks were used to record in-depth field observations on ecological characteristics for each species, such as blooming phenology, habit, and Raunkiaer's life form [48,49]. The native phytogeographical distribution of the plant species gathered from the study region was obtained using secondary sources such as floras, manuals, and recently published research papers [45] and specialized internet web pages of Germplasm Resource Information Network (GREEN) https://www.ars-grin.gov/ accessed on 8 May 2022 and www.efloras.org/ accessed on 8 May 2022. Based on data sources that were accessible, plant species were divided into native and alien species. The diversity indices i.e., Shannon–Wiener index [50], Simpson diversity index [51], and Evenness index [52] were calculated using the standard formula.

The Venn diagram evaluates the unique and common species among cropping systems by using Bioinformatics & Evolutionary Genomics tool (available at http://bioinformatics. psb.ugent.be/webtools/Venn/ accessed on 5 June 2022). Principal Component Analysis (PCA) was done to visualize the weed associations between crops using the package "vegan" in the software R 4.0.0 [53,54]. In order to compare the β-diversity of weeds at the habitat and landscape levels, we also calculated β-diversity, which was produced as a ratio of the regional and local diversity. We used the spatial turnover (Simpson pairwise dissimilarity) and nestedness-resultant components (nestedness-fraction of Sorensen pairwise dissimilarity) of β-diversity analysis applying "Sorensen" as family of dissimilarity index while the Dissimilarity analysis was conducted in the package "betapart" [55]. In addition, indicator species analysis was used to identify the key weed species for each habitat type using PAST software [56]. Indicator values were computed according to Dufrêne and Legendre [57], and the statistical significance of the maximum indicator value was determined using the Monte Carlo Test of significance [58].

Figure 7. The photograph depicts the studied agriculture and horticulture systems in the study area, 1. (**Top left**): paddy field in winter and summer season, 2. (**Top right**): mustard field in rainy and summer season, 3. (**Middle left**): maize field in summer and rainy season, 4. (**Middle right**): apple garden in rainy and summer season, 5. (**Bottom left**): vegetable garden in winter and summer season, 6. (**Bottom right**): paddy field in rainy season.

5. Conclusions

This study provides a comprehensive understanding of the distribution and diversity of weeds under the various management interventions that distinguish cropping systems from agriculture and horticulture systems. The findings are expected to assist farm managers to develop effective management plans to control and eradicate weeds infestation in the paddy, maize, mustard, apple, and vegetable agro-ecosystems of the IWH region. Such studies provide essential information to manage weeds where lack of sufficient data on weeds is one of the major constraints for formulating an effective weed management plan. The flowering phenology of weeds will help to establish an effective time for the application of herbicides or to implement manual weeding operations (i.e., removal of weeds) in the field, as the most appropriate weeding time is before the flowering of weeds. The dominant weeds and their clustered flowering phenological timings in a year for each of the dominant cropping systems will further help to identify an ideal time in a year for implementing weeding operations to achieve optimum results of weed eradication.

Supplementary Materials: The following supporting information can be downloaded at: https://www.mdpi.com/article/10.3390/plants12061222/s1, Table S1: Database of importance value index (IVI) of recorded species from different cropping systems of western Himalayas.

Author Contributions: Conceptualization, S.M.H. and F.A.L.; methodology, S.M.H. and M.K.; software, E.S.C. and M.W.; validation, S.M.H., R.C. and M.W.; formal analysis, M.W. and M.K.; investigation, S.M.H. and F.A.L.; resources, E.A.M.; data curation, M.W.; writing—original draft preparation, S.M.H.; writing—review and editing, H.O.E., F.A.L. and M.K.; visualization, F.A.L.; supervision, H.O.E.; project administration, R.C.; funding acquisition, H.O.E. All authors have read and agreed to the published version of the manuscript.

Funding: Researchers Supporting Project number (RSP2023R118), King Saud University.

Data Availability Statement: All obtained data are provided in the research article.

Acknowledgments: The authors would like to thank Researchers Supporting Project number (RSP2023R118), King Saud University, Riyadh, Saudi Arabia.

Conflicts of Interest: The authors declare no conflict of interest.

References

1.   Kumar, M.; Kalra, N.; Ravindranath, N.H. Assessing the response of forests to environmental variables using a dynamic global vegetation model: An Indian perspective. *Curr. Sci.* **2020**, *118*, 700–701.
2.   Kumar, M.; Kalra, N.; Khaiter, P.; Ravindranath, N.H.; Singh, V.; Singh, H.; Sharma, S.; Rahnamayan, S. PhenoPine: A simulation model to trace the phenological changes in *Pinus roxhburghii* in response to ambient temperature rise. *Ecol. Modell.* **2019**, *404*, 12–20. [CrossRef]
3.   Kumar, M.; Rawat, S.P.S.; Singh, H.; Ravindranath, N.H.; Kalra, N. Dynamic forest vegetation models for predicting impacts of climate change on forests: An Indian perspective. *Indian J. For.* **2018**, *41*, 1–12.
4.   Rawat, A.S.; Kalra, N.; Singh, H.; Kumar, M. Application of vegetation models in India for understanding the forest ecosystem processes. *Indian For.* **2020**, *146*, 99–100. [CrossRef]
5.   Richards, C.L.; Bossdorf, O.; Muth, N.Z.; Gurevitch, J.; Pigliucci, M. Jack of all trades, master of some? On the role of phenotypic plasticity in plant invasions. *Ecol. Lett.* **2006**, *9*, 981–993. [CrossRef] [PubMed]
6.   D'antonio, C.; Meyerson, L.A. Exotic plant species as problems and solutions in ecological restoration: A synthesis. *Rest. Ecol.* **2002**, *10*, 703–713. [CrossRef]
7.   Holzner, W.; Numata, M. *Biology and Ecology of Weeds*; Springer Science & Business Media: Berlin/Heidelberg, Germany, 2013; Volume 2.
8.   Einhellig, F.A. Interactions involving allelopathy in cropping systems. *Agron. J.* **1996**, *88*, 886–893. [CrossRef]
9.   Weston, L.A.; Duke, S.O. Weed and crop allelopathy. *Crit. Rev. Plant Sci.* **2003**, *22*, 367–389. [CrossRef]
10. Zimdahl, R.L. *Fundamentals of Weed Science*; Academic Press: Cambridge, MA, USA, 2018.
11. Sinden, J.; Jones, R.; Hester, S.; Odom, D.; Kalisch, C.; James, R.; Cacho, O.; Griffith, G. *The Economic Impact of Weeds in Australia*; Technical Series 8; CRC for Australian Weed Management: Glen Osmond, Australia, 2004.
12. Derpsch, R.; Friedrich, T.; Kassam, A.; Li, H. Current status of adoption of no-till farming in the world and some of its main benefits. *Int. J. Biol. Eng.* **2010**, *3*, 1–25.
13. Chmielewski, F.M. Phenology in agriculture and horticulture. In *Phenology: An Integrative Environmental Science*; Springer: Berlin/Heidelberg, Germany, 2013; pp. 539–561.
14. Kumar, M.; Padalia, H.; Nandy, S.; Singh, H.; Khaiter, P.; Kalra, N. Does spatial heterogeneity of landscape explain the process of plant invasion? A case study of Hyptis suaveolens from Indian Western Himalaya. *Environ. Monit. Assess.* **2019**, *191*, 794. [CrossRef]
15. Norsworthy, J.K.; Bond, J.; Scott, R.C. Weed management practices and needs in Arkansas and Mississippi rice. *Weed Technol.* **2013**, *27*, 623–630. [CrossRef]
16. Haq, S.M.; Hamid, M.; Lone, F.A.; Singh, B. Himalayan hotspot with Alien Weeds: A case study of biological spectrum, phenology, and diversity of weedy plants of high altitude mountains in District Kupwara of J&K Himalaya, India. *Proc. Natl. Acad. Sci. India Sect. B Boil. Sci.* **2021**, *91*, 139–152.
17. Ward, S.M.; Cousens, R.D.; Bagavathiannan, M.V.; Barney, J.N.; Beckie, H.J.; Busi, R.; Davis, A.S.; Dukes, J.S.; Forcella, F.; Freckleton, R.P.; et al. Agricultural weed research: A critique and two proposals. *Weed Sci.* **2014**, *62*, 672–678. [CrossRef]

18. Nunez, M.A.; Pauchard, A. Biological invasions in developing and developed countries: Does one model fit all? *Biol. Invas.* **2010**, *12*, 707–714. [CrossRef]
19. Monaco, T.J.; Weller, S.C.; Ashton, F.M. *Weed Science: Principles and Practices*; John Wiley & Sons: Hoboken, NJ, USA, 2022.
20. Forbes, J.C.; Forbes, J.C.; Watson, D. *Plants in Agriculture*; Cambridge University Press: Cambridge, UK, 1992.
21. Khan, M.; Hussain, F.; Musharaf, S. Floristic composition and biological characteristics of the vegetation of Sheikh Maltoon Town District Mardan, Pakistan. *Annu. Res. Rev. Biol.* **2013**, *31*, 31–41.
22. Ul Haq, Z.; Gul, B.; Shah, S.M.; Razaq, A.; Raza, H. Ecological characteristics of weeds of onion crop of University of Peshawar Botanical Garden, District Nowshera, Pakistan. *Pak. J. Weed Sci. Res.* **2016**, *22*, 263–267.
23. Ali, A.; Badshah, L.; Hussain, F.; Shinwari, Z.K. Floristic composition and ecological characteristics of plants of chail valley, district Swat, Pakistan. *Pak. J. Bot.* **2016**, *48*, 1013–1026.
24. Zeb, U.; Ali, S.; Li, Z.H.; Khan, H.; Shahzad, K.; Shuaib, M.; Ihsan, M. Floristic diversity and ecological characteristics of weeds at Atto Khel Mohmand Agency, KPK, Pakistan. *Acta Ecol. Sin.* **2017**, *37*, 363–367. [CrossRef]
25. Iqbal, M.; Khan, S.M.; Khan, M.A.; Ahmad, Z.; Ahmad, H. A novel approach to phytosociological classification of weeds flora of an agro-ecological system through Cluster, Two Way Cluster and Indicator Species Analyses. *Ecol. Indic.* **2018**, *84*, 590–606. [CrossRef]
26. Nafeesa, Z.; Haq, S.M.; Bashir, F.; Gaus, G.; Mazher, M.; Anjum, M.; Rasool, A.; Rashid, N. Observations on the floristic, life-form, leaf-size spectra and habitat diversity of vegetation in the Bhimber hills of Kashmir Himalayas. *Acta Ecol. Sin.* **2021**, *41*, 228–234. [CrossRef]
27. Chhetri, N.B.K.; Shrestha, K.K. Floristic diversity and important value indices of tree species in lower Kanchenjunga Singhalila ridge Eastern Nepal. *Am. J. Plant Sci.* **2019**, *10*, 248–263. [CrossRef]
28. Haq, S.M.; Malik, A.H.; Khuroo, A.A.; Rashid, I. Floristic composition and biological spectrum of Keran-a remote valley of northwestern Himalaya. *Acta Ecol. Sin.* **2019**, *39*, 372–379. [CrossRef]
29. Pichi-Sermolli, R.E. An index for establishing the degree of maturity in plant communities. *J. Ecol.* **1948**, *36*, 85–90. [CrossRef]
30. Saxena, A.K.; Pandey, P.; Singh, J.S. Biological Spectrum and other structural functional attributes of the vegetation of Kumaun Himalaya. *Vegetatio* **1982**, *49*, 111–119. [CrossRef]
31. Sahu, S.C.; Dhal, N.K.; Datt, B. Environmental implications of Biological spectrum vis-a-vis tree species diversity in two protected forests (PFs) of Gandhamardan hill ranges, Eastern Ghats, India. *Environmentalist* **2012**, *32*, 420–432. [CrossRef]
32. Waheed, M.; Haq, S.M.; Arshad, F.; Bussmann, R.W.; Iqbal, M.; Bukhari, N.A.; Hatamleh, A.A. Grasses in Semi-Arid Lowlands—Community Composition and Spatial Dynamics with Special Regard to the Influence of Edaphic Factors. *Sustainability* **2022**, *14*, 14964. [CrossRef]
33. Mamariani, F.; Joharchi, M.R.; Ejtehadi, H.; Emadzade, K. Contributions to the flora and vegetation of Binalood mountain range, NE Iran: Floristic and chorological studies in Fereizi region. *JCMR* **2009**, *1*, 1–18.
34. Vakhlamova, T.; Rusterholz, H.P.; Kanibolotskaya, Y.; Baur, B. Effects of road type and urbanization on the diversity and abundance of alien species in roadside verges in Western Siberia. *Plant Ecol.* **2016**, *217*, 241–252. [CrossRef]
35. Rashid, I.; Haq, S.M.; Lembrechts, J.J.; Khuroo, A.A.; Pauchard, A.; Dukes, J.S. Railways redistribute plant species in mountain landscapes. *J. Appl. Ecol.* **2021**, *58*, 967–1980. [CrossRef]
36. Asim, Z.I.; Haq, F.; Iqbal, A. Phenology, life form and leaf spectra of the vegetation of kokarai valley, district swat. *J. Biol. Environ. Sci.* **2016**, *9*, 23–31.
37. Rahman, I.U.; Afzal, A.; Iqbal, Z.; Ijaz, F.; Ali, N.; Asif, M.; Alam, J.; Majid, A.; Hart, R.; Bussmann, R.W. First insights into the floristic diversity, biological spectra and phenology of Manoor Valley, Pakistan. *Pak. J. Bot.* **2018**, *50*, 1113–1124.
38. Rahman, I.U.; Afzal, A.; Iqbal, Z.; Abd Allah, E.F.; Alqarawi, A.A.; Calixto, E.S.; Ali, N.; Ijaz, F.; Kausar, R.; Alsubeie, M.S.; et al. Role of multivariate approaches in floristic diversity of Manoor Valley (Himalayan Region), Pakistan. *Appl. Ecol. Environ. Res.* **2019**, *17*, 1475–1498. [CrossRef]
39. Joly, M.; Bertrand, P.; Gbangou, R.Y.; White, M.C.; Dubé, J.; Lavoie, C. Paving the way for invasive species: Road type and the spread of common ragweed (*Ambrosia artemisiifolia*). *Environ. Manag.* **2011**, *48*, 514–522. [CrossRef]
40. Khan, W.; Khan, S.M.; Ahmad, H.; Alqarawi, A.A.; Shah, G.M.; Hussain, M.; Abd_Allah, E.F. Life forms, leaf size spectra, regeneration capacity and diversity of plant species grown in the Thandiani forests, district Abbottabad, Khyber Pakhtunkhwa, Pakistan. *Saud. J. Biol. Sci.* **2018**, *25*, 94–100. [CrossRef]
41. Hussain, F.; Shah, S.M.; Badshah, L.; Durrani, M.J. Diversity and ecological characteristics of flora of Mastuj valley, district Chitral, Hindukush range, Pakistan. *Pak. J. Bot.* **2015**, *47*, 495–510.
42. Rosenzweig, M.L. A theory of habitat selection. *Ecology* **1981**, *62*, 327–335. [CrossRef]
43. Cornwell, W.K.; Schwilk, D.W.; Ackerly, D.D. A trait-based test for habitat filtering: Convex hull volume. *Ecology* **2006**, *87*, 1465–1471. [CrossRef]
44. Olokeogun, O.S.; Kumar, M. An indicator based approach for assessing the vulnerability of riparian ecosystem under the influence of urbanization in the Indian Himalayan city, Dehradun. *Ecol. Indic.* **2020**, *119*, 106796. [CrossRef]
45. Andrabi, S.M.; Reshi, Z.A.; Shah, M.A.; Qureshi, S. Studying the patterns of alien and native floras of some habitats in Srinagar city, Kashmir, India. *Ecol. Processes* **2015**, *4*, 2. [CrossRef]
46. Haq, S.M.; Khuroo, A.A.; Malik, A.H.; Rashid, I.; Ahmad, R.; Hamid, M.; Dar, G.H. Forest ecosystems of Jammu and Kashmir state. In *Biodiversity of the Himalaya: Jammu and Kashmir State*; Springer: Singapore, 2020; pp. 191–208.

47. Curtis, J.T.; Mcintosh, R.P. The interrelations of certain analytic and synthetic phytosociological characters. *Ecology* **1950**, *31*, 434–455. [CrossRef]
48. Raunkiaer, C. *The Life Forms of Plants and Statistical Plant Geography; Being the Collected Papers of C. Raunkiaer*; Clarendon Press: Oxford, UK, 1934.
49. Perez-Harguindeguy, N.; Diaz, S.; Garnier, E.; Lavorel, S.; Poorter, H.; Jaureguiberry, P.; Bret-Harte, M.S.; Cornwell, W.K.; Craine, J.M.; Gurvich, D.E.; et al. New handbook for standardised measurement of plant functional traits worldwide. *Aust. Bot.* **2013**, *61*, 167–234. [CrossRef]
50. Shannon, C.E. A mathematical theory of communication. *Bell Syst. Tech. J.* **1948**, *27*, 379–423. [CrossRef]
51. Simpson, E.H. Measurement of diversity. *Nature* **1949**, *163*, 688. [CrossRef]
52. Pielou, E.C. *Ecological Diversity*; No. 574.524018 P5; Wiley: Hoboken, NJ, USA, 1975.
53. Oksanen, J.; Blanchet, F.G.; Friendly, M.; Kindt, R.; Legendre, P.; McGlinn, D.; Minchin, P.R.; O'Hara, R.B.; Simpson, G.L.; Solymos, P.; et al. *vegan: Community Ecology Package*, R package Version 2.5-2; 2018. Available online: https://cran.r-project.org/web/packages/vegan/vegan.pdf (accessed on 8 June 2022).
54. Team, R.D.C. A Language and Environment for Statistical Computing. 2009. Available online: https://www.R-project.org (accessed on 8 June 2022).
55. Baselga, A.; Orme, D.; Villeger, S.; De Bortoli, J.; Leprieur, F.; Baselga, M.A. *Package 'betapart'. Partitioning Beta Diversity into Turnover and Nestedness Components*, Version 1.5.6; 2018. Available online: https://cran.r-project.org/web/packages/betapart/betapart.pdf (accessed on 8 June 2022).
56. Waheed, M.; Haq, S.M.; Fatima, K.; Arshad, F.; Bussmann, R.W.; Masood, F.R.; Alataway, A.Z.; Dewidar, A.F.; Almutairi, K.; Elansary, H.O.; et al. Ecological Distribution Patterns and Indicator Species Analysis of Climber Plants in Changa Manga Forest Plantation. *Diversity* **2022**, *14*, 988. [CrossRef]
57. Dufrêne, M.; Legendre, P. Species assemblages and indicator species: The need for a flexible asymmetrical approach. *Ecol. Monogr.* **1997**, *67*, 345–366. [CrossRef]
58. Haq, S.M.; Amjad, M.S.; Waheed, M.; Bussmann, R.W.; Maćków, J. The floristic quality assessment index as ecological health indicator for forest vegetation: A case study from Zabarwan Mountain Range, Himalayas. *Ecol. Indic.* **2022**, *145*, 109670. [CrossRef]

MDPI AG
Grosspeteranlage 5
4052 Basel
Switzerland
Tel.: +41 61 683 77 34

Plants Editorial Office
E-mail: plants@mdpi.com
www.mdpi.com/journal/plants